The Moral Austerity of Environmental Decision Making

The Moral Austerity of Environmental Decision Making *Sustainability, Democracy, and Normative Argument in Policy and Law*

Edited by John Martin Gillroy and Joe Bowersox

DUKE UNIVERSITY PRESS Durham & London 2002

Printed in the United States of America on acid-free paper ∞
Typeset in Trump Mediaeval by Keystone Typesetting, Inc.
Library of Congress Cataloging-in-Publication Data appear
on the last printed page of this book.

Contents

Figures and Tables

Figures

Tables

Preface

This book is a product of three years of lectures, workshops, and roundtables held at Bucknell University in 1997, 1998, and 1999 and sponsored by the John D. MacArthur Foundation. In the summer of 1997, at the annual meeting of the American Political Science Association in Washington, D.C., the eight discussants from these pages first got together to discuss the state of moral argument in environmental politics and policy. From these arguments (published in 1997 in *Policy Currents* 7:1–13) a collective decision was made to create the four issue areas of concern set out in this book and proceed to analyze the current state of moral argument in general and the ethical implications of the concept of sustainability in particular. The short arguments in Part 1 of the book were then written as a point of departure. Next, two sets of individual lectures were given at Bucknell between September 1997 and June 1998 on the subject of sustainability and policy for the twenty-first century. It is from these lectures that the case studies in Part 2 were drawn. Finally, in May 1999, the eight discussants came to Bucknell for two days of workshops and a final roundtable on sustainability and ethical discourse. For this event all of the discussants were given the case studies in Part 2 and integrated their reactions to them in their comments. All of the debate and discussion portions of Part 3 and the conclusion are drawn from transcripts of these meetings.

We wish to thank all of these individuals for taking the time to come to Bucknell and Washington, D.C., and for taking the time to prepare their arguments and cooperate in the discussion of environmental policy and ethics that took place over this three-year period. In addition, we wish to thank all those who facilitated the meetings at Bucknell, and the many colleagues, students, and citizens who participated as audience members in the discussions.

We also would like to thank Willamette University undergraduates and gifted research assistants Lee Che Leong, Johnny Vong, Magda Micholowicz, and Devadatta Gandhi, who completed their tasks with great alacrity. Final preparation of the manuscript could not have been done without the patience and technical wizardry of Karen Bond, administrative assistant to the Department of Politics at Willamette; we owe her greatly! Finally, we wish to thank our partners through all this, Margaret and Carrie, for their humor, patience, and support.

The Moral Austerity of Environmental Decision Making

Introduction: The Roots of Moral Austerity in Environmental Policy Discourse

An inherent austerity exists in the discourse and analysis of environmental policy. This austerity is not of empirical data, nor does it lie in the dearth of quantitative models or economic formulas with which we measure and trade off the costs and benefits of alternative public choices. The austerity is a moral austerity, that is, a lack of complexity in the terms of ethical debate over what is right, what is good environmental policy, and what responsibilities, duties, and obligations we have to both humanity and nature as we enter a new century of public policy decisions.

Our moral austerity is not the result of ill will or malicious souls; in fact, the field of environmental studies is filled with scholars and practitioners passionate about improving the relationship among individuals, society, and ecosystems. Rather, this normative austerity is inherent in environmental policy discourse itself, having two principal roots. First is the separation between politics and morality that has occurred in modern times, as the empirical world of the scientist and social scientist has parted ways with the normative concerns of the philosopher. These two professions speak less and less of the same language and cross-pollinate on increasingly infrequent occasions. As a result, policy analysis avoids the critical normative questions about the role of science, social justice, and theories of value, focusing instead on more and more intricate empirical casework. Policy analysis, as a relatively new "science," has chosen to concentrate its principles and assumptions in an effort to build an empirical inquiry from the foundation of microeconomics and behavioral/quantitative political science. Choosing this road, the amount of philosophical discourse over core principles, normative assumptions, and ethical terminology has decreased exponentially. What we are left with is a divorce between the empirical and the normative, between science and philosophy, between is and ought, that leaves policy discussion devoid of serious wrestling with such fundamental philosophical issues as what duties we owe to one another, to the future, and to nature. Ultimately, we believe that both political and ethical debate must share a common and explicit focus on first principles.

Radically separating the political and the ethical as subjects of debate has not always been standard practice. When Thomas Hobbes wrote *Leviathan* (1651) and John Locke wrote *Two Treatises of Government* (1698) there was an assumption that one had to link "is" and "ought." Even when Kant wrote his *Metaphysics of Morals* (1797) and Hegel the *Philosophy of Right* (1821), one was expected to integrate the normative and the empirical, to argue about the intersection of science and philosophy, and to answer the tough moral questions about who we are, how we cooperate, and to whom we ought to be responsible. Now policy is discussed as if politics and morals were mutually and necessarily exclusive subjects.

The second source of moral austerity in political argument is the overbearing presence of the market paradigm in public policy education, analysis, and discourse. If one examines any curriculum for a school of public policy or takes time to read policy journals, one is immediately confronted with the assumptions, principles, and methodology of the market paradigm, proffered as the conventional basis for policy formulation, recommendation, evaluation, and implementation. Specifically, with the separation of politics and morals just examined, an ever-growing gulf between the normative and the empirical exists. But with the need for quasi-normative standards with which to judge "socially better" policy, the modern "policy scientist" chooses a paradigm to frame his or her arguments that is based on the norms of classical economics. Rather than debate fundamental principles or generate a number of alternative paradigms based on different assumptions and ethical principles, the policy professional finds solace in the individual consumer, the invisible hand, the state that mimics market allocations, and the core principle of Kaldor efficiency. These assumptions are the bread and butter of evaluative structure. Even critics of this tradition in policy analysis assume its basic premises as a point of departure. By doing so they avoid the full discussion of the various moral dilemmas that lie at the heart of present and future environmental law.

The moral austerity of environmental policymaking is a dilemma, for the cause lies not necessarily in the individuals themselves, but in the language and frameworks they share. How do we breach the established gap between politics and morality when we have but one established paradigm, created by this gap, with which to analyze policy? How do we progress when every new movement in environmental criticism, from political ecology to sustainability, implicitly genuflects to the standard of efficiency?

This dilemma is especially acute as we enter a new century with *sustainability* as a "new" core principle for environmental policy. The principle of sustainability has taken the environmental community by storm. It is touted by economists, activists, businesspeople, philosophers, and political scientists as a new departure and a fitting standard with which to revolutionize future environmental law. But is it? Can the principle of sustainability reintegrate politics and morality? Will a sustainable future leave behind our dependence on the market paradigm and generate a new worldview where nonefficiency values have priority? Or is sustainability just another era of neoclassical economic analysis in environmental policy, with efficiency and market analysis at its core, that will widen the chasm between politics and ethics?

This book is an effort to begin the process of integrating politics and morality in policy discourse about the natural world. We begin in this introduction by looking at how we have arrived at the point we now occupy: How did this moral austerity come about? We first argue that moral argument is fundamental to public policy choice and that any normative standard being used to distinguish ought from is should be defined as a moral core of its corresponding policy argument. Next, we examine what seems to be the root cause of both sources of austerity (i.e., the separation of politics and morality as well as the dominance of the market paradigm) with an examination of the modern interpretation of Locke's political philosophy. We discuss Locke's creation as the father of modern liberalism and how his political argument has been used (or misused) by current environmental theorists to condemn liberal institutions and values, separating moral from political concerns and solidifying the market paradigm in the core of environmental debate. We then proceed to suggest that, even if the ecological critics are correct about Locke, within the Enlightenment tradition there are alternative principles and sets of assumptions that might better integrate politics and ethics while providing alternative paradigms for environmental policy analysis and choice. Throughout, we define four categories of essential issues that must be debated to enrich the level of normative discourse in policymaking and reintegrate politics and ethics: the role of science in making collective environmental decisions; the connection between environmental and social justice; the character and limitations of a strictly instrumental value theory; and the prospects for a theory of intrinsic value as an alternative for future environmental policy decision making.

Specifically, for environmental policy in the twenty-first century to be responsible to both humanity and nature, it must transcend its present reliance on science as the final arbiter of reasonable law (Issue 1); it can no longer ignore the connection between social and environmental justice (Issue 2); it should move past its "religious" faith in instrumental economic principles (Issue 3); and it needs to consider that intrinsic values may be of primary importance in making good public choices, and integrate them into policy argument (Issue 4). In this book, a panel of eight environmental discussants utilize these four issue areas as lenses through which to examine the ethical requirements of twenty-first–century environmental law and consider the ethical shortcomings of current policy discourse.

In Part 1 the panel members develop normative arguments regarding the necessity of change in each of the four issue areas, illustrating diverse and often contentious positions on the ideal relationship between policy and moral argument. In Part 2, to examine the current state of environmental decision making, seven cutting-edge analytic case studies in a variety of subjects written by recognized scholars in the fields of policy and law argue for a more sustainable environmental future. In Part 3, the eight panel members revisit their issue areas in light of the seven case studies and analyze both the lack of moral argument in the debate over what a sustainable policy is and how such ethical matters might be better integrated into environmental decision making. In the conclusion, the panelists discuss the problems and dilemmas involved with operationalizing the demands of moral argument in defining environmental law and policy in the twenty-first century, given the requirements of our democratic society.

But first, we state our assumptions about the role of moral argument in policy and examine in greater detail the fundamental causes of moral austerity in environmental policy discourse.

Moral Argument and Public Policy

Is moral principle but one possible basis for making public policy choices? It is axiomatic that normative standards must be set for any policy to be judged and analyzed and for success or failure to be evaluated. However, it might also be argued that moral principle is but one such standard, with collective ideology, individual self-interest, scientific postulates, and natural law being others. However, the essential point we wish to make is that any of these alternatives, if it plays the role of normative standard, plays the role of moral principle in policy discourse and action. Our concern is to concentrate on the diversity of normative standards or moral principles in policy studies and make these implicit assumptions clear, so that they might be defined, defended, and challenged in environmental policy debate. When we speak of moral principle in policy argument, we are therefore speaking of that core normative standard that defines a particular policy and determines its idea of the good, the right, and the responsibilities of humanity to itself and to nature.

If environmental policy decision making were entirely an empirical matter, then adopting one policy rather than another would be a matter of scientific evidence or a simple shift of legal focus that then formulates new policy, makes different choices among recommendations, and implements a new paradigm in place of an old one. However, the policy process is more complex than superficial change can accommodate. In addition to context programs and administrative activity, the formulation and implementation process relies on something deeper and more fundamental: a core moral or normative principle.

The most fundamental concept in defining policy is the idea of metapolicy. The genesis of this concept can be traced to Giandomenico Majone (1989, 146–49), who attempted to lend policy a dynamic and self-critical capacity by claiming that "our understanding of a policy and its outcomes cannot be separated from the ideas, theories and criteria by which the policy is analyzed and evaluated" (147).

Inherent in a metapolicy is the ideal of the foundational moral principle within each policy that sets its internal standards of evaluation and justification. Whatever principle defines a metapolicy will limit the options for policy formulation, recommendation, and implementation by setting the standards by which "persuasive" argument and "reasonable" policy will be judged. As long as this principle finds actualization in the institutional application of policy through law, and these results cause what "is" to measure up to what "ought to be" (as defined by the principle), then the "dominant" line of argument within the metapolicy holds its persuasiveness and law remains dynamic but consistent.[1] That is, the law is consistent in that the rules and expectations remain true to the central principle but is dynamic in that the ongoing debate has established options and distinct competitive moral standards that can be brought to the metapolicy as time and need require.

The fundamental core principle that defines the metapolicy cannot be sacrificed without the wholesale change of the character of that metapolicy. Although deliberation about a metapolicy may involve the consideration or "balancing" of many principles, public choice means the domination of one of these principles. When values conflict, this core principle will be decisive in the determination of law and policy. The "trump" principle is therefore critical to decision making and must be repeatedly applied over time to set expectations and establish consistency in the public administration of the metapolicy.

Change in policy and law occurs when we are no longer persuaded that the dominant line of argument for a metapolicy and its inherent moral standard is the best one for the issue. We examine the alternative lines of argument within the metapolicy for a better definition of what is at stake, how it should be handled, and how the ends of the policy should be defined. As a competitive argument gains in persuasiveness, its core principle will also gain more power in the deliberation process until it trumps the status quo principle and changes the conventional metapolicy. We might call this process the operationalization of a paradigm shift (Gillroy 2000, ch. 1). This process of change can be gradual, as one line of moral argument is found, over time, to be progressively more persuasive and the dominant standard fades. Change can also be sudden, as the dominant line of argument is abruptly replaced[2] by a competitor. One might define the former as evolution and the latter as revolution (Majone 1989, 147).

The institutions and administrative structures that translate metapolicy into law are themselves the product of policy arguments. Institutions that we accept as authoritative (Hart 1961) or legitimate (Fishkin 1979) encapsulate those processes and practices that are persuasive to us. In this way, lines of argument within a metapolicy have an effect on the evolution of the institutions and process by which they are transformed into law. The development of metapolicy as well as its ongoing dynamic quality are therefore the result of an interactive process between competitive theories about what ought to be and the institutions and political processes that write and interpret these arguments as law: where ought becomes is. "The relationship between policy and its intellectual super-structure, or meta-policy, is a dialectic one" (Majone 1989, 166).

Metapolicy contains the elements of continuity and change. To conceptualize the components that provide for both, it is useful to make the distinction, as Majone does, between core and periphery (1989, 150–54). Majone defines these two elements of a metapolicy to distinguish between "the relatively stable and rigid part of a policy and its more changing and flexible components" (150).

For our purposes, it is useful to describe the core as that part of the metapolicy that contains its fundamental principle. Although not unchanging, the core is resistant to change, and successful change represents a fundamental shift in the core normative principle and therefore the basic character of the metapolicy (Majone 1989, 150–51).

In contrast, the periphery can be described as containing those parallel principles and arguments[3] that compete for dominance within any metapolicy as well as the institutional apparatus that implements the dominant core argument. The idea of alternative principles in the periphery supports the dynamic quality of our definition of metapolicy and also creates the image of a persistent

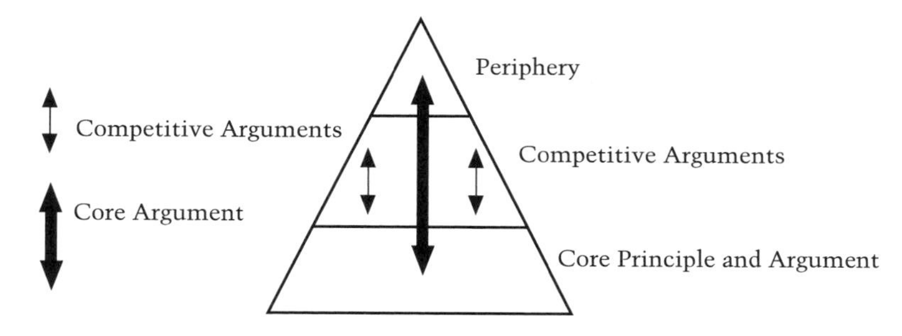

Figure 1. Metapolicy model. Source: Gillroy, 2000.

discourse about proper ends and the standards by which these should be judged within metapolicies. A metapolicy model can summarize this structure.[4]

In Figure 1, the top tier, the *periphery,* contains the law and institutional and administrative apparatus, with their rules and regulations, that define the status quo metapolicy and that operationalize the dominant or persuasive policy argument. At this level we find the public rhetoric associated with a debate over how a metapolicy is legislated, executed, and evaluated by experience. Here, law is codified and tested in light of the dominant metapolicy argument and its core moral principle.

The second tier, which is the *deliberative* section of the periphery, is larger than the first and is the reservoir of alternative moral principles and arguments for the policy that do not define it as a status quo metapolicy but that represent opportunities for future change and redefinition. Given the effects of the status quo metapolicy, it is here where a debate about the "reasonable" or "proper" ends of policy takes place. At this level, over time, judgments are made about the difference between what is and what ought to be. This deliberative zone contains the competitors for the core and the backup arguments that are called on when the core no longer persuasively serves the metapolicy or its constituents.

The third tier, containing the *core* principle and its policy argument, is foundational and defines the character of the metapolicy, providing the stable standard for the pyramid. The core principle determines the moral context and constraints for the deliberation in the second tier and the proper tools and legal arguments for the public process, debate, and institutional framework of the first tier.

Employing this model as a starting point, but before framing what normative or moral debate ought to take place within environmental decision making, we need to describe the current state of normative argument in environmental policy decision making. Here we examine the role Lockean philosophy has played in the debate over environmental policy and politics. Our modern interpretation of Locke has both solidified a separation between politics and ethics and created a core market paradigm for both policymakers and political ecologists as they debate the declining state of the environment and the need for substantial change in the way humanity lives within nature.

The Legacy of Locke: A Normative Point of Departure

Louis Hartz (1955, 140) argues that the philosophy of John Locke is practically an American cliché: we seem to live and breathe, walk and talk Locke's seventeenth-century liberalism. Meanwhile, prominent political writers and theorists and moral philosophers of an ecological persuasion continue to blame Lockean or quasi-Lockean justifications of individualism, property, accumulated wealth, and limited government for our environmental problems.

Political scientists often begin with Locke simply because of the apparent familiarity of the terrain: Locke deals with concepts of state formation, the legitimacy of government, its effect on private and public life, and the justification of economic activity. However, environmental scholars often seem to consider Locke's moral philosophy to be a trivial backdrop to the property concepts and liberal economic convictions they indict. It appears that Locke's only major contribution to Western society was the fifth chapter of the *Second Treatise,* "On Property." Apparently, the cause and potential effects of environmental degradation are most exquisitely exemplified here (Ophuls and Boyan 1992, 191, 204).

We argue that ecological interpretations of Locke and Lockean liberalism demonstrate the post-Enlightenment bifurcation of ethics and politics (see Bowersox 1995, chs. 1, 2; 1998). They also illustrate the influence that "demi-Lockean" presumptions against noneconomic paradigms may have had on modern political ecologists (Holsworth 1979; Kassiola 1990).[5] We perceive a uniquely determinist reading in the way various analysts utilize Locke or allude to "Lockean liberalism" in their environmental critiques of contemporary Western society. Lockean liberalism is purportedly dependent on a postulation of natural abundance. This unlimited cornucopia is then available for self-interested humans to appropriate in expression of their agency and morality. Such liberal beliefs as individual liberty and limited government are therefore *dependent* on access to unlimited resources: remove abundance and a Hobbesian "war of all against all" inevitably may result. Thus, in a state of scarcity, avoiding environmental tragedy may necessitate institutional change and political authoritarianism.

This Lockean cliché of unlimited natural abundance justifies unlimited human desire. Susan Leeson believes one may find Western society's ecological failing specifically in Locke:

> Lockean thought legitimated virtually endless accumulation of material goods; helped equate the process of accumulation with liberty and the pursuit of happiness; helped implant the idea that with ingenuity man can go beyond the fixed laws of nature, adhering only to whatever temporary laws he establishes for himself in the process of pursuing happiness; and helped instill the notion that the "commons" is served best through each man's pursuit of private gain, because there will always be enough for all who are willing to work. In short, Lockean philosophy led to a strong ideology of man's relationship to man and the earth, in which autonomous individuals seek comfort and enjoyment through hard work and material acquisitions. (1978, 303)[6]

Like Leeson, Ophuls argues that Locke's basic assumptions about the plenitude and ease of human exploitation of nature are necessary to facilitate the liberal polity: "Economics in general and private property in particular are the basis of the Lockean political order. Not only is material satisfaction the end of politics . . . but the process of acquiring property fulfills a vital political function. Instead of fighting over religion, the people unite to conquer nature and enrich themselves" (1997, 37).

Some others are less explicit in their use of Locke but clearly have him in mind. Robert Heilbroner depicts a Lockean political-economic system in the West dependent on continued economic expansion and a continuously rising standard of living to justify and protect liberal political institutions and values (1980, ch. 3). William Leiss attacks the rampant commercialism purportedly legitimated by Locke and his compatriots because it alienates humans from their natural surroundings and gives the illusion of unlimited potential wealth and perpetual social change through expanding development (1972, 183–85).

The temptation to utilize this Lockean legacy as an ecological scapegoat has also enamored William Catton, who chastises a "naïve liberalism" for ignoring human tendencies toward violent conflict and dependence on continued expansion of geographic and productive frontiers to avoid social chaos (1980, 103). Paul and Anne Ehrlich, in their recent rebuttal of brownlash literature, portray their opponents as demi-Lockean cornucopians (1996, chs. 2, 10). Deep ecologists Bill Devall and George Sessions discuss the environmental effects of "the dominant social paradigm," a paradigm characterized by discrete individualism, rationalism, and human dominance of nature readily linked to Lockean liberalism (1985, 42–45). Ecocentrist Robyn Eckersley, while critical of authoritarian responses, also assumes that a very economistic and cornucopian Lockean liberalism lies at the heart of modern environmental ills (1992, 23). Though they now span three decades and numerous ideologies, classic analyses of the "roots of the environmental crisis" thus often share an emphasis on the apparent economic determinism in Locke and Lockean liberalism: "It was precisely this that allowed John Locke, whose political argument is essentially the same as that of Hobbes in every particular except scarcity in the state of nature, to be basically libertarian where Hobbes is basically authoritarian. Thanks to the Great Frontier, Locke and [Adam] Smith found that there was so much abundance in the state of nature that a Hobbesian war of all against all was unlikely; every person could take away some prize, and competition would be socially constructive rather than destructive, with the 'invisible hand' producing the greatest good for society as a whole" (Ophuls and Boyan 1992, 204).

Recently, environmental ethicist Peter Wenz reiterated cornucopian Lockeanism, suggesting its centrality to the development of commercialism, liberalism, and the rational exploitation of nature for human liberty and protection of individual rights (1997, 207–11). Myrl Duncan explicitly blames the cornucopian "Chapter V on Property" for the persistence and vigor of antienvironmental individualism generally, and recent antienvironmental "takings" decisions in particular (1996, 1095).

The portrayal of a determinist Locke by environmental analysts is critical for their essential premise: *if* one can disprove the proposition of abundance, *then* one can question the logical validity of the derivative consequences of the proposition, that is, individual autonomy, liberal toleration, the limited state, and the legitimation of limitless accumulation/exploitation (Ophuls 1997, chs. 4, 5). Certainly these empirical projections of abundance or scarcity should not be ignored or trivialized. But the real concern is whether such sublimation of debate into a conflict over resource numbers is an effective strategy for addressing a primarily normative dilemma about fundamental moral principle and alternative paradigms for policy choice.

It seems counterproductive and ill conceived to confront an economistic argument with simply another economistic model. An argument attacking sociopolitical and economic beliefs and institutions that promote reliance on technical and scientific rationalism to satiate our consumption patterns may become suspect if that argument itself relies simply on the same strategy. People may question the utility and efficacy of an "antiscientific" argument that tends to dominate its audience with its own (equally suspect?) scientific data (Fowler 1991, ch. 7). Although several authors have attacked the apparent authoritarian arguments of Ophuls, Heilbroner, and others, they themselves employ the same rhetorical strategy (Leeson 1978; Kassiola 1990).[7]

The debate over the biophysical fact of abundance or scarcity and economic assumption about the individual and society exacerbate the conflict between politics and morals. In *Ecology and the Politics of Scarcity,* Ophuls (1977) argues against ethics as a sufficient strategy to curb environmental destruction. Rather than rely on individual or collective moral reflection and debate, we must turn to politics, which he conceives as increasing governmental authority imposing an environmentally oriented ecological aristocracy. Such steps are necessary to ensure planetary and human survival. Ophuls clearly sees change in individual or collective principles and values as secondary and politically useless: "Individual conscience and the right kind of cultural attitudes are not by themselves sufficient to overcome the short-term calculations of utility that lead men to degrade their environment. Only a government possessing great powers to regulate individual behavior in the ecological common interest can deal effectively with the tragedy of the commons" (154).

Meanwhile, environmental ethicists often see little connection between their philosophical recommendations and political reality. While defending the attribution of rights to animals, for example, Tom Regan faced persistent questions from a University of Wisconsin audience about the impact such rights might have on existing political structures. Ultimately, he shrugged and threw up his hands, uttering, "I am only a moral philosopher. I don't specifically worry about those things."[8]

Herman Daly recognizes the conflict between economic and moral frameworks for structuring society. If a growth economy is a way to structure social stability by providing potentially unlimited access to one's desires, then moral strictures (another way to provide for stability) get in the way and potentially threaten stability by placing ethical limits on growth. Yet if scarcity is imagined within the economic paradigm, the only solution to potential strife may be the imposition of order. Daly's comments tellingly illustrate the ongoing displacement of a wider ethical debate with an austere proto-Lockean preference utilitarianism, while also implicitly questioning the stability provided by the "victorious" paradigm: "So the growth economy fosters the erosion of the values upon which it depends, such as honesty, sobriety, trust, etc. On the supply side, the 'infinite' power of science-based technology is thought to be capable of overcoming all biophysical limits. But even if this erroneous proposition were true, the very world view of scientism leads to the debunking of any notion of transcendental value and to undercutting the moral basis of the social cohesion presupposed by a market economy. As internal moral restraint is eroded, then external police power is substituted" (1991, 238).

This conflict between implicit economistic means and explicit normative ends lies at the heart of the debate over growth, throughout the field of environmental studies. Both the abundance and scarcity postulates of the economic paradigm or worldview beg the ethical question, then deem it irrelevant. But are not the values themselves determinant of whether rain forest destruction or species extinction is deemed "catastrophic"?[9]

Environmental analysts may be severely undermining their own normative goals by formulating their logical argument on mainly empirical and positivist grounds. Empirical projections of future shortages tell us *nothing* about how we should interpret them. As J. S. Alper notes: "If biological facts are to be used in moral decisions, it must be assumed that biological facts are relevant in making ethical decisions. This premise itself is of course a value judgment. Biological facts have relevance in ethical decisionmaking only if we choose to allow them to have relevance" (1981, 99). Ecocatastrophe becomes catastrophic only when one assumes scarcity and places value in future generations or ecosystem integrity. If abundance is assumed as a truth, then we can continue exploiting and accumulating without ceasing, for, to use Locke's criterion, one could expect that there will always be "enough and as good" for others.[10] But is this really the issue?

Environmental and policy analysts such as Ophuls and Daly may have directed our attention to Locke, yet distracted us from the more fundamental conflict among different ways of structuring our normative understanding of human nature, our relations to the surrounding world, and the means and ends of a "sustainable" society. The empirically quantifiable and verifiable argument for and against scarcity is not a fit substitute for genuine argument about such intangible and "incommensurable" ideas as "the good" and "the just." Environmental arguments ought not to be presented as biophysical inevitability; such arguments will ultimately undermine and delegitimate the larger, more important lessons to be learned from environmentalism, such as the social, political, and ethical benefits that can be achieved by living in greater harmony with nature. It is this conflict between the need for normative complexity and the progressive marginalization of normative debate and conflict in environmental studies that concerns us in this book.

The self-defeating nature of environmental argument has to do with the nature of the argument itself. The political ecologist's argument has the aura of a scientific "law of nature." Due to a combination of our Lockean characteristics as individuals and the limited "fact" of scarcity or abundance, we must prepare ourselves for, and indeed execute a departure from, our attachment to the liberal economy, polity, morality to which we have grown accustomed. We are then commanded to change our institutions and (only then) consider our values: "Accordingly, the individualistic basis of society, the concept of inalienable rights, the purely self-defined pursuit of happiness, liberty as maximum freedom of action, and laissez faire itself all become problematic, requiring major modification or perhaps abandonment if we wish to avert inexorable environmental degradation and eventual extinction as a civilization. Certainly democracy as we know it cannot conceivably survive" (Ophuls 1977, 152).

The same tendency to employ an economic and biophysical fait accompli to dismiss moral argument is evident in the writing of Murray Bookchin, philosophical architect of "social ecology," who utilizes a "Lockean" economistic picture of eventual ecological devastation to suggest the necessary abandonment of present economic, social, and political institutions: "It is crucially important, in social ecology, to recognize that industrial growth . . . stems above all from harshly objective factors churned up by the expansion of the market itself, factors that are largely impervious to moral considerations and efforts at ethical persuasion . . . the most driving imperative of the capitalist market . . . is the need to grow. . . . A society based on 'grow or die' as its all-pervasive imperative must necessarily have a devastating ecological impact" (1981, 367–68).

However, in their argument for scarcity rather than abundance, environmental analysts make presumptions strikingly similar to those they attribute to Locke. They rely on "law-like" evidence that resource scarcity (and not abundance) is real, preferring not to argue on the basis of values or perceived "goods" that, regardless of the amount of a resource, we have certain responsibilities considering use. This undermines their own normative argument by relying on (unfalsifiable, potentially subjective) imperatives of science, and exhibits a profound distrust of political and ethical debate. This distrust stems from an assumption of the same economistic paradigm they attribute to Locke, which sees morals and values as ultimately incommensurable. Morals and values are therefore a threat to social and ecological stability, and thus logically subordinate to an economistic understanding of human nature and society, whether or not the paradigm is based on abundance or scarcity. In suggesting this, environmentalists fail to recognize a more fundamental problem in the way we conceptualize and comprehend values, good, right, and the nature of public morality and immorality in the post-Enlightenment age. They fail to adequately reflect on the consequences of relegating values to the "facts" of self-interest and subjective preference.

But Locke offers more than just an affirmation of natural abundance. In fact, whereas many theorists and political ecologists alike tend to disregard Locke's moral theory (Laslett 1963, 79; McPherson 1962), it is clear that Locke himself hoped to place morality on a level of certainty equivalent to that of the physical sciences (1959 [1690], bk. 4, ch. 3, §18). Locke's moral philosophy is not a mere afterthought. Many scholars argue that Locke's moral philosophy is coequal with if not central to the rest of his political and economic writings (Tully 1980; Riley 1982; Grant 1987; Rapaczynski 1987). Locke himself argues that moral knowledge is a central concern in the *Two Treatises* and the primary focus in the *Essay*. In the introduction to the *Essay* he makes clear that "our business here is not to know all things, but those which concern our conduct" (*Essay*, intro, §6). Similarly, Locke states in the *First Treatise* that his primary concern is not to discuss the existence of political power but rather its moral legitimacy and ethical contours: "The great question which in all ages has disturbed mankind, and brought on them the greatest part of those mischiefs . . . has been, not whether there be power in the world, nor whence it came, but who should have it" (1T, 106).

Locke's moral theory also stands at the heart of his politics and property theory. In the *Two Treatises* Locke argues for the construction of a political society through consent and bounded by the universally and perpetually applicable law of nature (1T, 41; 2T, 26). Paralleling his argument in the *Essay*, Locke states that the preservation of the self and (all) others is that first law of nature: "The state of nature has a law of nature to govern it, which obliges everyone: and reason, which is that law, teaches all mankind . . . that being all equal and independent, no one ought to harm another in his life, health, liberty, or possessions. . . . Every one as he is bound to preserve himself . . . so by like reason . . . ought he, as much as he can, to preserve the rest of mankind. . . . And that all men may be restrained from invading others rights, and from doing hurt to one another, and the law of nature observed, which willeth the peace and preservation of all mankind" (2T, 6–7).

Locke rejects a purely self-centered, market view of the moral universe. The preservation of society and humanity are not mere afterthoughts to one's own preservation (Tully 1980, 40–46).[11] Locke's justification of property and the development of political society can be viewed within this original purposive framework of natural and moral law, not just independently of it, as Ophuls and others have chosen to do (2T, 25; *Essay* II, 21, §67–72).[12]

A critical problem does surface in fulfilling our moral requirements to preserve others and ourselves in connection with Locke's theory of property. For if property rights are exclusive, then it would appear that preservation of others (in accordance with natural law) would be jeopardized as resources are in short supply. If property rights are inclusive, they would place heavy burdens on individuals and the polity, requiring regulation and restriction of property (and perhaps its redistribution) to assure access to the means of subsistence by all (Ophuls 1977; Ophuls and Boyan 1992; McPherson 1962; Rapaczynski 1987).[13]

However, it is not a question of infinite *Lebensraum*, as claimed by ecological critics, though in Locke's own time one could legitimately believe that there were still plenty of open spaces in the world for expansion, cultivation, and natural appropriation (2T, 36–45). The real Lockean keys to abundance are the fundamental human qualities of reason, industry, and the transformational, productive capacity of labor. Together these can stave off the rather burdensome and problematic requirements for state regulation (2T, 95). In this alternative, labor is central to Locke's account of retaining the spirit of the fundamental law of nature: preserving oneself and others.[14] When we labor, we mix our personhood with the objects of our labor; what was once purportedly "waste" now becomes something of value in which we have property right (2T, 27–35). It seems both plausible and within the spirit of Locke for him to assign a moral dimension to labor as the means for fulfilling the fundamental law of nature. Moreover, Locke specifically notes the productivity of human labor and its ability to make "more" from less: "To which let me add, that he who appropriates land to himself by his labor, does not lessen but increase the common stock of mankind" (2T, 37).[15]

The English commons, once thought of as a physical limit, becomes an expanding source of plenty as it is appropriated. Scarcity is not a function of biophysical limits; rather, it is the catalyst of human inventiveness and productivity. Indeed, Locke boldly asserts that "this shews, how much numbers of men are to be preferd to largenesse of dominions" (2T, 42). Humans, as the seat of productive labor, are truly the "ultimate resource": "For 'tis not barely the plough-man's pains, the reaper's and thresher's toil, and the baker's sweat, is to be counted into the bread we eat" (2T, 43).[16] Similarly, human activity on available resources, not merely the extent of natural resources at hand, secures the wealth and security of a people (2T, 41–42). Reason relates the duty of each to fulfill the fundamental law of nature through labor, transforming the commons into a greater expanse of potentially appropriable property for other individuals.[17] Locke also integrates morality and politics by connecting natural abundance to civil liberties: all share in the abundant natural wealth available for appropriation, and do so without bumping into other individuals' claims or values. Universal economic activity—not executive and juridical authority—is the societal conservator.

Our position throughout this book is that, like Locke, we must worry about both moral argument and politics. Neither can avoid the implications of the other. In fact, to do so may delegitimize both. Hence, the contemporary political scientist and moral philosopher must realize that their subjects are interrelated in theory and practice and that moral argument must be included in policy decision making as the foundation for scientific and political "facts." Surely, the history of the twentieth century teaches that politics without morality is dangerous, and morality without politics is empty and without effect.[18]

Therefore, the fundamental problem in making public choices about the environment lies in the way decision makers conceptualize values and combine them with facts in the post-Enlightenment age. Policy is rooted in a theoretical paradigm of assumptions and moral principles that define the normative concepts of "socially better" and "public interest" while they also set the standards by which "good" or "optimal" environmental policy is judged. Decision makers, whether consciously or unconsciously, start from a theoretical paradigm with which they attempt to understand human nature and human conflict, the nature of society, and humanity's place in the natural world. Such paradigms help in diagnosing societal problems and prescribing suitable remedies.

Selection of a paradigm should depend on a discourse on values combined with one's philosophical predispositions as well as historical evaluations of comprehensiveness, accuracy, and effectiveness. But up to now the market paradigm, with the principle of efficiency and the assumption of self-interested maximization of wealth, has held traditional pride of place as a point of departure for policy analysis. Similar to Kuhn's 1962 paradigms, the market paradigm has become all-encompassing and internally consistent. But this internal coherence comes at a price: the market paradigm excludes rival or purportedly historically discredited paradigms and also ends moral argument about the "proper" normative principles and assumptions that ought to be part of the environmental decision-making process. Meanwhile, both traditional cost-benefit analysts applying the market paradigm and political ecologists who criticize such neoclassical methods avoid the fundamental questioning of formative values. They opt instead for the more accessible, less problematic discussion of the principle of efficiency, its visible socioeconomic structures, and the factual constraints of the natural world.

In summary, we contend that the normative assumptions behind neoclassical environmental policy, assumed not only by "Lockeans" but also by many contemporary ecological theorists and practitioners,[19] are as follows:

> Assumption 1. Humans are utility-maximizing individuals seeking their own interests, which are primarily defined by and dependent on economic accumulation.
> Assumption 2. Conflict and social disturbance will result unless self-interested, maximizing humans are provided with sufficient access to economic space.
> Assumption 3. Social disturbances may then be handled through augmenting distribution or, better yet, increasing access to appropriative wealth without disturbing the appropriation patterns of other members of society. This may become increasingly difficult as scarcity increases.

In remaining within an economic or market paradigm, environmental analysts implicitly agree that any paradigm based primarily on noneconomic ethical or moral principles should be discredited. Thus, the argument against growth and continued accumulation, though proffered to advance normative goals, dismisses those same values as irrelevant (even counterproductive) to the logic of the argument itself.[20] From our point of view, this dismissal involves several additional assumptions on top of the previous three:

> Assumption 4. Historical paradigms that have tried to explain human nature and human conflict and ensure social stability have been largely based on objective hierarchical moral/religious value systems and are therefore not pluralistic.
> Assumption 5. Such value systems, if maintained as objective, will be a threat to social stability because of the incommensurability of their claims and the potentially violent rivalry that results.
> Assumption 6. Therefore, in the interest of the preservation of society, we must transform all value claims into economic interest, scientific fact, or market preference claims, which, like other interest claims, are universal, fungible, and attainable through market choice. (Bowersox 1995)

But can we change policy without adopting assumptions that provide a foundation for environmental values rather than Lockean or neo-Malthusian market concerns (Paehlke 1989; Gillroy 1993)? One is reminded here of the honest admission of an earlier proponent of scarcity, John Stuart Mill. In his discussions of the necessary advent of the stationary state required by diminishing returns and marginalized resources, Mill "digresses" to ponder an alternative justification based not simply on the laws of the physical world or the laws of economics, but on values: "If the earth must lose the great portion of its pleasantness which it owes to things that the unlimited increase of wealth and population would extirpate from it, for the mere purpose of enabling it to support a larger, but not a better and happier population, I sincerely hope, for the sake of posterity, that they will be content to be stationary, long before necessity compels them to it" (1904 [1884], V4, 339).

Unlike many present analysts, Mill indicates a keen awareness of the "logical" limitations of an economistic and scientific necessity argument for future scarcity and acknowledges the role values play in motivating any such analysis. The switch in his argument to an open declaration of opposition to economic determinism invites public debate about what is ultimately good for humans, human society, and the world. Unfortunately, Mill is an exception.[21] Too often, contemporary analysts hide behind empirical data and market assumptions, giving only modest reference to values. Ultimately, they may fear the step taken by Mill, as they fear losing in an open forum.

Overall, political ecologists show a typical Enlightenment proclivity to distrust or completely discount the efficacy of moral dialogue or debate about fundamental values. They seek the sheer "inevitability" of scientific facts and those same market forces they so validly critique by altering the supply curve and reshaping demand.[22] These political ecologists uncritically take as their starting point the same "Hobbesian" view of human nature they attribute to Locke: we cannot expect any more of human nature than a bluntly economic view of self-interest. Not arguing for an imposition of morals in a world unaccustomed or overtly hostile to open moral debate, they opt for realpolitik. The Ehrlichs suggest simply that no one is "exempt from the laws of nature" (1996, 231). They then proceed to quote from a statement by fifty-eight scientific academies that "underscores the need for government policies and initiatives" that will help achieve "zero population growth within the lifetime of our children" (233). Fearful of incommensurable debate over individual "preferences," they seek a biophysical fact to supplant moral argument. But the role of moral argument in policy decision making about the environment is the subject of this book, and policymakers may have to analyze and replace dominant paradigms, however daunting that task may be (Meadows et al. 1992, xvi; Ehrenfeld 1993, 121–22; Kassiola 1990, 181).

The apparent implication of Locke's system for environmentalism is that to fulfill political obligations in the most efficient and practical manner, morally substantial debate over individual and societal goals has to be subordinated to an instrumental exploitation of the natural environment. But Locke's moral and political philosophy seeks to reconcile such liberal values with the moral responsibilities perceived as logically entailed by the law of nature. He properly diagnoses the dilemma of reconciling property accumulation with the preservation of humanity and provides a political and moral prescription with which to cure that ill. For Locke, human collective action can be subordinated to ethical principle and moral argument (Bowersox 1995, ch. 2). This is the critical point that environmental analysts have not considered. Ultimately, we may wish not to wade too deeply into the ramifications of public debate and conflict over values and normative principles. But if we wish to critique the market paradigm, and its separation of morals and politics, we may have to create complete alternative systems of principle and move beyond the implicit assumption that "the market" is the only legitimate realm for political disagreement.

The Possibility of New Normative Frameworks within the Enlightenment Tradition

Our contemporary use of Locke allows economistic and scientific assumptions to play the role of moral principle in environmental policy choice. It also exacerbates the struggle between those who wish to keep democratic "liberal" institutions and therefore forsake moral argument, and those who want what is "morally right" for the environment and therefore recommend abandoning the political philosophy of the Enlightenment and its democratic institutions for authoritarian administrative structures. This dichotomy, however, is fundamentally a false one. Even if the current arguments about Locke are correct, and a closer reading does not show him arguing for the coordination and integration of political and moral discourse beyond simple market frameworks, alternative strains of non-Lockean Enlightenment thought may exist that are not market-based and that may integrate moral argument about humanity and nature more fully into the environmental policy debate. Returning to Ophuls's latest book suggests that there are possibilities for Enlightenment argument within policy that transcend economic determinism and the debate over abundance or scarcity.

A Requiem for Modern Politics and Morality?

Ophuls makes two arguments in his book *Requiem for Modern Politics* (1997).[23] First, he contends that the liberal-democratic paradigm that has determined the evolution of our politics, economics, and social life is inherently self-contradictory. This dominant "Lockean" paradigm has produced an antiecological (43), despotic (236), and narcissistic (53) reality that exhibits moral entropy (45) and nihilism (194) while it produces a mass of dead citizens (75) or amoral individuals (27) who are slaves (170) rather than free persons. Second, Ophuls traces this evolution of totalitarianism and despotic democracy, inevitably, to the principles and philosophical assumptions of the Enlightenment: "The Enlightenment Project Has Failed—and failed badly, exposing humanity to an unprecedented planet-wide catastrophe-in-the-making that is the ironic product of its highest ideals" (279).

In the first argument, although Ophuls overreaches in places (making literacy/writing [1997, 179] inherently imperialist and tyrannical), he makes some essential points about the inherent contradictions in market assumptions about freedom, equality, and private property. For example, his observations about the inherent failings of self-interest as a definition of reason are persuasive, whether applied within an economic context, where means and wealth maximization bury moral ends, positive freedom, and the value of the person, or within a political context, where personal preference and desire satisfaction promote interest and ideology over public spirit.

In his second argument, Ophuls defines the Enlightenment as a single set of "Lockean" preference utilitarian principles and market assumptions. In this way, he continues to avoid a debate over normative values and dismisses the greater Enlightenment dialectic that could be utilized to provide different principles and assumptions that might effectuate the paradigm shift toward "ecological maturity" (1997, 272), a "politics of consciousness" (274), a "recovery of morality" (274), and a "restoration of governance" (276) that he maintains we need to navigate into the future. This is the fundamental oversight of Ophuls's argument.

If one studies the Enlightenment and the various actors and arguments that it produced (Schmidt 1996), one is on sounder ground to state that the market-based paradigm that Ophuls describes and condemns is but one argument, albeit a dominant one, within an ongoing dialectic that is our true legacy from the Enlightenment. Using an Enlightenment dialectic as a point of departure, and agreeing with Ophuls that "to know the root ideas or governing paradigms of a culture or civilization is to understand its behavior" (1997, 178), the task is to find, within Enlightenment discourse, the antithesis and synthesis to Ophuls's thesis in order to determine the possibilities of a better, and more sustainable, future.

In his book, *The Lincoln Persuasion,* the late David Greenstone describes the essential dialectic within American liberalism between what he calls "humanist liberalism" and "reform liberalism" (1993, 50–63). Humanist liberals are concerned with instrumental values and individual preferences and approach government with a utilitarian calculus that parallels what Ophuls argues is the sole legacy of the Enlightenment. In addition, however, Greenstone describes reform liberalism as concerned with positive rather than negative freedom and with the moral capacities of individuals and their development rather than with preferences and desire satisfaction. What is most valuable about Greenstone's dialectic, for our task, is its insistence that the Enlightenment Project is concerned with a discourse among alternative paradigms seeking core status. If the Enlightenment is a multiheaded creature with competing principles and assumptions about who people are, how they act collectively, and what the proper role of the state is, then the entire Project cannot be condemned for the failures of but one of its lines of argument.

Let us suppose that the Enlightenment is not just the evolution of a preference utilitarian market paradigm that originated with Hobbes and evolved through the thought of Locke, Smith, and Madison. In the same way that Greenstone paints a picture of competing paradigms, one essentially Jeffersonian and one beginning with the thought of John Adams, we could argue that another strand of Enlightenment argument begins, for example, with Rousseau and evolves through Kant and Hegel. This second strand of Enlightenment thought may promote positive freedom and the intrinsic value of humanity and nature over instrumental values and the exploitation of nature. It might focus on essential capacity as the core of practical reason rather than preference and choice. It may be concerned with quality rather than quantity, with organic growth rather than exponential, with ends rather than means, and with politics as a check on avarice rather than a facilitator of desire satisfaction. Perhaps this alternative Enlightenment paradigm, as the core assumptions of a new political and economic order, could reconfigure republican government so that instead of involving itself in the affairs of individuals, it "administers justice among men who conduct their own affairs" (Ophuls 1997, 262).

Conceivably, within the dialectic that is the Enlightenment, we have alternative assumptions, principles, and priorities that will allow us to move from "pioneer" to "climax" conditions without destroying ourselves in the process (Ophuls 1997, 12–15). To dismiss the entire Enlightenment Project because we condemn one of its products is to ignore principles and historical context that are available for adoption as the core of a new paradigm or paradigms that might take us past the pioneer stage and to the next level of sociopolitical evolution. Fundamentally, it may be possible to debate the environment within the Enlightenment tradition that supports our liberal-democratic tradition, but move beyond the dominance of market assumptions and instrumental values while integrating alternative but codependent and more complex definitions of the political and the moral into policy analysis and evaluation.

The Key Issues: What Is at Stake in Sustainability?

The principle of sustainability is moving the current debate about public environmental policy. If the roots of our liberal-democratic society do not require us to separate moral from political argument, and the possibility of alternative nonmarket paradigms for environmental policy analysis exists, then how do we start the process of making these possibilities *probabilities?* How might we create a more vibrant and complex moral argument around the fundamental issues of a sustainable future?

The first thing we should do is identify and discuss those fundamental issues that will dictate the normative complexity of future policy argument about the environment and the role of morality in sustainable policy. Specifically, we need to begin to discuss the role of science in environmental policy, the connection between social justice and environmental law, and the distinctions and possibilities of moving beyond a strictly instrumental value theory and toward a new evaluation of humanity and nature as having intrinsic value, independent of use. We can then proceed to the exposition of case studies on sustainable policy and, finally, reexamine these issues with the case studies in mind.

At that point we may be better informed about the actual state of normative austerity or wealth in environmental policy and better able to at least identify the problems inherent in moving past conventional market valuations and the separation of politics from ethics in policy discourse.

So, let us begin.

PART I

Moral Principles and Environmental Policy: Basic Issues and Dilemmas

Issue 1: Science as a Substitute for Moral Principle?

Science as a Substitute for Moral Principle

Susan Buck

To ask if science can be a substitute for moral principle in determining environmental policy is to assume that moral principle is the basis for a significant number of environmental policy decisions. This is not the case.

A few decisions *are* made with a substantial moral component, for example, whether wilderness areas should be preserved, and to what extent the racial balance in nearby communities should influence landfill siting. Certainly, moral principle has a strong, at times even dominant, effect on positions taken by groups that are influential political players in the policy system, such as Greenpeace and the Sierra Club. Moral principle motivates some elected officials in their responses to particular policy problems, although few elected officials dare to follow moral principle when it runs counter to political, economic, or social preferences of the voters. When moral principle *is* explicitly considered, it is most likely to be found in the early stages of the policy process (agenda setting, problem definition, formulation, and legitimation) and less likely in the final stages (implementation and evaluation). During the implementation stage, government agencies do not indulge in broad moral principle-based decision making because that is the job of elected officials. The focus of this essay is the proper role of scientific knowledge and practical experience in the implementation stage of the environmental policy process.

In this essay, I argue first that public administrators are charged with implementing legislative policies because they have the requisite technical and scientific expertise to do so, and because the administrative process is faster and more flexible than the legislative process. Legislative guidelines often give the bureaucrats substantial discretion in implementation; bureaucrats are *supposed* to use their scientific expertise in shaping environmental policy. Second, I argue that explicit policy-relevant consideration of moral principles does not and should not shape bureaucratic decisions on a routine basis. The essay is written from the perspective of public administrators in the trenches of environmental policy; its concern is not with the inputs into policy decisions but rather with administrative outcomes: What are the policy goals set by the legislative process? How well have they been met? Below are some examples of implementation directions from Congress to the agencies (emphases added):

> It is the intent of Congress that the Administrator [of the Environmental Protection Agency] shall carry out [the Toxic Substances Control Act] in a *reasonable and prudent manner,* and that the Administrator shall consider the *environmental, economic, and social impact* of any action the Administrator takes or proposes to take under this chapter. (15 U.S.C. 2601[c] [Toxic Substances Control Act, 1976])

> The purposes of [the Endangered Species Act] are *to provide a means* whereby the ecosystems upon which endangered species and threatened species depend may be conserved, *to provide a program* for the conservation of such endangered species and threatened species, and *to take such steps as may be appropriate* to achieve the purposes of the treaties and conventions. (16 U.S.C. 1531[b] [Endangered Species Act of 1973])

> The Congress finds that . . . to serve the national interest, the renewable resource program must be based on a *comprehensive assessment of present and anticipated uses, demand for, and supply of renewable resources* from the Nation's public and private forests and rangelands, through *analysis of environmental and economic impacts, coordination of multiple use and sustained yield opportunities* . . . and *public participation* in the development of the program. (16 U.S.C. 1600[3] [Forest and Rangeland Renewable Resources Planning Act of 1974])

Through the legislative process Congress has decided the larger moral questions: Should government regulate toxic substances? Should we strive to preserve endangered species? Are forest resources primarily a source of commercial timber? The public administrator is now charged with implementing these decisions, translating them into governmental actions that affect other branches of government or the citizens. Bureaucracy plays an important role throughout the policy process,[1] but it is in implementation that bureaucracy shows its greatest influence. Although implementation may be hampered by poor policy design or a lack of commitment by policymakers, once responsibility for the policy passes to the hands of the administrators, other factors come into play; the most important of these for our purposes are administrative expertise and administrative discretion.[2]

Administrative expertise may be defined as policy-relevant scientific information, technical know-how, and administrative experience. Scientific and technical expertise is helpful during implementation, which is difficult because it involves a number of interdependent actions that must be accomplished almost simultaneously (Mazmanian and Sabatier 1983, 22). Implementation actions include planning programs, promulgating rules and regulations, acquiring resources (e.g., money, land, personnel, equipment), monitoring, and enforcement. Administrative expertise is necessary to accomplish all of these tasks; the scientific and technical aspects of administrative expertise are especially important in crafting rules and regulations, monitoring, and enforcement. For example, bureaucrats use their technical backgrounds to judge threshold levels of lead toxicity (*Connecticut Coastal Fishermen's Association v. Remington Arms Co.*, 989 F. 2d 1305 [1993]), or to decide if ornamental cedar trees must be destroyed to protect the local apple crop (*Miller v. Schoene*, 276 U.S. 272 [1928]).

Because bureaucrats have administrative expertise, Congress delegates broad quasi-legislative authority to administrative agencies. Although Congress may delegate very specific guidelines to the agencies and may even attach "hammer clauses" to encourage regulatory compliance (e.g., Resource Conservation and Recovery Act Amendments of 1984), usually Congress gives considerable discretion to environmental agencies to promulgate rules and regulations that have the force of law. Except for the handful of agencies that are required by statute to use formal rule-making procedures, most agencies issue rules under a relatively simple procedure (Administrative Procedure Act [APA], § 553).

Administrative discretion is essential for effective administration. Legislators cannot possibly draft legislation in the detail that is necessary for implementation. First, they lack the technical expertise that is such a powerful resource for the bureaucracy. Second, the time pressure on members of Congress is immense; even if they had the personal expertise or professional staff to hammer out the details, they cannot spare the time: they must delegate to the agencies. Third—and this is especially true in environmental administration—the rapid pace of scientific discovery and technological change would make legislation obsolete almost as soon as it was passed. Discretion compensates for changing technology. As scientific data accumulate, or drought endangers national forests, or new species are discovered, discretionary powers of administrators can accommodate the changes. The legislative process is cumbersome;

it was designed that way to protect the United States from crisis management. The administrative process adjusts to new conditions more rapidly because rule making is more flexible than legislating: it rarely requires formal negotiation, administrators are not subject to the same lobbying pressures as elected officials, and there are fewer "veto points" (points at which the legislative or administrative process is permanently derailed). Discretion also gives administrators leeway to fit policy decisions to individual cases, to "humanize" the governmental process. Of course, the devil is in the details, and Congress is often willing to lay the blame for unpopular, expensive, but necessary decisions on executive agencies.

Knowledge of scientific information and practical experience not only shape agency rules and regulations, but they also succor the agencies when the agencies are drawn into federal courts to defend their actions. In fact, the courts expect agencies to make decisions on the basis of administrative expertise, and traditionally the courts defer to the agencies when their decisions are based on their scientific knowledge and administrative experience. Justice Stevens wrote for the Court in *Chevron U.S.A. v. Natural Resources Defense Council, et al.* (467 U.S. 837, 865): "In these cases the [EPA] Administrator's interpretation represents a reasonable accommodation of manifestly competing interests and *is entitled to deference:* the regulatory scheme is technical and complex, the agency considered the matter in detailed and reasoned fashion, and the decision involves reconciling conflicting policies" (emphasis added, notes omitted).

There is little room for large debates over environmental ethics and philosophy in the discretionary activities of environmental professionals. For example, we can argue both sides of whether social justice issues ought to be part of environmental policy, but what practitioners need to know is how incorporating social justice issues will affect their existing programs' goals as well as whatever new policy goals may emerge. Practitioners are also concerned about their constituencies, their budget and staffing level, their interactions with Congress, and their long-term career prospects if the administration changes. These are not questions about moral principles.

This is not to say that public administrators are operating in a moral vacuum or that they are ethically bankrupt. They indeed have rigorous and even noble ethical principles that inform the process by which their work is done. The most persuasive approach to analyzing bureaucratic ethics relies on the Constitution, which symbolizes core regime values of the American polity such as equality, liberty, and property (Rohr 1989). Although these regime values are certainly not absolute, nor indeed are they the only possible regime values we might explore, they are salient values for the American political system. Public administrators take an oath to uphold the Constitution, not *Walden*, however much they may approve of the sentiments in the latter. They are free in their private lives to pursue environmental goals that differ from their professional obligations, but if they use their position as public servant to undermine government policies or effectiveness, they are being unethical.

Under the Constitution, the president is charged to "take Care that the Laws be faithfully executed" (U.S. Constitution, Art. 2, Sec. 3). To be perfectly constitutionally accurate, that charge is to the president and not to his subordinates. However, as far as we know, the Founding Fathers did not envision the administrative state. Because courts routinely invalidate administrative actions that exceed the agencies' statutory guidelines, I think it is fair to place the obligation, if not the constitutional imprimatur, on agency behavior as well as presidential actions. In the *Chevron* case cited above, Justice Stevens wrote: "An agency to which Congress has delegated policymaking responsibilities may, within the limits of that delegation, properly rely upon the incumbent administration's views of wise policy to inform its judgments. While agencies are not directly accountable to the people, the Chief Executive is, and it is entirely appropriate for this political branch of the Government to make such policy choices—resolving the competing interests which Congress itself either inadvertently did not resolve, or intentionally left to be resolved by the agency charged with the administration of the statute in light of everyday realities" (457 U.S. 837, 865–66).

An executive agency that substituted its own broad policy decisions for statutory ones would be *ultra vires*, that is, "in excess of statutory jurisdiction, authority, or limitations" (APA §706[2][C]). See, for example, *Calvert Cliffs' Coordinating Committee v. United States Atomic Energy Commission*, 449 F. 2d 1109 (D.C.Cir. 1971) (Atomic Energy Commission must fully integrate requirements of NEPA into its construction and operating permit procedures), or *American Textile Manufacturers Institute, Inc. v. Donovan*, 452 U.S. 490 (1981) (OSHA is within its statutory authority to promulgate a cotton dust standard that will impose significant economic losses on the industry, but OSHA has no authority to require wage protections for workers unable to comply with interim safety measures).

Philosophical debates may inform legislators (although I doubt this happens with any significant regularity), and they certainly drive much of the lobbying and other political pressures on the legislative, executive, and judicial branches of government. *However, public administrators do not decide issues primarily on their personal policy preferences, however morally principled they may feel them to be, nor should they if the administrators are doing their jobs properly.*

In democratic systems, it is difficult to distinguish between the public will and the public good. In an increasingly technological world where environmental problems are scientifically complex, the general public is not competent to assess risks and to propose solutions (W. Rosenbaum, 1998, 128–30). Usually, the "public" voices we hear are those of policy specialists who have the capacity to make their views known and to apply political and economic pressure to elected officials. Elected officials rarely have the scientific training to make technical assessments, but in our system of government, they have the primary responsibility for framing broad environmental policy. They may still make decisions on environmental issues without or against scientific information, but they have surrounded themselves with specialized personal and committee staff who can provide sound advice, and they have access to administrative expertise that is found in the executive agencies charged with implementing environmental decisions. Moral principle has a valuable role in many stages of the policy process, but in the implementation stage, scientific and technical information coupled with bureaucratic experience are the preferable, and ethical, basis for policy decisions.

Science Is No Substitute for Moral Principle

Robert Paehlke

Science is essential to environmental policy decision making, but it is not in itself a sufficient basis for consistently appropriate (and correct) decisions. The world, for better or worse, is much more complicated and interesting than that. There are many reasons for this: (1) many environmental policy decisions involve extrascientific, explicitly moral dimensions; (2) there are many significant gaps in our scientific knowledge and decisions cannot always wait on filling those gaps; and (3) science (especially perhaps that science that ventures into the always morally active realm of human activity) is itself vulnerable to powerful political forces that are best understood, in part, on a nonscientific basis. For these reasons and others, environmental policy, and the environmental movement itself, are best understood through—and arise out of—both scientific and moral and/or political analysis.

The environmental movement itself always has been ambivalent about science—needing it, arising out of it, but critical of some of the technologies that also arise out of science and their environmental consequences. Environmentalism is, in a sense, the sociopolitical arm of the science of the environmental impacts of technological "advance" (much as conventional economics, as an ideology, is bound up with and mutually supportive of technological advance and economic expansion). Where everyday science is ordinarily specialized, environmental science has more by way of integrating dimensions and effects: its principle focus is on the unintended consequences of human activities, a research objective that demands comparisons and judgments regarding net benefits.

Judgments regarding net benefits in turn require a sorting out of the relative merits of the many and various things that we humans value (positively or negatively). The environmental sciences—ecology, toxicology, epidemiology, and (perhaps) environmental economics—provide information that sometimes forces societal (essentially moral) choices between health and wealth, between the well-being of humans and the well-being of other species, between resources and "nature" (or resources now versus resources later). Neither the science nor the moral analysis are alone a sufficient basis for effective environmental policy decisions. This is best seen in any number of concrete examples, grouped here under the three categories identified in the first paragraph above.

Some Issues Involving an Extrascientific (Moral) Dimension

Here I am not concerned with the technically difficult questions that science has not yet answered, but with those (mostly value-laden) questions that it cannot in principle answer. My sense of this territory has always been guided by the wisdom of Arnold Brecht (1959). Some environmental policy questions simply cannot escape this extrascientific territory, and when we pretend that they can (a frequent occurrence), we do a profound disservice to both science and democracy. The disservice to democracy exists because those with power prefer, of course, that issues be publicly perceived to be technical questions whose answers are beyond political contestation (and lie within the territory where the views of small numbers of people are determining of outcomes). It is bad science because more is claimed than in fact follows from scientific findings.

Consider the case of nuclear power. Both those in the business and those opposed to the industry have always claimed to have science on their side, and there are yet many scientific questions that have not been entirely resolved. But the primary differences are with regard to the essentially nonscientific question of what risks are worth taking—a question with profound moral dimensions. All can agree that the risks of core meltdowns of Chernobyl-like proportions (or worse) are of low probability (though not, of course, how low). Science can say something about the probabilities of such an event (though the number assigned is as much scientifically informed judgment as scientific "finding"). Science can say very little about whether or not that risk is a risk worth taking. That conclusion has more to do with how highly one values the economic and social activities that are (or might be) met through the use of nuclear energy, how one imagines the long-term human future, and what other (most often, less catastrophic but virtually certain) risks one is prepared to endure via alternative means of generation to avoid the (perhaps) remote risk of an extreme nuclear catastrophe. Science can inform such a decision, but cannot make it.

Most environmental policy decisions are like that, though perhaps less starkly and dramatically so. Science can tell us that if we build dam A or cut forest B, there is a risk that one or several species will face possible extinction in the wild. But is the extinction of one species not in some circumstances tolerable? Moreover, is that extinction more or less important than a large loss of a viable ecosystem containing many (nonendangered) species (as might result if we do not cut the forest, but someone else extracts wood or other building materials somewhere else instead)?

Moreover, the choice as to which scientific knowledge is deemed relevant often depends on nonscientific (value) decisions. If a forest is seen as having value only, or primarily, as a source of wood or pulp, one will gather the scientific evidence that will lead to a conclusion that some forests are "overmature" and that more "resource" can be produced through "scientific" forest farming with genetically selected (or altered) tree species continuously replanted following repeated clear-cutting. If a forest is valued primarily in terms of its biological diversity, scientific data are readily available to demonstrate that forest farming (with single-age, single-species monocultures) is ecologically far and away the wrong choice; the right choice is to gently remove selected individual trees or to simply leave the forest alone. Either choice is, again, a scientifically informed, value-based decision.

Science can also tell us much about the human response to toxic substances in the environment and the pathways by which toxins in the environment might reach (or have reached) our vital organs, but it cannot answer the question of where one ought to place the burden of proof: on environmental scientists or on (possible) polluters. That is, where is the presumption of guilt with regard to chemicals? Must safety or hazard be proven, and to what standard of proof? Nor can science tell us whether a three-exposures-in-a-million probability of childhood leukemia is worth $100 million in GDP (or the whole of the human future, if it is your child). All such matters require combining moral analysis, scientific analysis, and political analysis (e.g., what is the likelihood of any of the money generated through a decision not to regulate ending up in medical research or treatment expenditures?).

Gaps in Scientific Knowledge

As is well-known, environmental policy decisions must frequently be made on the basis of incomplete, ambiguous, or contradictory scientific evidence. Here we are in the realm of what we can in principle determine scientifically but as yet have not resolved (and may never fully resolve). Frequently, then, waiting for scientific consensus may involve waiting until it is too late to reverse or avoid a negative environmental outcome. The decision to adopt the prudence principle and the way that principle is interpreted and applied are not scientific matters; they are primarily decisions made within the realm of ethical analysis and judgment. So too, in part, are decisions regarding the allocation and management of funds for further scientific research.

What are some of the important things we do not know? The list is very long. We do not know all of the interactions that go on within complex ecosystems: What happens when top predator populations are diminished? How dependent are particular forests on micro-organisms for soil health, or small mammals or particular bird or bat species for regeneration? What will be the future rates and patterns of climate warming under varying scenarios of greenhouse gas emissions? How significant to human health are releases into the environment of endocrine disrupters? What are the human dose (or emission) response curves for any number of known toxic chemicals?

Science comes up short with regard to many dimensions of environmental policy. We are, of course, always gaining scientific ground, but are we gaining that ground as fast as new technologies are creating new problems? It is hard to say. To act on indications, suspicions, and partial results is not easy—especially if so acting is going to inconvenience large numbers of voters or cost someone a lot of money. "Sound science" in this context, as with Ronald Reagan and acid rain, can simply be an excuse for perpetual inaction. Needless to say, it is also true that acting solely on what could prove to be groundless fears is also a risky business, especially when the action involves significant forgone or diminished economic opportunities. In the end, though, choosing which way to lean—to the economy or to the environment—is frequently made necessary by the character of the data at hand. Such essentially moral choices face all environmental decision makers, from consumers to presidents. Science informs decisions and, guided by prudence principles and the like, can help to exclude some options. Often, however, it cannot make the difficult decisions for us.

Overwhelming Science with Power (and Vice Versa)

Science itself also operates within a moral-political context and is not immune to being influenced by power. Science has its own defenses (peer review, skepticism, the shared value of truth seeking, replicability, etc.), but it is hardly immune to corruption and to the influence of a lack of balance with regard to the availability of research funding. Science, like all human enterprises, is also prone to blind spots, intellectual arrogance, and especially to misinterpretation and misuse. A viewpoint rooted in an environmental ethic and in political sophistication uncommon (in my experience) among natural scientists can, I contend, help one to know when it is important that things might have "gone wrong" scientifically. There are numerous examples of this in the history of environmental policymaking. I discuss three here.

First, some of the early epidemiological science of asbestos statistically evaluated the health of then-current asbestos workers without seeking out for inclusion those persons who had previously worked in the industry but were no longer working. This systematically skewed the result in favor of the healthy. Other studies did not take fully into account the common twenty- (or more) year lag between asbestos exposure and illness onset. Even after Irving Selikoff's definitive 1963 study of 632 World War II shipyard workers, industry-funded studies purported to show that only one type of asbestos was problematic, others that the problem was the plastic bags that the asbestos came in, and yet others that once there was some venting of the workplace the problem would be radically reduced (though that reduction could not be confirmed for twenty years). All these assertions were partially or wholly false, but all were science used in aid of delayed policy action. In the 1970s, asbestos was still being used extensively in new school and office construction, only to be removed at great expense in the 1980s and 1990s. (This case is discussed in Paehlke 1979, 1996).

Second, despite several years of clear warnings from Canadian government scientists regarding overfishing of cod off Newfoundland, governments did not act to reduce allowable catches. The politics of employment reduction in one of North America's highest unemployment jurisdictions was politically difficult in the extreme, and there was always one scientific study that cast doubt on the growing consensus. Some scientists thought that water temperatures might be changing, others that some cod had caught a new disease, some that the perceived problem was cyclical rather than human-induced. There was always money for research that could find that difficult decisions were unnecessary. Fishing technology (sonar, larger boats, seagoing canneries, etc.) continued to improve and, seemingly suddenly in the 1990s, a bounteous resource that had enriched all who came for five hundred years was gone. Despite a moratorium, it has not returned. Politics overwhelmed science, and the (generally correct) government scientists (and inland fishers who also knew something was wrong) were limited in their willingness and capacity to be openly and effectively political (see Paehlke 1996).

Third, "conventional" science is limited both in terms of geographic reach (poorer nations and communities do not always fully know what it is they are importing, or will do anything for money and do not wish to know) and in terms of chronological perspective. As well, science, in seeking precision and relative certainty, has some tendency to miss or undervalue cumulative and long-term impacts. Specifically, for example, engineered landfills are pronounced safe because they are unlikely to leak leachate for fifty or one hundred years, or because the clay soils of their site will greatly slow groundwater movement. In either case, problems are being created for a century from now, when the source of the problem may be all but forgotten. Finally, one-project-at-a-time scientific evaluations can miss the most significant overall effects of activities.

All these scenarios suggest that science can be politically vulnerable in some contexts. Other forms of knowledge can help to check that vulnerability. On the other hand, in an era of "global competitiveness" and the rising power of large corporations relative to both governments and civil society, science can also be a strong part of the counterarguments to the advocates of short-term economic gains at all costs. Science, as all human activities and perspectives, may be error-prone and vulnerable to economic power, but it is also politically credible and potent: it is one of the few remaining potential rebuttals to the undiluted power of global corporations. Less-narrow science may well be better science. Science might well be more effective if more scientists were willing to acknowledge the political vulnerability of their collective enterprise. One way to improve science in this role is to work harder at integrating scientific findings into larger policy-relevant overviews—even if that involves "mixing" science with moral and nonscientific forms of policy analysis.

Conclusion

To put this last point somewhat differently, one partial solution for some of the problems identified in this essay is more frequent systematic and thoroughgoing attempts to integrate the multiple particular findings of the environmental sciences into broad environmental-economic policy strategies. Many scientists might be uncomfortable with such undertakings, as some of the conclusions would of necessity be outside their own realm of expertise and have clearer political and policy implications than would ordinarily be the case. Such integrative intellectual undertakings could help to develop a dialogue between scientific and policy inquiries at a more conceptual level, and actively seek to engage with environmental value analysis and assertion. Such engagement can only help scientists in setting priorities for their own, more particular research.

Such engagement between scientists and other scholars (and perhaps secondarily with activists, public servants, and/or private sector actors) might also help in sorting between the first and second sections above, and in recognizing the several variations on political and related pressures on effective science noted in the third. Four types of more comprehensive and integrating analyses that I see as promising in this regard are comparative risk assessment (Gillroy 1995), life cycle analysis (Paehlke 1999), footprint analysis (Wackernagel and Rees 1996), and, particularly, environmental space/sustainability analysis (Carley and Spapens 1998).

I do not discuss these possibilities here, except to say that all are broadly comprehensive, involve the integration of scientific work from a wide variety of disciplines and perspectives, and lead toward conclusions that have social scientific, policy, and value relevance. The hope is that moral knowledge and scientific knowledge will be genuinely engaged, policy acknowledged to be more than a technical exercise, science valued but not conferred with such special intellectual status that it is rendered more prone to blind spots and arrogance, and value analysis elevated from the realm of mere opinion and mindless relativism. In the end, all must contribute to a democratic environmental policy process.

Issue 2: Environmental Justice without Social Justice?

Why Environmental Thought and Action Must Include Considerations of Social Justice

Joel J. Kassiola

Political theory is affected by environmentalism, and environmentalism cannot do without political theory . . . they should be mutually enriching.—Andrew Dobson (1998, 11)

The crux of the problem is that the mainstream environmental movement has not sufficiently addressed the fact that social inequality and imbalances of social power are at the heart of environmental degradation, resource depletion, pollution, and even overpopulation. The environmental crisis can simply not be solved effectively without social justice.—Robert D. Bullard (1993, 23)

Earlier, I have written: "It is crucial to have contemporary environmentalism fully informed by ethical consciousness about the central issues of environmental ethics if this movement for social change is to be both effective and morally worthy of our support in the upcoming twenty-first century" (Gillroy et al. 1997, 3). I aim to clarify and elaborate on this general position by focusing on the specific moral and political value of social justice, as well as to examine the grounds for the claim of the essential connection among environmental thought, action, and social justice. I hope to convince the reader that a morally sound environmental political theory, an effective environmental social change movement, and the implementation of a desirable environmental public policy can be achieved only with a thorough consideration of social justice. In addition, like Dobson, I hope to show the mutual relevance of and benefits to environmentalism and political theory, and to stimulate more political theorists to recognize the important contribution they can make to understanding and responding to the environmental crisis.

Critics of this position fear the inappropriate applications of moral and political values in the making of public policy and the implementation of policy regarding the environment. One such critic in this volume is Susan Buck, who writes: "I argue that explicit policy-relevant consideration of moral principles does not and *should* not shape bureaucratic decisions on a routine basis. . . . There is little room for large debates over environmental ethics and philosophy in the discretionary activities of environmental professionals" (my emphasis). No less an authority in political theory than John Rawls is cited by my colleague in this project, Joe Bowersox, as holding the "politics" side of this "politics versus ethics debate" when Rawls asserts: "As a practical political matter no general moral conception can provide a publicly recognized basis for a conception of justice in a modern democratic state" (see Rawls 1985).

This restricted and devalued view of ethics, morality, and social justice is driven by several presuppositions that are rarely made explicit: (1) that alleged descriptions of what is true are prescriptive (see Buck's insertion of "should" in the above quotation); (2) that moral truth does not exist or is not knowable (a version of moral skepticism); (3) that ethical disagreement is not susceptible to rational adjudication; and, therefore, (4) that substantive moral views should be replaced with procedural rules (Rawls's conclusion in his influential theory of justice; see Dobson's classificatory scheme for conceptions of social justice [1998, ch. 3]). To argue against these profound and dominant contemporary propositions is beyond the scope of this essay; however, they need to be mentioned in order to understand the opposing view and to provide the moral and political context for this discussion. Furthermore, I recognize that the required steps from moral and political theory to environmental law, and from public policy to administrative decision making and implementation are large ones. Nonetheless I am persuaded that matters pertaining to social justice (and moral and political values in general) must not be put aside, *pace* Buck, Rawls, and the prevailing opinion today among environmentalists and the public as a whole, when it comes to thinking about the environmental crisis and implementing remedial policies.

To be sure, in our value-skeptical—some would say nihilist—modern culture, any explicit normative position is likely to be criticized merely for being deontological, or, as I understand the meaning of this term, relying on basic but contentious conceptions of and principles about values and obligations such as social justice. This essay is addressed to such value-skeptical critics who wish to exclude moral considerations from environmental policymaking and implementation; those analysts who maintain that environmental theory, policy, implementation, law, and enforcement practices, can, *and* should, all be created and executed while intentionally excluding social justice. I say "intentionally" here because I do not believe that social justice can actually be excluded in reality from the wide-ranging scope and important nature of the environmental problems confronting humanity at the beginning of the twenty-first century, as they involve issues of life and death, and the quality of life, of billions of people. So many individual lives, so many species, and so many ecological consequences are in the balance as a result of existing global and national injustice that we must not, I argue, deny the importance of social justice to environmentalism.

The world, and the United States as an illustration, is characterized by an extreme extent of social inequality, racism, sexism, and other forms of discrimination that produce egregious global injustice. In such a perverse world, the social injustice experienced by groups and individuals based on class, race, sex, and other marks for societal discrimination such as religion, ethnicity, national origin, and disability, will be expressed in all public policy and implementation decisions, including, of course, environmental ones. This is a practically significant issue and not some abstruse theoretical exercise, as the antideontology advocates would have it. An environmental justice proponent captures several very concrete manifestations of such unjust environmental policy when discussing the nature of environmental racism in the United States: "Environmental racism is racial discrimination in environmental policy-making. It is racial discrimination in the enforcement of regulations and laws. It is racial discrimination in the deliberate targeting of communities of color for toxic waste disposal and the siting of polluting industries. It is racial discrimination in the official sanctioning of the life-threatening presence of poisons and pollutants in communities of color. And, it is racial discrimination in the history of excluding people of color from the mainstream environmental groups, decisionmaking boards, commissions, and regulatory bodies" (Chavis 1993, 3). When these forms of environmental racism are also applied to the poor and other systematically disadvantaged social groups as a result of widespread social injustice, one can perceive the close link between social injustice and the environment, environmental policymaking, and efforts to improve the environment within environmental groups committed to social change. Clearly, space limitations here do not permit a presentation of supporting data for the claim about the pervasiveness or extreme degree of global social injustice today; although there is room for debate over specific, purported instances of such mistreatment, I shall have to assume agreement in general about this proposition. Nonetheless, the moral hollowness of a global order wherein some nations have annual incomes per person of $300, the constant threat of famine, and an average lifespan of fifty years, while other nations have an annual per person income over a hundred times this and an average lifespan of eighty years, should be obvious.[1]

If such an assumption is made about the existence of worldwide injustice adversely affecting billions of victims of such mistreatment, I pose the following question: *How can morally acceptable environmental policy, in its efforts to address environmental problems, be made and carried out without considering the fundamental social fact of our world regarding social injustice or its absence?*[2] To think about environmental policy without considering the point of the moral concept of social justice, knowing what we do about the extent of racism, classism, sexism, and other forms of discrimination in the world, seems to deny social reality, reveal an incredible naïveté and insensitivity, or to constitute yet another expression of ubiquitous discrimination itself: environmental injustice. This occurs even when the intended environmental policy goal is laudable, for example, preserving wilderness, saving endangered species, cleaning up the air and water, and protecting the ozone layer.

Just as all policymaking has an ethical (and normative political) component (even if low-level administrators charged with the enforcement of policies are unmindful of this component, as Buck seems to think), all policymaking that addresses the environment within the context of pervasive worldwide social injustice must include, I contend, the resulting horrible consequences from such injustice if it is to be just itself. Part of what it means to live in a socially unjust community comprises not only inadequacies in diet, medical care, education, transportation, police protection, job opportunities, and housing, but also environmentally based expressions of the fundamental unjust social structure. Therefore, my first reason for linking environmentalism and social justice involves the moral and political acceptability of public policy; both ordinary citizens and political leaders are morally and politically obliged to eradicate all forms of social injustice, including environmental ones, within their community, notwithstanding the fact that a moral consensus does not exist. The reply to Rawlsian concerns about the lack of a moral consensus leading to chaos and war is participatory democracy or its current manifestation within the environmental movement: "stakeholder meetings" (see Bowersox below).

My second reason for arguing for the essential inclusion of social justice within environmental public policy is not a moral or normative political reason but an environmental one. It involves the inescapable sharing of environmental fate for all humans—rich or poor, advantaged or oppressed—as well as for all living creatures as a result of being inextricably environmentally connected. Indeed, one may consider the definitive component of the ecological outlook to be this theme of universal ecological connectedness; there are no environmental "gated communities" where the consequences of the environmental crisis can be warded off permanently. Global injustice endangers *all* living species on Earth because of the dual extremes of the urgent basic needs of the billions of poor people on the planet *and* the excesses in profligate material consumption and resulting wastes produced by the billion or so affluent citizens worldwide. If poor peasants are driven by survival needs to "slash-and-burn" food production methods, thereby worsening the deforestation problem with implications for global warming and other concomitant climatic changes, *all* inhabitants on Earth are adversely affected, including the rich (although they, by definition, will have more resources to utilize in reducing these effects than the poor). If a billion Chinese purchase CFC-loaded refrigerators or run their cars on lead-based gasoline because it is cheaper, the consequences for the stratospheric ozone layer and global warming will be ominous for all living creatures on Earth.

Human poverty is bad for the environment because people in a crisis state become desperate and cannot care for the environment given the brute immediacy of their needs. On the other end of the economic and justice spectrum are the rich (although they are often overlooked, distorting our understanding of the relation between social justice and environmentalism); global injustice permits them the means by which to inappropriately overconsume and pollute nature because of their wealth. *My point here is that addressing and improving social justice will not only be morally beneficial by ameliorating current social injustices but will be environmentally beneficial as well by reducing the causes of environmental abuse produced by both the rich and the poor.*

If the consequences of social injustice for the environment unite all people in a common desire to improve our natural surroundings or, alternatively, prevent environmental disaster, then concern for the environment can indeed help bring about social justice. If moral arguments do not move the rich and other beneficiaries of societal inequality to work to eliminate grinding poverty and debilitating discrimination—and examination of the world today indicates that they seem not to be moved—then perhaps recognition of the undesirable environmental effects for *them* of such poverty and injustice might awaken them to remedial action. The worldwide attention and efforts to save the Brazilian rain forest as well as understanding the relation between the impoverished conditions of that nation's landless peasants and the global environmental significance of their desperate actions to eke out a living through deforestation (along with the relatively rich ranchers who destroy forests for profit) could be used as an illustration of this point. Do the world's elite seek to intervene on behalf of the poor of a country with little global environmental impact, such as Guinea, Haiti, and others? Increasing the industrial elite's understanding of planetary ecological relatedness and the importance of environmental sustainability could improve the degree of social justice in ways that moral arguments do not. This result is as practically important as any administrative outcome of interest to Buck.

If environmental racism and classism are eradicated and communities of color, the poor, and the politically powerless are not used as dumping grounds for toxic wastes, the environmental dangers and unsustainability of, say, nuclear power, pesticide use, or the entire endless and unsatisfying[3] consumption lifestyle and values of the global rich may become more vivid to the practitioners themselves of this perverse and dangerous worldview. As a result, the industrial and capitalist culture's *environmental* undesirability as well as its moral and political injustice may become socially recognized, and thereby increase the chances for the transformation of this flawed and environmentally dangerous social order. The sectors of the nations and planet currently being unjustly treated socioeconomically and environmentally will no longer absorb wastes and have their natural wealth pillaged, becoming "sacrifice zones" so that one-quarter of the world's human population can live the industrial capitalist ideal of limitless consumption, a manner of living that is clearly not sustainable environmentally across the entire human population. Thus, both the undesirability and impossibility of this worldview and lifestyle can be better impressed on the group of people who must abandon it: the global elite and those aspiring to such status. That is why, like Dobson (1998; epigraph above), I too claim that social justice and environmentalism are essentially linked.

This claimed relation between environmental thought and action and social justice can work in both directions (or are "mutually enriching," according to Dobson): not only from the social justice side, producing greater environmental justice, as noted above, but from environmental concerns producing greater social justice. Environmental justice should, in addition, help humans and nonhumans. In short, sensitivity to environmental justice can help us overcome the myopic anthropocentrism of the industrial worldview, which I take to be the central implication of the disputed but important concept of "environmental sustainability." We tend to value only current human species and overlook the moral and environmental relations and obligations we have with other living species and future generations. The environmental and moral principle of biological egalitarianism or interspecies equality leads to the conclusion that to cause unnecessary pain to a member of a different species for "sport" (as opposed to survival) is unjust. The recognition and acceptance of this principle could also make humans more aware of intraspecies equality such that all humans should be treated with respect for their rights, promoting efforts toward achieving a world where this ideal is approached, even though we are far from it today.

Morally sound environmental public policy should lead to the reduction of the many manifestations of social injustice, both globally and nationally. Such policies should also highlight these forms of mistreatment and in that way educate the public as a step in making policy to ameliorate the mistreatment. Moreover, morally desirable environmental public policy will also be environmentally desirable. In this instance, moral and environmental requirements will converge or mutually reinforce each other, helping to eliminate environmental expressions of poverty, political weakness, poor education, and nutritional deprivation, among many forms of social injustice. *Social injustice is not only morally bad but environmentally bad as well.* This, in part, explains how the moral and political power of the grassroots environmental justice movement in the United States improved the mainstream, reform-oriented environmental organizations that were focused on mainly white, middle-class, recreational values and issues. This was accomplished by bringing previously underrepresented groups such as women, people of color, and poor people into the environmental movement for social change, raising important but previously ignored issues such as hazardous waste siting and lead and pesticide poisoning (see Bullard 1993; Cable and Cable 1995).

My aim in this essay is to show that a good, clean environment in an unjust society is impossible and, furthermore, to demonstrate the essential link between environmental thought and action and considerations of social justice. Sound environmental policymaking and implementation must include considerations of social justice for both moral and environmental reasons.

Environmental Justice: Private Preference or Public Necessity?

Joe Bowersox

In an earlier piece on the relationship between environmental policy and environmental ethics, Susan Buck noted two specific problems of linking environmental policies with particular images of social equity. First, Buck claims that policymakers and administrators act "unethically" if they base their professional, public actions on "personal policy preferences"—their subjective ethical evaluations. Second, she argues that linking a "new or controversial policy issue" such as social equity to the "established issue" of environmental protection ultimately undermines fragile public consensus on environmental problems. Such a consensus apparently is necessary to pass and implement policy initiatives addressing domestic and international problems, such as endangered species mitigation or ozone protection (Gillroy et al. 1997, 9–10). Ultimately, Buck seems to suggest that the apparent Enlightenment separation of the public "political" from the private "moral" realm is indeed not only preferable but necessary for environmental protection: ruminations on justice, Socrates, and "the good" should be left for parlor games, not policy debates.

In this essay I wish to argue otherwise, suggesting that environmental dilemmas demonstrate categorically the centrality of "public" and "private" discourse over conceptions of "justice" to environmental policy formation. In addition to this central concern, I illustrate two tendencies in moral thought that hinder such debate, endangering our ability to develop effective and lasting public policies for environmental protection. These two tendencies are the problematic consequences of voluntarism and the axiological primacy of the self.

Voluntarism, Justice, and the Democratic State

People commonly assume that ethical evaluations are merely individual whims masquerading as categorical imperatives—radically subjective opinions incommensurable with the ethical evaluations of others. Because "everyone is orthodox to himself" (Locke 1955 [1689], 13), we often follow Locke's strategy, reducing objective truth claims to personal preferences (28–30). Buck is not alone in her desire to bifurcate the political from the metaphysical. Indeed, as Rawls noted years ago, this seems to be one of the foundational political commitments of the Enlightenment: "As a practical political matter no general moral conception can provide a publicly recognized basis for a conception of justice in a modern democratic state. The social and historical conditions of such a state have their origins in the Wars of Religion following the Reformation and the subsequent development of the principle of toleration, and in the growth of constitutional government and the institutions of large industrial market economies. . . . A workable conception of political justice . . . must allow for a diversity of doctrines and the plurality of conflicting, and indeed incommensurable, conceptions of the good affirmed by members of existing democratic societies" (1985, 225).

Like Rawls, many contemporary policy analysts fear that the linkage to and use of substantive standards of justice would doom already contentious and fragile policy deliberations to endless conflict over first principles (Rawls 1985, 223, 249; Buck in Gillroy et al. 1997, 10). Like church and state, policy and substantive moral philosophy are best left to their separate and exclusive realms. However, even Rawls recognizes the need for at least minimal and contingent consensus on "familiar intuitive ideas and principles" that, in a democratic society, justify "a system of fair social cooperation between free and equal persons" and "narrow the range of public disagreement" (1985, 229, 228). Rawls calls this a "political" result, yet clearly, these ideas and principles are the metaphysical struts of democratic pluralism laid bare. Thus, even this most paradigmatic proceduralist formulation of public justice by the past century's most constructivist contractarian cannot avoid a certain "metaphysical" fait accompli.

Many have taken issue with Rawls's result (Okin 1989; Sandel 1984; Walzer 1983, 79), but his formulation *does* demonstrate a fundamental fact about contemporary political societies, and one directly relevant to environmental policy: a democratic society that is in principle dedicated to minimizing coercion is primarily dependent on voluntary compliance with its laws and policies. Though public sanctions for noncompliance may exist, in most industrial democracies the surveillance and disciplinary limitations of the state apparatus mean that most citizens are expected to comply *for reasons other than likelihood of fines, imprisonment, and so on.* Although some may fear a speeding ticket, most drivers obey in part because they affirm the importance, in fact the *fairness,* of laws that ask them to refrain from speeding in a school zone.

Peter Wenz (1988) articulates this point precisely. Because most environmental policies require *someone* to sacrifice previously permitted or subsidized behaviors or entitlements (e.g., clear-cutting a forest or driving a single-occupant vehicle) that now endanger human health, a common pool resource, or an individual species, these "someones" must at minimum perceive that they are treated fairly: "Because social solidarity and the maintenance of order in a relatively free society require that people consider their sacrifices to be justified in relation to the sacrifices of others, environmental public policies will have to embody principles of environmental justice that the vast majority of people consider reasonable" (21).

In environmental dilemmas the importance of addressing justice issues to develop effective policies to which citizens nearly universally consent is clearly visible: Bullard (1993), Collin and Morris Collin (1994), and others illustrate the particularly disturbing need of minority communities for environmental justice. But one simply has to look at just about any local environmental dispute to see how quickly "metaphysical" justice issues bubble to the surface; though some analysts tend to dismiss NIMBY sentiments, the concerns of displaced loggers, or the "rights" complaints of property owners as simply "parochial and emotion-laden" posturing (C. Davis 1993, 104), a few do see articulated within these statements legitimate fairness concerns that must be considered for policies to succeed (Rabe and Gillroy 1993; B. Brown 1995).

Given a hypothesis of unlimited coercive power and an unbending political will to utilize such, it is *possible*, perhaps, to imagine an environmental regime that governed without concern for justice and placed the cost of environmental protection on the backs of the poor, minorities, women, farm workers, rural loggers, or whatever suspect class said regime sought to scapegoat. Though theoretically possible, I doubt that it is practically possible (witness the failure to completely control social behavior of the most totalitarian regimes of the twentieth century, Nazi Germany and Stalinist Russia), and certainly not desirable or effective.

Ultimately, I am suggesting that the equity concerns of both minorities and property rights advocates must be engaged and not dismissed as either "too controversial and divisive" or "superficial and self-interested." And if they are to be engaged, that conversation will necessarily involve claims to first principles, as interlocutors justify the exchange of one "fairness claim" for another. Hence, justice (metaphysical *and* political) must be a part of any environmental policy solution.

Justice, Certainly, But What Type of Justice?

Having addressed the necessity of incorporating "metaphysical" justice claims into environmental policy debates, the question now arises regarding what standard of justice ought to be employed and what entities it ought to cover. Many writers on environmental justice, whether consciously or unconsciously, have employed some deontological claims in their analysis, often in the form of rights (Bullard 1993; Gibbs 1995; Fisher 1994; Lazarus 1992). However, few have considered the potential ecological difficulties that justice concerns introduce into environmental dilemmas (Tarlock 1992). On the other hand, many environmental ethicists of the holist sort have endorsed consequentialist methodologies, which, in their conception of "equity" for nature, show little regard for human justice in general, let alone for particular communities or minority groups that may inordinately suffer the costs associated with environmental degradation *or* environmental protection (Callicott 1989, 20–27, 46–47). Although I understand the attractiveness of both strategies (one seeks respect for humans, the other seeks respect for nature), I believe both are incomplete, even fundamentally flawed.

The underlying problem lies in their formation of valuational reference in the self. Often referred to as the axiological primacy of the self, such has been an assumption of ethics generally since Hume (1888, 468–69). It is prevalent in deontological rights theories (human and nonhuman), in that most consider the paradigmatic rights holder to be an entity that can express its interests and exercise or set aside its own rights, without the epistemological difficulties associated with having a guardian (whether a person or a government) speak on behalf of the entity's rights (Rodman 1977; Flathman 1976, 87–88). An entity must "know" itself, articulate its presence, and therefore participate in a community of mutually respecting, yet subjectively oriented selves. To the extent one allows various human and nonhuman entities to participate in this community of rights holders via guardianship, one risks inconsistency and hypocrisy by suggesting that the guardian knows and will articulate the needs, interests, and desires of the entity. But does the guardian really speak for the entity, or for the guardian's own desires? Clearly, we are back where we started: the axiological primacy of the self.

Among environmental holists, the situation is not much better. Callicott (1989) is just the most honest when he suggests that environmental holism overcomes the problem of the axiological primacy of the self by folding everything into the larger "Self." This is done via the insights of holistic ecological science and a particular reading of quantum theory: "If . . . both imply in structurally similar ways in both the physical and organic domains of nature the continuity of the self and nature, and if the self is intrinsically valuable [given the axiological primacy of the self], then nature is intrinsically valuable. If it is rational for me to act in my own interest, and I and nature are one, then it is rational for me to act in the best interest of nature" (173).

Although ingenious and attractive in its consequentialist appeal for us to "do good things" in the name of our greater self, Callicott's formulation returns us to the pressing epistemological problem that underlies the axiological primacy of the self: the self is primary because it is the only thing we can thoroughly know. Yet, clearly, we do not know everything about the function, value, and needs of all constituent parts of this "greater self" we call the ecosphere—we haven't even identified all the parts (J. Diamond 1990). Furthermore, we, a mere "part" of that entity, must make decisions regarding the other parts. The situation is akin to the left hand deciding that the bump on the right forearm is a cancerous mass and not simply a benign mole: besides basic ignorance, there is not the metaphysical distance between valuer and valued, subject and object, necessary for ethical judgment. "Intuited" raw self-interest may overshadow judgment: Is it better for the body and my (smaller) self, to amputate the right forearm? A subjectivist axiology brings us to the verge of inconsistency and hypocrisy once again.

Though I have explored tendencies in only one deontological theory and one consequentialist theory, the axiological primacy of the self lies at the heart of most problems plaguing theories of justice today, whether individualist or holist. No wonder, then, that Buck and others seek to escape this metaphysical quagmire, where we clearly tend to become private moral authoritarians seeking to impress upon a necessarily relativistic public sphere our own moral predilections via artful reasoning. One can see Locke and Madison spinning in their graves: no matter what Joel Kassiola, I, or others may say regarding justice, to many we are simply dangerous demagogues, convinced of our own righteousness. Nevertheless, we are back to that fundamental premise: in a democratic society "justice" must be done to maintain the consensus necessary to limit coercion.

Fortunately, I think there are both procedural and substantive solutions that can meet this necessity for justice and guide us, both politically and metaphysically, between the Scylla of voluntarism and the Charybdis of the axiological primacy of the self.

In recent years interest in discourse theory has reappeared in both philosophical circles (Dryzek 1990; Habermas 1987) and public policy circles (Andranovich 1995; Painter 1988; Dryzek 1987). Whereas most have focused on its procedural benefits for gaining a greater variety of information (Dryzek 1990, 205–6) or resolving local conflicts (Andranovich 1995; Finney and Polk 1995), communicative discourse, consensus decision making, stakeholder negotiation, or whatever you call it can provide a practical forum for discussing fundamental truth claims about what is right and what is fair, promoting examination of previously unconsidered and "unseen" humans and nonhumans, new scientific information, and claims regarding the outcomes of various actions or policies (Apel 1980; Painter 1988; Dryzek 1990). Based on insights from critical theory and hermeneutics, discourse theory in its most useful forms suggests that, given the axiological primacy of the self, the best contingent truth claims don't result from some abstract mindgame of reason (à la Rawls) but from the nonexclusive, participatory, noncoercive deliberations of both silent and vocal members of a discursive community. Though similar to the ideal scientific community criticized as an unreachable and idealistic model of policy formation in the past (Lindblom 1959), these discursive forums operate under the assumption that though such an ideal never may be achieved fully (i.e., some will always be more persuasive because of their eloquence and not their correctness), if structured adequately and containing a multiplicity of voices with sufficient empowerment, decisions—both practical and metaphysical—can be reached (Painter 1988; Finney and Polk 1995).

Although not a silver bullet for all problems, the continued success of such forums—whether called watershed councils, stakeholder meetings, or issue forums—does suggest some procedural and substantive breakthroughs in the "politics versus ethics" debate. First, it suggests that there may be a procedurally superior alternative to traditional pluralist conceptions of policy formation: though not always appropriate and certainly not always as "efficient" (Raab 1995), more open, participatory, and deliberative processes are better at overcoming adversarialism and parochialism associated with interest group mobilization, lobbying, and "hearing and comment" public input processes (Finney and Polk 1995; Dryzek 1987). Furthermore, they demonstrate that in most cases, citizens can meaningfully and substantively deliberate fundamental truth claims, perhaps leaving behind their original unexamined and uncontested evaluations. When presented with new information, new participants, and new alternatives, people can reach tentative agreement on first principles, moving from subjectivity to intersubjectivity. That is, I believe, a major substantive leap from Buck's individual government employee acting on a private moral whim.

Though problems remain with discourse theory (e.g., assuring adequate participation, promoting deliberation and mutual evaluation rather than confrontation and obstructionism, see Painter 1988; Dryzek 1990; Finney and Polk 1995), there is more here than simply a well-meaning, green version of Rodney King's "Why can't we all just get along?" Communicative discourse theory assumes that we must indulge in substantive metaphysical debate, risking disagreement, harsh words, and divisiveness over the contingent nature of truth. We cannot afford to sweep it under the Lockean rug of "toleration." There is, of course, no guarantee that discursive procedures will produce greener and more just policies. There also is no guarantee that they will produce a systematic and workable deontological or consequentialist theory of environmental justice protecting human and nonhuman communities. However, if the process provides greater and broader access to information, interests, and concerns, I, like Mill (1986 [1849]), think there is reason to be a bit more optimistic: given time and patience, more voices, more positions, and more deliberation can produce better, more complete, and more considered options.

Having thus outlined a response to why justice is necessary and how justice might be approximated in discursive practice, let me briefly suggest a substantive standard of justice that serves well in discursive settings, addresses the strengths of deontological and consequentialist standards, avoids the fundamental weaknesses of a subjectivist axiology, and is fit for environmental contexts. It is a standard of justice that has been around for a while yet largely forgotten: it is the early Socratic standard of justice as temperance.

In the *Gorgias,* Socrates responds to the claim of Polus that the happiest person is the tyrant who achieves all his desires and constantly searches for new pleasures, even at the expense of doing injustice to others (Plato 1941, 470a–e). Socrates argues that, like a leaky jar (493–94), the tyrant mistakes pleasure for happiness and never is filled. Ultimately, Socrates suggests that the only truly happy person is the temperate person, who, unlike the tyrant, governs his or her desires and recognizes that his or her own "profit" lies in the mutual satisfaction and friendship of others. Tyrants are never secure, never at peace, and never among friends—their own injustice is their prison (507a–511b).

I find some very appealing aspects in Socrates' idea. First, Socratic justice as temperance assumes as its starting place a subjectivist axiology: first and foremost, we understand and acknowledge our own desires and beliefs and act rationally to fulfill them. Second, the concept of temperance as discipline in one's appetites fits well with ecological and political necessities: every act of consumption, every act of political power, has its consequences for ourselves and other entities, eliminating or exhausting some opportunities while constructing others. Third, whereas the first two points assume a consequentialist and utility-maximizing calculus on the part of individuals, Socrates' rebuke of hedonism (a pleasure principle) in favor of eudaemonianism (a happiness principle), which maintains that an individual's happiness requires the happiness and fulfillment of others, promotes a deontological maxim of mutual recognition and respect, even of other species and entities. Ultimately, I think all three aspects make such a conception of justice amenable to discursive communities engaged in environmental policymaking: democracy requires justice; environmental protection and restoration require temperance; human happiness requires the happiness and fulfillment of others. Socrates has something to teach environmental policy analysts after all.

Issue 3: Nature Has Only an Instrumental Value

Sustainability: Descriptive or Performative?

Bryan Norton

I have devoted the past several years to trying to understand the much used but little understood concept of sustainability. This term is politically attractive to environmentalists because it seems to offer a "big umbrella" goal that many environmentalists can embrace, without emphasizing divisive disagreements. For example, adoption of sustainability as an environmental goal has reduced earlier tension between advocates of development and advocates of environmental protection by including both under a general rubric of "sustainable development." Speaking politically, then, the definition of sustainability suggested by the World Commission on Environment and Development (WCED; 1987), sometimes referred to as the Brundtland definition after the Commission's chair, achieved an important breakthrough.

In this essay I look not just at the past, but also at the future of the concept of sustainability. For reasons explained in "Why Not Foxy Hedgehogs?" (this volume), I see the problem of sustainability in prospective terms. I do not think we will "find" the correct definition of sustainability existing somewhere. A more tractable goal would be to "construct" a conception of sustainability that is useful in discussing and perhaps also in explaining, judging, and prioritizing environmental goals. Given the emphasis most advocates of sustainability place on making the concept measurable, one might assume that sustainability is a straightforwardly scientific concept. One might assume, that is, that when we finally succeed in formulating an adequate definition of sustainability, the main feature of that definition will be some descriptive measure that can be truly or falsely applied to paths of development. I refer to this assumption, for reasons that will become clear later, as the descriptivist assumption, and my purpose here is to examine and call into question this assumption. Now, I do not mean to argue that the term sustainability has no scientific or descriptive aspect at all; I do argue, however, that if a useful concept of sustainability were to be constructed it would not be *primarily* a *scientific* concept. For reasons I explain presently, I prefer to think of sustainability as *primarily a policy term with scientific content.*

I argue that, if we are to achieve a truly satisfactory and useful definition of sustainability as a value-laden term in policy discourse, which it must be if it is to be a substantive guide to policy, we must pay much more explicit attention to questions of social value. I think we cannot overlook the inevitable role of social values in any argument that we *should* live sustainably, and it is hard to see how this important value aspect could be reflected in a simply descriptive/empirical term. An important implication of taking the policy role of the term sustainability seriously is that the construction of a useful definition of sustainability will have to involve much broader scientific issues and much deeper philosophical ones than has been evident in the current debate about the meaning of sustainability. I illustrate this point by examining the debate of mainstream economists and ecological economists regarding weak and strong sustainability.

Strong vs. Weak Sustainability

In a series of eloquent lectures and elegant papers, Robert Solow has defended the view that sustainability can be fully defined, characterized, and measured within the neoclassical theory that shapes the mainstream economic tradition of resource analysis. Here, I focus on his brief summary paper, "Sustainability: An Economist's Perspective" (1993). Solow's basic idea is that the obligation to sustainability "is an obligation to conduct ourselves so that we leave to the future the option or the capacity to be as well off as we are." He doubts that "one can be more precise than that." A central implication of Solow's view is that, although to talk about sustainability is "not empty," "there is no specific object that the goal of sustainability, the obligation of sustainability, requires us to leave untouched" (181). Thus, although Solow clearly sees sustainability as an obligation of people in the present, we will see that this aspect fades into the background as he attempts to make sustainability a measurable concept.

The effect of this approach to defining sustainability is to place the term within the general theoretical structure that economists have used to understand economic growth. Solow's goal, then, is to measure sustainability in terms of economic growth patterns. He makes two important arguments to favor his approach. One argument is the view, widely shared by economists, he says, that no resource is irreplaceable, that every natural resource has an adequate substitute. In the jargon of economists, all resources are "fungible" (1993, 181). Solow also argues that we cannot know what people in the future will prefer, so we cannot know what, specifically, will improve their welfare (181). For simplicity, we can refer to these as the fungibility premise and the ignorance premise, respectively. Given this complex of related ideas, Solow also, and validly, draws the conclusion that the problem of obligations to the future can be understood as simply the task of defining a rational and intergenerationally equitable *investment policy.* In this view, monetary capital, technology, labor, and natural resources are interchangeable elements of capital. Sustainability, within this complex of principles and assumptions, is a matter of balancing consumption with adequate investment so that the future faces a nondeclining stock of total capital.

This nifty simplification of the intergenerational problem apparently avoids any need to specify particular elements or processes of natural systems that should be protected for the future. Sustainability, as defined by Solow, can therefore be tracked simply by measuring and projecting highly aggregated measures of economic growth. As long as we do not impoverish the future, we will have fulfilled our obligations to it. Solow's argument, starting with the fungibility and ignorance premises, thus concludes that the best we can do is to provide capital for the future. Note that implicit in Solow's argument is another assumption: that the way to measure fairness to the future is to measure some characteristic of the people that live at different times (such as "welfare," or "income," or "capital") and to compare these across time. And, by using growth theory to propose a general approach to such a comparison, Solow in effect turns what began as a normative question—What do we owe the future?—into an empirically measurable trend that can be read from data recording economic growth and savings rates.

Ecological economists have challenged this complex of assumptions and economic models. In particular, they have questioned Solow's principle that all forms of capital can be aggregated together and compared across generations. They argue that certain elements, relationships, and processes of nature represent irreplaceable resources, and that these resources constitute a scientifically separable and normatively significant category of capital: natural capital. We therefore face what is described by Herman Daly and John Cobb as the clash between "strong" and "weak" senses of sustainability (1989, 72).

In the text of their book, Daly and Cobb describe the disagreement between mainstream and ecological economists as a disagreement regarding which "paradigm" is appropriate for analyzing economic trends. This idea evokes the distinction between "normal" and "revolutionary" science of Thomas Kuhn (1962), who suggested that most science is undertaken within a paradigm, which is a constellation of assumptions, norms, concepts, and methods that give unity to a scientific discipline. Occasionally, when practitioners of a paradigm are unable to solve important anomalies, the paradigm is challenged by an emerging, competing paradigm, and a "revolution" ensues. By describing their criticism of the mainstream economic approach as attacking the mainstream paradigm, they suggest that they are rejecting not only the conclusions but also the basic concepts and measures of mainstream economics.

In fact, however, the "green" accounting system proposed in an appendix to their book hardly challenges the standard assumptions of mainstream economic modeling. Daly and Cobb propose that, having identified certain resources as natural capital, these particular assets should be protected because they are essential to the welfare of future generations. Or, to put the same point differently, they understand losses of natural capital as having an impact on people's income in the future: losses of natural capital, beyond a certain point, will make people in the future poorer. In their argument, they dispute the conclusion of economists William Nordhaus and James Tobin on the point of substitutability, concluding that resource depletion is already affecting income, and that "since 1972, the stagnation of productivity for about a decade are signs of the effect of rising real resource costs, particularly energy resources" (Daly and Cobb 1989, 410). They say, "We have thus deducted an estimate of the amount that would need to be set aside in a perpetual income stream to compensate future generations for the loss of services from nonrenewable energy resources (as well as other exhaustible mineral resources). In addition, we have deducted for the loss of resources such as wetlands and croplands. . . . This may be thought of as an accounting device for the depreciation of 'natural capital' " (411). The value of resources for the future is equated to an "income stream" from a trust fund that will compensate the future for losses in income due to our destruction of natural capital today.

But despite their emphasis on natural capital, Daly and Cobb never question the mainstream assumption that all values must be measured in the common currency of human welfare: present dollars. And this reduction of all values to a single measure allows them to protect the marginalist assumptions of welfare economics. They never question the comparability of natural and other forms of capital with respect to their impacts on welfare as experienced at different times, so trust fund compensation is equivalent to actually saving natural capital. Their analysis thus retains the main features of marginalism and compensability so familiar in welfare economics.

At this point in the argument, however, the differences between strong and weak economic sustainability become less and less tangible, even as their similarities—especially the willingness to compensate the future at the present exchange rate for a lost future opportunity—become more evident. Daly and Cobb's attack on the mainstream paradigm thus seems less radical and revolutionary than they suggest. Their alternative accounting system hardly requires a new paradigm, and the system they suggest seems to find it morally acceptable to destroy natural capital provided some form of compensation is set aside in each case. Indeed, their disagreement with Nordhaus and Tobin is apparently a difference of empirical fact: Daly and Cobb believe that shortages of natural resources cause real and significant price increases, increases large enough to reduce future welfare: Nordhaus and Tobin deny this. This is hardly a paradigmatic disagreement likely to cause a "revolution" in economic theory and conceptualization.

Furthermore, underlying the apparent similarity is another shared idea: that sustainability is or should be a term measurable in economic terms, as a comparison across time of the aggregated welfare of people who live at different times. In short, once one considers the object of concern to be capital—natural or otherwise—the economic camel has its nose under the edge of the tent of social values, and it is inevitable that it will occupy the whole tent. One suspects that, though Daly and Cobb do sincerely question Solow's strong fungibility assumption, on another level they have no answer to the ignorance argument. Because they lack an independent, noneconomic definition of natural capital, I doubt whether Daly and Cobb can give substantive guidance regarding what to save for the future. If the idea is that, to be fair to the future, we must save what will prevent a decline in the welfare of future people, we must in some way respond to Solow's challenge that we cannot know what people in the future will prefer and that we cannot therefore know what will promote their welfare. Their proposal to set up a trust fund seems no more likely than Solow's weak sustainability requirement to tell us what, exactly, to save for the future.

Daly and Cobb, no less than Solow, see the problem of sustainability—being fair to the future—as a matter of comparing measurable amounts of a descriptive variable across time. On this approach, our ability to say anything specific about what to save in the present to support future welfare requires knowledge of what people in the future will want. If we cannot know this, then we are reduced, as are Daly and Cobb and Solow, to insisting on some form of compensation package for the future. Whether we specify that package through the models of growth theory, requiring nondeclining wealth, as does Solow, or by measuring income losses due to losses of natural capital, as do Daly and Cobb, compensation will be considered a morally acceptable "substitute" for specific elements of natural capital. Neither line of reasoning, apparently, can identify some resources or some processes that are so important that, if lost, people of the future will be worse off, even if their level of consumption is no lower than that of future generations.

At this point, it seems to me, we face an important decision regarding which way to go to be more specific about what we should do for the future. One possibility, one that I have experimented with (with varying success), is to pursue what we might call ecological sustainability. If one were to go in this direction in search of a definition of sustainability, one might argue that both strong and weak sustainability apply only to one aspect of sustainability: that of maintaining wealth or "economic" resources. Because it would no doubt be wrong to impoverish the future, we can thus adopt weak sustainability as a *necessary,* but not sufficient, condition of fairness to the future. Then, in addition, one might set out to specify certain physical characteristics—elements or processes—of the ecological systems that we should not destroy or damage. We could say that, for example, if our generation irreversibly destroys the great reefs of the world's oceans, *people of the future will be worse off than they would have been had we not destroyed those reefs, regardless of the fact that they may be better off, in terms of income or wealth, than we are.* Regardless, that is, of whether we maintain a fair savings rate and place enough wealth in trust to compensate for lost income due to the destruction of the reefs.

One example of this line of reasoning can be found in the writings of those who insist that we use natural systems so that they maintain their "integrity," their "health," their "resilience," or some other descriptive characteristic of ecological systems (see, e.g., Costanza, Norton, and Haskell 1992; Arrow et al. 1995; Holling 1996). One might interpret these authors as stating a two-condition (or, perhaps, n-condition) requirement of sustainability. One requirement would be that we ought not to impoverish the future (and this condition can be measured according to methods chosen by mainstream economists); the second is to maintain some measurable physical characteristic of ecosystems, such as resilience.

I think this third approach (which, speaking broadly, we could call strong ecological sustainability) presents a very attractive line of argument. Its strengths are that it incorporates economic concerns into a broader criterion that measures other features of the environment across time; it makes sense of the feelings and attitudes of many environmentalists, who think saving the environment is more than an economic calculation, and it also provides an important role for ecological information about the impacts of our actions on ecological systems. It might also be considered an advantage, by those who are committed to treating sustainability as a scientific, measurable concept, that it "keeps the faith" by proposing measurable ecological sustainability requirements as well as economic ones. It may be that a consensus will emerge behind this approach.

Evaluating Descriptivism

Despite the interest and the advantages of an approach such as this, I believe it still rests on a confusion. To get an inkling of this, consider the following questions: On what basis does strong ecological sustainability *justify* a given policy? Do terms such as health, integrity, and resilience represent variables that can simply be measured and compared across generation-length times? Are they, that is, pure descriptive terms? If so, how can they represent the normative obligation to protect health, integrity, or resilience that seems at the heart of calls to live sustainably? If not, if they embody important social values, how can they be represented as purely scientific? Interestingly, much the same can be asked of the two types of economic sustainability. To the extent that sustainability is represented as a pure description, it cannot recommend a course of action or policy direction. It is at least possible, even after one has asserted that people of the future will be better off than we are, to continue, despite this, to propose other constraints based on noneconomic values. So, if the economists would have us believe that their measures of nondeclining capital/wealth and nondeclining income entail, with no further argument, a policy recommendation, then their descriptive measures must include also some value judgments. As it stands, Solow has hidden his own high value placed on wealth and capital within his definition of what is fair. A more transparent discussion of how to construct a sustainability concept would bring these values out into the open for discussion.

It may now be clear why I insist that the term *sustainability*—as well as similar terms such as *health* and *integrity*—is best thought of as a term in policy discourse. Once social values are characterized in the value-laden discourse of policy, it may be possible to choose one or more descriptive criteria that express those social values. But no descriptive criterion can stand, naked and alone, as a definition of sustainability. I am suggesting that, to define sustainability for a community—sustainability as a full-fledged policy goal, not just as an arbitrarily chosen descriptive variable—the community must in some way articulate broad values and social goals. It would then be possible to seek and, perhaps, to choose measures that are likely to track and usefully represent these values and goals. With this approach, the driving force in choosing sustainability goals is thus a community process of value articulation and goal setting. The choice of a measurable characteristic, though unquestionably important, now arises derivatively: What measure will help us to track how well we are doing in protecting the social values our community holds dear? Searching for indicators now becomes a search for indicators that reflect social values.

Let us now return to Solow's argument based on ignorance, an argument that is not really challenged by Daly and Cobb. Solow says we have no obligation to save any particular thing because we cannot know what people of the future will want or need. We can ask, however, why it is so important that we know what people of the future will prefer. Solow's implicit answer would apparently be: *Because the goal of sustainability is to maintain a fair savings rate and a nondiminished opportunity to consume, and because we cannot know what things will be preferred by future people—what consumption will promote their welfare—the best we can do is to maintain capital.* But this argument simply spins out the consequences of Solow's value system, which measures all value in terms of ability to consume; it does not provide any independent justification for the overweening value he has placed on wealth and capital. If we accept Solow's economistic and exclusive valuing of wealth and capital, we will already have bought into the idea that cross-generational fairness is a matter of comparing the opportunities of people in different generations to enjoy utility (consume). But if we reject Solow's reduction of the good of the future to a measure of aggregated capital, if we believe there are important community values that are not reducible to economic measures and calculations of growth rates, then we should reject his reduction of our obligations to the future to merely measuring wealth.

Description or Performance?

Knowledge and ignorance of future preferences is Solow's problem, but we need not adopt it as our own, provided we can pose the intergenerational question in a fresh way, a way that makes it less important to predict future preferences. If we reject Solow's formulation of the question, we find that the question to ask is not what people of the future will prefer, but rather what we, in the present, are willing to commit to as important expressions of our values as a community. Those who have adopted protection of environmental goods and special places as a goal recognize the special role that the natural world plays in the development of human communities. The problem at hand is not to predict what people in the future will want, but to articulate those values that are sufficiently important to present people that they are willing to make an effort to project these values into the future. This can be accomplished by focusing on the present and requires no prediction of what people in the future will want.

If specifying our obligations to the future depends on predicting in detail what individuals in the future will want or need, then assertions of obligations to the future will, at best, be plagued by unavoidable uncertainties; if we can be fair to the future only if we can predict their needs in detail, then there will always be an impossible task at the heart of all specific ("strong") sustainability requirements. By insisting that intergenerational moral obligations be measured in terms of comparisons of aggregated welfare, economists have formulated the problem of intergenerational fairness so as to require information that cannot be available at the time crucial decisions must be made. The collapse of sustainability into weak sustainability on the basis of ignorance is therefore preordained by the theoretical scaffolding chosen to express sustainability as a measurable characteristic of economic systems.[1] At its deepest level, economists' argument for weak sustainability rests not on the *fact* of our ignorance about future values, but rather on a deep and unquestioned commitment to reduce all moral questions to descriptive questions, to questions that can be fully resolved on an empirical basis.

There is a name for the mistake committed by economists and many other utilitarians: it has been called the "discriptivist fallacy" by J. L. Austin, the Oxonian analyst. In his *How to Do Things with Words,* Austin (1962) argues that many of our sentences that look like ordinary statements have purposes other than to describe. As examples and illustrations, Austin mentions "I do" (when uttered in the context of a marriage ceremony); "I name this ship the *Queen Elizabeth*" (while striking the ship's bow with a bottle of champagne); and "I bequeath this watch to my brother" (in the context of a will). He then says that, in these examples, to utter the sentence in question (in the appropriate circumstances) is not to *describe* the doing of what is being done, but rather, to *do* it. Austin proposes that we characterize such uses of language as "performatives," and mentions that they can be of many types, including "contractual" and "declaratory" (5–7). Later, Austin says: "A great many of the acts which fall within the province of Ethics . . . have the general character, in whole or in part, of conventional or ritual acts" (19–20).

Applying Austin's idea, and based on the analysis of this essay, a new way of thinking about intergenerational morality emerges. If we see the problem as one of a community making choices and articulating moral principles—a question of which moral values the community is willing to commit itself to—then the problems of ignorance about the future become less obtrusive. The question at issue is a question about the present; it is a question of whether the community will or will not take responsibility for the long-term impacts of its actions, and whether the community has the collective moral will to create a community that represents a distinct expression of the nature-culture dialectic as it emerges in a place. We do not then ask what the future will want or need; we ask by what process a community might specify its legacy for the future. If one wishes to study such questions empirically, there is important information available. One might, for example, study how communities engaged in watershed management processes or community-based watershed management plans achieve, or fail to achieve, consensus on environmental goals and policies. Empirical studies such as these may contribute to the process of community-based environmental management, but I am suggesting that the foundations of a stronger sustainability commitment lie more in the community's articulated moral commitments to the past and to the future than in any *description* of welfare outcomes.

This basic point makes all the difference in the way information is used in defining sustainability, and it changes the way we should think about environmental values and valuation. If the argument of this essay is correct, then the problem of how to measure sustainability, though important, is logically subsequent to the prior question of commitment to preserving a natural and cultural legacy and to the question of what we deem important enough to save and at what cost. So, we face the prior task—and I admit it is a difficult and complex one—of developing community processes by which democratic communities can, through the voices of their members, explore their common values and their differences and choose which places and traditions will be saved, achieving as much consensus as possible and continuing debate to resolve differences. These commitments, made by earlier generations, represent the voluntary, morally motivated contribution of the earlier generation to the ongoing community.

Whereas choosing measurable indicators is logically subsequent to commitment to moral goals, the tasks of choosing measurable indicators can, and must, proceed simultaneously with the articulation of long-term environmental goals. It cannot be otherwise, because the choices that are made by real communities regarding which indicators are relevant to their moral commitments represent, in effect, an operationalization of moral commitments. Although scientists and technicians must play an important role in formulating and validating these measures, they cannot be chosen without ongoing input from the public. The task of choosing community values, similarly, cannot be sharply separated from the specification of certain indicators that would track the extent to which actual choices and practices achieve those commitments. The specification of a legacy, or bequest, for the future must ultimately be a political problem, to be determined in political arenas. The best way to achieve consensus in such arenas is to involve real communities in an articulation of values, in a search for common management goals, and to include in that process a publicly accountable search for accurate indicators to correspond to proposed management goals.

The advantages of this shift in perspective are now evident: this approach suggests that the key terms sustainable and sustainable development are not themselves abstract *descriptors* of states or societies or cultures in general, but rather refer to many sets of commitments of specific societies, communities, and cultures to perpetuate certain values, to project them into the future, ones that include a strong sense of community and a respect for the "place" of that community. The problem of how to measure success and failure in attempts at living sustainably is now the problem, for each community, of choosing a fair natural legacy for the future, in a fair and democratic way.

Here, it is undeniable, as the economists will be quick to point out, that, ultimately, people in the present must balance their concern and investments in the future against real needs today. There may be situations in which setting aside special places or protecting traditional relationships between cultures and their natural settings will compete with other values. But now the question is transformed. If we see the problem as one of commitment of people in the present to not allow certain of their values and commitments to be eroded, the fact of our (partial) ignorance of future people's preferences, though a limitation in some ways, is not really relevant to the advocacy of sustainable communities. The case the sustainability advocates must make is that, to the extent the community has committed itself to certain values and associated management goals, these goals are deserving of social resources and "investments" in the future. Certainly, the task for the protectionist is a daunting one, given the competing demands on society's limited resources. To the extent that a community and its members see the creation of a legacy for the future as a contribution to an ongoing dialectic between their culture and its natural context, and to the extent that they accept responsibility for their legacy to the future, they have embraced a commitment that gives meaning and continuity to their lives, that, in some deep sense, affects their sense of self and community (see Holland and Rawls 1993, 14–19; Ariansen 1997). I believe that that commitment, not the ciphers of economists and other scientists, must ultimately represent the core idea expressed by the terms sustainable and sustainable development.

Are Environmental Values All Instrumental?

Mark Sagoff

Are environmental values always or even primarily instrumental? Are environmental goods and services, in other words, to be valued always or even primarily as means to an end? Environmental economists propose that the exclusive or at least the principal goal of environmental policy is to increase or maximize human well-being or welfare. As welfare economists Edith Stokey and Richard Zeckhauser have written, "The purpose of public decisions is to promote the welfare of society" (1978, 275). In this approach, the value of the environment is instrumental, that is, to enhance welfare, and the purpose of environmental policy is to maximize the long-term benefits nature provides human beings.

This essay argues that environmental economists confront a fundamental theoretical dilemma. On the one hand, many of the most important values and commitments that concern people in their views about the environment often have nothing to do with their welfare, well-being, or benefits they may seek to enjoy. Rather, they reflect religious, ethical, aesthetic, or political judgments these people defend as matters of principle, argument, or faith. Environmental economists, in contrast, believe that the goal or purpose of environmental policy is to maximize human well-being or welfare, that is, the net benefits the environment offers society over the long run. To admit nonwelfare and noninstrumental values into consideration—even to concede they may be relevant to public policy—would be to abandon the central scientific assumption of environmental economics that only instrumental values count. From the scientific point of view, values that are not instrumental—that are not related to well-being—cannot enter the social welfare calculus on which environmental policy must be based.

A Dilemma for Environmental Economics

Environmental economists believe the purpose of environmental policy is to maximize well-being or welfare; therefore, only those values that colorably relate to human welfare or well-being can count in scientific policymaking.

This commitment to social welfare separates environmental economics, which is a science, from the unscientific processes of democratic decision making. In the sloppy and nonquantitative Sturm und Drang of democratic political life, citizens may defend or oppose environmental policies on many grounds other than how they believe those policies will affect them. People can and often do introduce into political deliberation religious, moral, aesthetic, and cultural judgments, not just considerations related to their well-being as individuals.

It is the consideration of noninstrumental values—the presence of moral, aesthetic, cultural, and religious judgments in their own right—that makes democracy unscientific. These kinds of judgments have to be weighed on their merits in the context of deliberation, and that is not a scientific project. Welfare-related instrumental values, in contrast, can be "priced" at the margin, because we are used to trading or measuring well-being in terms of dollars. Economists can measure and aggregate values or preferences that are related to welfare, that is, all those that are related to welfare as means to an end.

From the scientific perspective, only one normative judgment is objectively true: the goal of environmental policy must be to maximize well-being or, technically, the present discounted value of future benefits associated with the environment. If people have other views or opinions they think bear on policy, one must remember that they are not scientists. Their rejection of the scientifically correct view of environmental policy may reflect ignorance or willful irrationality—or possibly an effort to eke out some advantage.

Environmental economist A. Myrick Freeman III explains the moral premise of his science: "The basic premises of welfare economics are that the purpose of economic activity is to increase the well-being of the individuals that make up the society, and that each individual is the best judge of how well off he or she is in a given situation" (1993, 6). Eban S. Goodstein makes this commitment explicit in his textbook on environmental economics: "The ethical foundation of economics is utilitarianism, a philosophy in which environmental clean-up is important solely for the happiness (utility) that it brings people alive today and in the future" (1995, 31). Tom Tietenberg, in his popular textbook on environmental economics, concurs: "In economics the environment is viewed as a composite asset that provides a variety of services" (1996, 16). Economic science strives to discover how society can allocate natural resources efficiently, so that "the net benefit from the use of those resources is maximized by that allocation" (19–20).

Pick up any textbook on or introduction to the science of environmental economics, and you will find clearly stated the fundamental thesis that the purpose of environmental policy is to increase welfare or well-being. It follows, then, that all those values—even if they dominate political discussion—that assume or propose a different normative foundation for environmental policy simply oppose science. Goodstein underscores the point: "Economic analysis is concerned with human welfare or well-being. From an economic perspective, the environment should be protected for the material benefit of humanity and not for any strictly moral or ethical reasons. To an economist, saving the blue whale from extinction is valuable only insofar as doing so yields happiness (or prevents tragedy) for present or future generations of people. . . . Economists adopt a hands-off view toward the morality or immorality of eliminating species" (1995, 24–25).

On the other hand, to rule out of consideration religious, ethical, aesthetic, and cultural judgments—to say that these do not count because they are not related to welfare—would be like trying to play *Hamlet* without the Prince of Denmark. In many areas of environmental policy—the protection of endangered species is an example—almost all discussion and debate turns on moral, cultural, and even religious considerations. Few if any of us believe that we could benefit in any way as a result of the protection of Furbish's lousewort, for example, although a majority of us may agree that society has a moral duty of some kind not to cause or tolerate the extinction of a species. Thus, the prevalence of moral, religious, and aesthetic judgments in debates about environmental policy might pose a problem for environmental economists. These nonwelfare values may be scientifically irrelevant, but they are very important to those who hold them.

The science of environmental economics tells us that all values pertaining to the environment are instrumental values, with human welfare or well-being as the single end to which these values are means. Religious, moral, and related aesthetic attitudes, on the other hand, teach that personal advantage or benefit is not the end of the good life or the good society. Rather, we stand for a great many goals and principles that cannot be embraced by, or may even conflict with, social welfare as the goal of social policy. Should environmental policy rest on science, or express or reflect popular morality or religion? Which is appropriate in a democracy?

What Is Welfare and Why Is It Good?

To answer questions such as these, we should learn what environmental economists mean by welfare or well-being. The first thing to note is that these concepts have no relation, either conceptual or empirical, to human happiness. In other words, when economists today speak of welfare or utility, they do not refer to the same things that the classical utilitarians meant by these terms, namely, pleasure (in the instance of Bentham) or happiness (in the instance of Mill). Economists avoid references to pleasure and to happiness for good reason: pleasure can be despicable; consider, for example, the pleasure fiends take in the suffering of others.

As a goal for social policy, happiness may seem less controversial than pleasure, but although chemists can produce or cause pleasure, nobody knows the formula that leads to happiness. If anyone knew where to find the bluebird of happiness and how to capture it, he or she would be as a god, and we might worship that knowledge. Actually, it is hard enough to decide what might work in one's own case—an iffy business at best—much less how to increase the happiness of others. Hence, the first basic postulate of environmental economics may be this: each individual is the best judge of how well off he or she is in a given situation. The other basic postulate asserts that the preferences of individuals are to count in the allocation of resources (Samuelson 1953, 223; Quirk and Saponsik 1968, 104). "In this framework," economist Alan Randall notes, "preferences are treated as data of the most fundamental kind. Value, in the economic sense, is ultimately derived from individual preferences" (1981, 7).

We now have before us two crucial scientific postulates. The first asserts that the individual is the best judge of what will benefit him or her, that is, how well off he or she is in a given situation. The second holds that the satisfaction of the preferences of individuals, insofar as resources allow, constitutes the chief aim or goal of environmental policy.

A problem arises because not everyone bases his or her views, opinions, beliefs, and positions about environmental policy on judgments about what will benefit him or her. For example, when environmental economists support particular policies, their advice presumably reflects judgments they make on grounds other than their own welfare or utility. Environmental economists often propose that efficiency should be the standard for environmental policy. They make this suggestion because they believe that good arguments support it. They are not trying to benefit themselves; for example, they are not promoting efficiency because they hope to be hired as consultants to determine which outcomes are the most efficient. Their policy preference reflects judgments on issues other than what these scientists believe will benefit them.

Environmental economists make it quite clear that they speak from objective perspective of what is right or good, whereas the rest of us seek only our own advantage. Indeed, the rest of us never think of anything but how well off we may be in a given situation. That is what ties preference to welfare. Speaking with the authoritative voice of objective science, economist Freeman explains:

"Since the benefits and costs are valued in terms of their effects on individuals' well-being, the terms 'economic value' and 'welfare change' can be used interchangeably. Society should make changes . . . only if the results are worth more in terms of individuals' welfare than what is given up by diverting resources and inputs from other uses" (1993, 7).

Now, environmental economists have often shown that markets can almost never be trusted to make the efficient or welfare-maximizing allocation. Whatever the market does, an economist can always second-guess its result by referring to some unpriced cost or benefit. After all, there is always someone who would have favored a different outcome and might have paid something for it. Resource economists Alan Kneese and Blair Bower of Resources for the Future observed that by the 1960s, it had become clear to them and their colleagues that market failures are so pervasive, at least with respect to environmental "public goods," that "the pure private property concept applies satisfactorily to a progressively narrowing range of natural resources and economic activities" (Kneese and Bower 1972, 66).

Given the pervasive failure of markets to allocate resources to the highest bidder—and thus to maximize welfare—we need scientific managers to allocate resources efficiently. To be sure, in principle one would prefer a market allocation, but given that externalities are ubiquitous, public goods pervasive, and unpriced costs and benefits preponderant, there is no scientific alternative to centralized planning. The upshot of environmental economics as a theory is that professional economists know better than the market. They should then allocate resources to benefit not themselves but society as a whole. Individual consumers make the basic economic choices by determining how much they are willing to pay for different outcomes given the expected impacts of those outcomes on them. Environmental economists scientifically transform those individual preference orderings into a larger cost-benefit calculus on which policy may be scientifically based.

What Does Willingness to Pay Measure?

Environmental economics is not only a normative science but also a quantitative one. It is normative because it proposes social welfare as the basis of environmental policy; it is quantitative in that it proposes willingness to pay as a measure of welfare. For example, Freeman defines "the benefit of an environmental improvement as the sum of the monetary values assigned to these effects by all individuals directly or indirectly affected by that action" (1979, 3). Accordingly, preference and the willingness to pay that measures it have become the foundations of economic science for the environment. Willingness to pay measures the amount of welfare the individual expects to gain or lose as a result of a particular allocation of resources.

A problem arises, however, in defining the relationship between preference (and therefore willingness to pay) and welfare. If people formed their preferences with their welfare in mind, this connection would be apparent. Theoretical problems arise for environmental economics, however, because people so often consider matters other than their own welfare or well-being when forming judgments about environmental policy. As long as preference reflects what the individual believes is good for him or her, then willingness to pay may measure the extent of that good. But if one's preference concerns some other value—such as the belief that a policy is efficient, or consistent with cultural commitments, or religiously mandated—what difference would it make how much the individual is willing to pay to satisfy it?

Indeed, the reasons people offer for adopting one or another environmental policy—including the reasons economists offer—are not to be traced to the expected welfare of those individuals. Of course, economists offer their views as reflections of important objective arguments, not as indications of their own utility orderings. The funny thing is that economists are not the only ones who think they do this. Virtually everybody else is more than willing to defend his or her preference or position with arguments that refer to what society should do, rather than what that individual wants for himself or herself.

Consider, for example, a person who believes that society should protect endangered species because this is God's will, or because it is the right thing to do, or because this goal expresses the highest aspirations of our society. Such a person may consistently hold that society ought to implement the Endangered Species Act, and that this would not positively affect his or her own welfare. In other words, it is conceivable that people can form political, social, moral, aesthetic, and religious views and opinions about the environment, and, in doing so, they will consider values and concerns other than their own welfare or advantage. If so, these people have strayed from the sphere of authority economists have assigned to them: they are not basing their preferences or positions on judgments about their own well-being.

To what extent do the preferences people express reflect moral, religious, aesthetic, or other such principles rather than considerations of self-interest or personal well-being? The answer intuitively would be: almost always. People in general—not just environmental economists—advance views and positions on the basis of objective arguments and considerations as distinct from private wants. Just as economists present their views about efficiency as objectively correct, so others present positions about "sustainability" or "moral responsibility," or "God's will," or "obligations to future generations," or "community values" as objectively correct. Hardly anyone defends a policy on the grounds that he or she expects to benefit as a result. Welfare considerations, in fact, may have relatively little to do with what economists call preferences.

Nonuse Values

The empirical data support the idea that individuals form judgments about controversial environmental policies primarily on the basis of considerations that have nothing to do with what they believe will benefit them. Rather, these values reflect views about what they believe is right or correct *simpliciter,* or is consistent with the culture of the society to which they belong. Social scientists who interviewed respondents to contingent valuation (CV) surveys in 1992, for example, found "a large range of strategies for constructing stated WTP [willingness to pay] that had little or nothing to do with respondents' expected utilities" ("Ask a Silly Question" 1992, 1986). Reviewing several such protocols, three economists concluded that "responses to CV questions concerning environmental preservation are dominated by citizen judgments concerning social goals and responsibilities rather than by consumer preferences" (Blamey, Common, and Quiggin 1994).

A careful empirical study found that ethical commitments often dominate economic or welfare considerations in responses to CV surveys: "Our results provide an assessment of the frequency and seriousness of these noneconomic considerations: They are frequent and they are significant determinants of WTP responses" (Schkade and Payne 1994, 88). In general, responses to CV surveys, as commentators repeatedly point out, reveal "social or political judgments rather than preferences over consumer bundles" (Blamey, Common, and Quiggin 1993).

These findings become particularly consistent with respect to what environmental economists call nonuse, passive, or existence values. Tietenberg describes nonuse values as "those that derive from motivations other than personal use." He notes: "The evidence that existence values exist seems quite persuasive. Certain behavioral actions of people reveal strong support for environmental resources even when those resources provide no direct or even indirect benefit. One example is provided by the millions of dollars contributed each year by large numbers of people to environmental organizations" (1996, 62–63).

Tietenberg is plainly right in thinking that people care about environmental assets or resources—endangered species, for example—that provide no direct or even indirect benefit to them. People care about a lot of things besides their own welfare. Yet, if welfare is the sole desideratum of environmental policy, then these other values must go by the board. Nobel laureate economist Amartya Sen drew this conclusion elegantly in 1977. He wrote that moral, aesthetic, religious, and other values important to human beings reflect commitments to concerns other than one's personal well-being. These kinds of moral and ideological commitments drive a "wedge between personal choice and personal welfare," although "traditional economic theory relies on the identity of the two" (1977, 317, 344).

Environmental economists, as scientists, assert that the purpose of environmental policy is to maximize the welfare of individuals. Economists Michael Bowers and John Krutilla have announced the goal of environmental policy as a settled question: it is to allocate environmental resources "to provide the greatest discounted net present value from the resulting flow of goods and services" (1989, 32). What, then, to do with all opposing views and positions, for example, the political, ideological, and moral values of those who consider nature to be worth preserving for its own sake or for its intrinsic qualities rather than for some benefit that they expect? Perhaps these other views are just wrong. Yet they crop up again and again in the values people express about the environment and the reasons they give for preferring particular policies.

If existence or nonuse values have no clear relation to judgments about welfare—if they cannot be tied to any benefits direct or indirect—why do economists engage at great public expense in scientific surveys intended to elicit these values and measure them in welfare terms? These CV surveys generally misrepresent respondents' judgments and convictions about the public interest as consumer preferences about their own personal "satisfactions." These surveys, as commentators note, "are simply opportunities for individuals to comment, without very much opportunity for thought, on a hard issue of public policy. In short, they most likely are exhibiting off-hand opinions on the same policy issue to which the cost-benefit analyst purports to give his own answer, not private preferences that might be reflected in their market transactions" (Farber and Hemmersbaugh, 1993, 267).

Why Satisfy Preferences?

If welfare considerations have no clear relationship to preference, why should preferences be satisfied? To satisfy a person's preference would not necessarily improve the individual's welfare or that of anyone else, except in the rare or exceptional case that the person formed the preference wholly on the basis of judgments concerning his or her own well-being. Otherwise, preference satisfaction and welfare seem to be completely distinct and unrelated notions. It then becomes impossible to understand what willingness to pay measures or how this measurement can be tested.

One could argue that the reasons people form their preferences do not matter: it is the satisfaction of the preference, however formed, that creates utility, in an amount that willingness to pay measures. The reason to satisfy preference, then, is utility. And utility consists in or measures the satisfaction of preference. Thus, the reason to satisfy preference is utility, which is itself defined as the satisfaction of preference. This argument merely states a tautology, an identity of the form A is A. For some, this tautology is sufficient to secure a normative science.

Has the satisfaction of preference any relation to happiness or well-being as these concepts are commonly understood? Absolutely not. In fact, empirical research reveals that preference satisfaction is not correlated with perceived happiness. Studies show that once people meet their basic needs, they do not become happier or more content as their income increases and they are therefore more able to satisfy their preferences. Social research confirms what common wisdom suggests: money and with it preference satisfaction do not buy happiness (Campbell, Converse, and Rodgers 1976; Erskine 1964; Guring 1960). One review concludes, "Studies of satisfaction and changing economic conditions have found overall no stable relationship at all" (Argyle 1987, 144). For example, "those who win large sums of money in football pools or lotteries are not found to be on the whole more happy afterwards" (144).

Contentment depends more on the quality of one's desires and on one's ability to overcome them, than on the extent to which they are satisfied. The things that make one happy—friends, family, achievement, health—depend largely on virtue and luck; they are not available on a willingness to pay basis. Consumption does not produce contentment: "And this is virtually inevitable because the faster preferences actually *are* met, the faster they escalate" (Rescher 1980, 19).

It gets worse. One may plausibly argue that there are no such things as preferences. No one has ever seen one, for example. People express views, beliefs, ideas, opinions, arguments, positions, inferences, findings, and commitments; they are moved in their behavior by all of these along with virtues, vices, memories, habits, mistakes, and all kinds of foolishness. Preferences rarely have much to do with anything. To be sure, Bartleby the scrivener in Melville's wonderful story kept saying, "I would prefer not to." The appeal to preference as a ground for action seems a little dotty. People are usually able (and expected) to come up with some better justification or explanation than that.

In fact, preference may best be understood as an abstract construct of economic theory. At best, the concept of preference refers to a private mental state. Because it is a private mental state, preference cannot be observed. To determine a person's preference, an economist describes a choice that person makes. But choice, too, because it is also a mental act, cannot be observed. What is observed are actions. But a person's actions or observable behavior can be interpreted in any number of ways, depending on the dispositions, beliefs, desires, and "opportunity sets" the observer assumes motivate that behavior. Thus, it is simply not true that choice reveals preference, as economists generally think. Rather, to describe bodily motions as indicating a choice of one sort or another, the observer has already to ascribe a preference to the agent, for without such a preference, that description of the behavior would not make sense. Thus, choice is really inferred from preference, rather than preference inferred from choice (Sagoff 1985).

In economic theory, preference functions as a conceptual construct, that is, an abstraction like utility. In fact, preference and utility are conceptually related. Neither can be seen, felt, touched, or even located in the world. The belief that preference satisfaction increases utility is not one that can ever be tested, ever be falsified. This is because one can always change one's mind about what preference a person "really" possessed or how much utility he or she "really" gained as a result of some outcome. If two observers disagree about the nature of a person's preference, how might they settle the disagreement? They could subject the individual to other tests on the usual assumption that preferences obey axioms such as transitivity and antisymmetry. But social psychologists have often shown that people's preferences do not usually conform to these kinds of axioms.

The science of environmental economics is like a pill: it is best swallowed whole. Chew it over a bit and one must spit it out. Ultimately, there is no way to improve, strengthen, amend, or chasten it: one must either embrace the theory as a whole or reject it as a whole. Many environmentalists, including those who call themselves ecological economists, try to "green" the theory by discovering more existence values, pumping up the shadow prices for certain environmental goods, and lowering discount rates. Like the woman who rode on the back of an alligator, they may have fun for a while. They will find, however, that they must either get off the alligator, or they will wind up inside it.[1]

Issue 4: Intrinsic Value Implies No Use and a Threat to Democratic Governance

A Practical Concept of Nature's Intrinsic Value

John Martin Gillroy

Basing policy on the intrinsic value of nature is both practical and necessary to moving away from the overuse of the environment by the market and the undersupply of environmental quality as a collective good. Taking the second point first, I contend that the preservation of the intrinsic value of the environment can become a critical and necessary distinction to policymakers when preservation of systems integrity is seen as a *principle* rather than a resultant *policy.* Second, I point to lines of argument within environmental law that can support preservation as a new and practical principle for twenty-first–century environmental policy.

Preservation as Principle

There are four basic arguments against the concept of intrinsic value as it is applied to nature. One can argue, first, that there is no such thing as intrinsic value, that the word either has no meaning or it is "inherent" and undefinable outside of the "conscious" subject itself (Thompson 1990). Second, one may argue that if nature has an intrinsic value, it requires that we apply human attributes such as "rights" to nature, which is seen as very problematic (Hargrove 1992). Third, one might contend that even if intrinsic value exists, it has no distinct meaning but conflates into instrumental theories of value. Last, one may admit that intrinsic value is possible but argue, as Taylor does in this volume, that it is undesirable, as it is inherently antidemocratic.

Being interested in moral concepts as they apply to public policy choice, I concentrate on the third argument in providing answers to the other three. By demonstrating, through the case of the reintroduction of wolves to Yellowstone, that intrinsic value has practical meaning independent of nature's use value for the policymaker, I simultaneously argue that we can define the environment's intrinsic value in its own terms as the integrity of functioning natural systems, and that, therefore, intrinsic value must have meaning as a philosophical concept. In the course of this argument I also claim that the recognition of intrinsic value is a prerequisite to democracy.

The most ambitious argument conflating intrinsic value as a subcategory of instrumental use value is made by Bryan Norton (1986). Norton does not attack intrinsic value directly in this essay; he identifies preservation with intrinsic or nonanthropocentric values and conservation with instrumental or anthropocentric value. He then argues, as a counterargument to Passmore (1974), that the difference between them is a "difference in emphasis" (195) and "a matter of degree" (200) rather than a fundamental distinction affecting the motives and product of policymaking. Norton argues that it is not the motives of a preservationist or conservationist that matter as much as the policy goals defined by these positions. He contends that preservation is best defined as a policy resulting in the persistence of "untrammeled"[1] wildness or "the exclusion of disruptive human activities from specified areas" (201). He concludes that "conservation and preservation are different activities, which might result from varied and complex motives" (201). In effect, he maintains that there are many conservationist reasons for "preserving" wilderness and biological diversity, and that because conservationism is more accessible to policy methods, we need only a theory of nature's conservation to adequately maintain the "health" of ecosystems (Norton 1992).

My concern is that there is a profound difference between valuing nature instrumentally and valuing it intrinsically. Motives do matter; in fact, they determine not only what standards will define "reasonable" and "persuasive" policy, but how and under what circumstances nature will be used by humanity. The history of human contact with natural systems has frequently had conservation attached to it, yet our overuse continues unabated and our efforts to consider nature a priority in our economic decision making has been piecemeal at best. This may be because we value nature merely as of instrumental value to us.

Take a human example. Can we treat one another with respect as human beings if we do not value one another intrinsically? To own other persons as slaves, we need to think of them not as persons but as "things" for our use. To objectify them is a prerequisite to their enslavement; it is very difficult if not impossible to recognize others as human beings and still enslave or exploit them. To free others it is necessary to acknowledge that they are of intrinsic value independent of our use of them, that they are not property but *people* being treated as *property.*

No matter how valuable a slave is to its owner, no matter how long slaves have been in one's "family" or how many generations of owners have depended on them, they are still slaves, valued instrumentally, and this determines the relationships of the people involved. Property has merely instrumental value, and in our decision making about slaves we will not respect them as persons or treat them as we would feel obliged to had they freedom and moral capacity like ourselves.

Civil rights and human rights movements are critical because they are mutual recognition that humans are moral agents with value independent of their uses. Rights separate our inherent worth from our utility and exist specifically to announce to the world that we, as human beings, have *intrinsic value.* Human rights invoke duties in and from others, including collective duties through public policy, that are not duties regarding mere property. We have a distinct consciousness and think differently about what we value intrinsically, and we think that the motives of others are extremely important in their treatment of us. No matter how comfortable our physical existence as slaves or property, we long to have our freedom or intrinsic value fully recognized both in the eyes of other humans and in the laws and policies of governments. Many have left comfort for discomfort and suffered much pain to claim their freedom and rights as persons. In terms of human interpersonal relationships, both public and private, motives count; it is how we tell respect from condescension.

Within the sphere of interhuman affairs, valuing someone intrinsically is also the most essential prerequisite to the promotion of democratic values. It is the "barbarians" and the "slaves" whom we value instrumentally and to whom we deny citizenship and participation in the democratic process. It is instrumental valuation of persons that fuels genocide, slavery, exploitation, and persecution and is the first and most fundamental prerequisite in defeating democratic values.

Why should nature be fundamentally different? First, we do not have to attribute human moral qualities to nature to define its unique capacity and purpose. We can define nature's functional integrity as an ecological capacity and evolution as its functional purpose. We need not hold nature as a moral agent in order to invoke human duties to its value independent of our use of it. But unless we recognize that it has independent status as functioning life, we will have neither a basis for respecting it as an end in itself, nor a policy standard that will respect its capacity and provide for its flourishing in our collective decision making.

Nor will nature ever receive equal baseline consideration in democratic decision making. I have argued that it is only those persons we value intrinsically that we allow to participate in democratic discourse; this is also true of nature. I am not recommending that we extend the franchise to gophers, but that, in our democratic decision-making process, the flourishing of whole natural systems be considered of value independent of use and therefore be given full status as an end in itself in policy deliberations. Only by conferring this status on nature will we move past our treatment of the natural world as a mere resource and grant basic consideration to the persistence and flourishing of natural as well as human systems in democratic discourse.

Norton's (1986) mistake is that he defines preservation as a policy result rather than a foundational moral principle; in this way, he misses its unique character by ignoring the distinct motives of those seeking to protect the functional integrity of nature. If we classify conservation and preservation in terms of consequential policy, for example, as either the "preservation" of pristine wilderness or the "resilience" of resource supply over time (216), then we have limited these concepts to the descriptions of narrow and specific outcomes, which Norton is correct to point out might come from a variety of motives and value systems. It is true that, from this approach, intrinsic or instrumental value matters less than achieving the policy goal.

However, when preservation is viewed as a moral principle informing humanity's relationship to the environment, preservation can mean a wider range of policy goals than just the setting aside or protection of wild places. As a moral principle, it affects our consciousness of nature, as it has affected our consciousness of ourselves and other people; it has unique explanatory power in describing nonuse value in a way utilitarian policy from a principle of conservation does not; and it is better able to suggest a range of policy alternatives aimed at respecting nature's integrity as a functional system.

Let us take the reintroduction of wolves into Yellowstone National Park (McNamee 1997) as an example. By Norton's definition, this cannot be a preservationist policy, as it does not leave the "untrammeled" park intact, but is a human perturbation that changes the biodiversity of the park and its ecology by introducing what can now only be called a foreign species to the present ecology of the park. The system is not left unspoiled or left alone to evolve undisturbed, but is being managed by humanity. It is not a conservationist policy either: there is no human "resource" interest served by the reintroduction. The local citizens are in more "danger," the tourists cannot see the aesthetic difference with or without wolves, even scientists could study wolves elsewhere and get all the information they needed. There is no resource value to the wolves in the long or short run; in fact, they will produce many resource, opportunity, and other economic costs to all involved. If we feel that this reintroduction is an important policy, how is it justifiable? Consider approaching this policy using preservation not as a policy outcome but as a principle for the policymaker trying to respect nature as a functional end in itself.

The reintroduction of wolves makes sense as a preservationist *act,* that is, if we define preservation as a *principle* rendering a *motive* to *act.* If we value nature independently of our use of it and recognize an intrinsic value in the persistent functioning of natural systems, then we can say that the ecological integrity of Yellowstone invokes a duty from us to its intrinsic value as a functioning natural system. This intrinsic capacity, in order to flourish, requires that we act by "removing unwanted species and favoring other," less "economically productive ones" which are necessary to restore a natural system to a former state of flourishing. Management is not wrong or right in and of itself; it depends on the reason for management.

Norton (1986) collapses preservation into conservation by focusing on the resultant policy goal and narrowing this policy to a single result: the set-aside of pristine wilderness. This does not allow preservation to involve active management or reclamation of natural systems and defines it not in terms of natural systems resilience (216) but in terms of "predictability" or "autogenic" change (217). Norton limits the preservationist to support of ecological systems on their natural course, with "limitations on the severity and persuasiveness of human management" (218), where "total management of whole areas" works counter to preservationist policy. He does not allow for a preservationist policy to restore species that we have hunted out of existence in order to promote the integrity of the system as an end in itself, independent of our use of it. Under these circumstances, the principle of preservation requires a human perturbation to reintroduce the wolves for the sake of the ecology and its flourishing, regardless of the effects on ranchers, tourists, or scientists or the resource value of the wolves.

Because the reintroduction of wolves is not a set-aside of wilderness and requires human intervention in nature's autogenic course, it cannot be preservation by Norton's definition. Reintroduction of wolves focuses on natural systems resilience, which Norton limits to conservation policy. However, this resilience is not to "maintain a constant flow of resources for fulfilling human demands," as Norton defines it (1986, 216), but for the integrity of the natural system itself, independent of our use of it. So it cannot be conservation policy under any definition. Therefore, wolf reintroduction is a policy that a decision maker functioning under Norton's paradigm could not conceptualize or justify as reasonable in either conservation or preservation policy.

For this policy choice, the distinction between intrinsic and instrumental value is important to the consciousness, first, that wolves have more than human use value, and second, that they are components of the intrinsic functional value of natural systems such as Yellowstone. Third, the distinction is important because we have a duty to restore the wolves to assure the persistence of natural systems integrity. The justification of this policy can be seen as reasonable only if a principle of preservation informs a variety of policy goals rather than a single "hands-off" policy that could be justified by a variety of motives, as Norton would have it.

Having nonanthropocentric reasons for action is critical to a full evaluation of what is important about nature. In the same way that failure to treat one another as ends in ourselves rather than means to the ends of others results in exploitation, enslavement, and waste, so not treating natural systems as functional ends in themselves leads to their overuse and exploitation. In the same way that freedom is only important to the individual as a moral end in itself, wolves in Yellowstone are only important to the natural system and its full empowerment as a functional end in itself. Their restoration to Yellowstone cannot be justified within anthropocentric conservationism.

My definition of preservation as principle can render a much wider range of possible policy in the name of empowering natural systems as ends in themselves. Ecological integrity might mean taking exotic species out of systems, replacing them with others necessary to the integrity of that particular ecology. It might require reclamation of both what we have destroyed and what nature itself has destroyed (e.g., reconstructing the White River in Vermont after a storm). Preserving systems integrity supports a definition of ecosystem health as resilience, which Norton limits to conservationism. It even allows human use while it also allows us to make important distinctions, as between old- and second-growth forests, common and unique ecosystems, and native and exotic species, which conservation fails to appreciate in its general classification of nature as a resource. In the same way that we can treat one another as means as long as we simultaneously treat one another as ends, so too we may use nature as long as we respect it as a functional end in itself.

Norton's argument for abandoning motives and theories of intrinsic value is based in his concern that policymakers do not listen to philosophers and can operate only within a framework that acknowledges the instrumental value of nature and its conservation from an anthropocentric perspective: "The Achilles heel of the discipline has been its inability to secure a hearing among environmental decision-makers" (1986, 202). In terms of the practicality of preservation, I contend that it is both already present within environmental law and accessible as an alternative in administrative decision making.

Preservation in the Law

First, support for administrative decision making by the principle of preservation, as I have defined it, is alive and well in environmental statute law. With the principle of preservation, a new "natural systems approach" to management might be fully integrated as a competitive argument to the resource efficiency and multiple-use reality of our environmental and natural resources law.

This competing argument finds life in a cross-section of statutes. For example, in the Clean Air Act (42 U.S.C.A. §§ 7401–7671q, 1992), we see a competitive argument for the inclusion of natural system research (§7403 (e)) and natural system risk assessment (§7408(g)) in the establishment of national air standards. In addition, there is a human intrinsic value/public health argument in setting air standards (§7409) that has recently been used by the Environmental Protection Agency to tighten National Ambient Air Quality Standards (NAAQS) for criteria pollutants. In the Endangered Species Act (16 U.S.C.A. §§ 1531–1543, 1973), we have an argument for the designation of threatened and endangered species that can be made only on the basis of ecological data (§1533 (b)), as well as an argument for critical habitat (§1533 (bB2)) that focuses on "the ecosystems upon which endangered species and threatened species depend" (§1531 (b)).

Current environmental law provides more than efficiency, multiple-use arguments. Two parallel arguments within these statutes are accessible to administrators. One, the currently dominant argument, concentrates on efficient use of nature as a resource. The second argument suggests an opening for natural systems management and a competitive principle seeking preservation of natural systems integrity.

In addition to the fact that provisions for preservation and ecosystem management already are provided in the statutes, there is further support for change in a series of Supreme Court decisions. These decisions establish that the Court will allow administrators to decide on the interpretation of statutes or which of the two parallel arguments they will use to write rules:[2] "If this choice represents a reasonable accommodation of conflicting policies that were committed to the agency's care by the statute, we should not disturb it unless it appears from the statute . . . that the accommodation is not one that Congress would have sanctioned. . . . Is it a reasonable policy for the Agency to make?" (*Chevron* 1984, quoting Shimer).

Conclusion

A value concept may be judged on the basis of the answers to three questions:

— Does the concept of value define something unique about humanity and/or nature?
— Does the value concept change the human consciousness of that to which it is attributed?
— Does this new consciousness cause different policy principles and priorities to be applied to collective decisions involving that which is so valued, resulting in unique outcomes?

I contend that my definition of intrinsic value is successful on all of these levels: it does describe the capacities of living things in a way not covered by other terminology; it changes our consciousness of the unique value both of human beings and of nature; and it suggests policy, such as the reintroduction of wolves into Yellowstone, that would not occur to policymakers without the principle of preservation and the goal of natural systems as valuable independent of our use of them.

Overall, being able to consider intrinsic value is important to the proper status of humanity and nature alike. By concentrating on preservation as principle and not policy outcome we can have a unique definition of environmental motives that inform policy not accessible to a conservationist or resource approach to nature. In addition, this definition of nature's intrinsic value already exists in the law; we need only the bureaucrats motivated to apply it, as the Supreme Court has already said they are free to do.

On Intrinsic Value and Environmental Ethics

Bob Pepperman Taylor

A few years ago, Eugene Hargrove suggested that after briefly attempting to apply the notion of "rights" to nature, environmental ethicists turned to the idea of "intrinsic value" to do the work that rights clearly failed to do for them: "To find a kind of intrinsic value that could trump instrumental value—in the way rights can—they [environmental ethicists] started looking for nonanthropocentric intrinsic value" (1992, 183). Although Hargrove himself promotes a "weak anthropocentrism" rather than a nonanthropocentric environmental ethic, his own view of the importance of intrinsic value is driven by concerns that he shares with many of his nonanthropocentric colleagues: to find values that will "trump instrumental values" and thereby avoid moral calculations in which natural values seem to be disadvantaged in relation to more conventional economic or utilitarian considerations.

Serious questions have been raised about the coherence of applying intrinsic value to natural objects and processes in the way Hargrove promotes. Janna Thompson, for example, has suggested that only conscious individuals are capable of embodying intrinsic value (1990, 159), and that even if the concept could be appropriately applied to nonconscious nature, it would fail to solve the problems it claims to address. For example, intrinsic value cannot explain why we should preserve *these* species instead of destroying them so new ones with their own intrinsic value can evolve over time, or, more generally, how we are to mediate competing intrinsic values throughout the moral universe (Thompson 1983, 91). Anthony Weston has raised similar issues in greater detail in a widely read article, and is perhaps the most articulate critic of intrinsic value (1996, 292–98). In one of his most rhetorically compelling claims, Weston has argued that appealing to intrinsic value is itself contrary to the lessons of ecology, which should make us suspicious of separating things off from one another and building hierarchies of value: "The appeal to intrinsic value in nature . . . is incompatible with the very ecological insights for which environmental ethics wanted to speak" (1992, 112). For Weston and Thompson, arguments about the intrinsic value of nature are philosophically incoherent and counterproductive.

These concerns and objections are common enough and, I believe, carry a great deal of power. It seems to me, however, that there is an even more basic disagreement, or at least difference of sensibility, between those who would use intrinsic values as trumps and those who distrust such a move. Hargrove criticizes pragmatism for confusing what, to him, is a perfectly obvious and useful distinction between instrumental and intrinsic values. Such a confusion threatens to subvert any serious attempt to articulate aesthetic values in nature: "When an aesthetic intrinsic value judgement is converted into instrumental terms, the person having the aesthetic experience is depicted as using natural scenery as a trigger for feelings of pleasure. When these feelings of pleasure are then compared with the other instrumental values that can be obtained, for example, by clear-cutting or stripmining, the value of the aesthetic experience then appears trivial, ridiculous, and indeed indefensible" (Hargrove

1992, 197). For Hargrove, instrumental values seem to inevitably reduce to economic values; thus, environmental values don't stand a chance because aesthetics or moral respect simply can't compete with other utilitarian goods in the marketplace. My concern here is not to criticize Hargrove's understanding of instrumental values (although his view does seem unreasonably, even cynically, restrictive). Rather, it is to point to the assumption behind the claim: that in the event that environmental concerns have to compete with economic concerns on an even footing, there is no reason to believe they will be taken seriously at all. This assumption leads Hargrove to develop his "weak anthropocentric intrinsic value" model of aesthetic appreciation of nature to trump instrumental (i.e., economic) values. Because the majority of citizens are not necessarily any more qualified to evaluate the aesthetic value of particular natural objects, formations, or processes than they are to evaluate the aesthetic value of fine art, these evaluations will have to be made and enforced by authoritative experts: "In real life . . . the value of a painting does not depend on the occurrence of particular emotional experiences in the general public. Rather it depends on the judgment of experts who interpret social ideals—the equivalent of the perception of Aristotle's 'good man.' Precise aesthetic judgments, comparable to those provided by art critics, can also be obtained by consulting professional nature interpreters, naturalists, and most environmentalists" (Hargrove 1992, 198). At the end of the day, the development of intrinsic value as a trump to other values is a way of bypassing democratic discussion and control of environmental decision making. The not quite hidden assumption is that "the people" don't really understand any language other than that of the market and are probably driven by only the most crude of utilitarian desires. Why else would Hargrove fear evaluating our environmental values in relation to clear-cutting and strip-mining? Only experts and environmentalists can be trusted to appreciate natural beauty and value it appropriately.

On this score, the contrast between Hargrove and Weston, for example, couldn't be greater. If Hargrove hopes to use intrinsic value as a trump to silence or avoid public debate on environmental policy and values, Weston's attack on intrinsic value is inspired by a belief in the absolute necessity of promoting such a debate: "We learned the values of nature through experience and effort, through mistakes and mishaps, through poetry and stargazing, and, if we were lucky, a few inspired friends. What guarantees that there is a shortcut? . . . What we have yet to accept is its [pragmatism's] inconclusiveness and open-endedness, its demand that we struggle for our own values without being closed to the values and the hopes of others. The search for intrinsic values substitutes a kind of shadowboxing for what must always be a good fight" (Weston 1996,

303). The problem, for Weston, is how to encourage the ongoing public discussion, debate, and development of environmental values in all the various particular contexts in which environmental problems and issues arise and with all the interested parties. The task for environmental ethics is not to produce trumps that preempt discussion and overdetermine policy outcomes. Weston's project, on the contrary, is much more like that of Bryan Norton, who writes, "The core problem is one of building human communities, including a more communal perception and more democratic decisionmaking process, that are capable of charting a course toward sustainability" (1995, 355). If Hargrove's presumption is that the democratic public cannot be trusted with environmental values, Weston's and Norton's presumption is that the public is the *only* appropriate location for decisions about such values.

The underlying distrust of, or faith in, democratic deliberation that seems to underlie Hargrove's and Weston's opposing views of intrinsic value is part of a more general story in contemporary environmental ethics. Intrinsic value is just one of the techniques or conceptual tools that are used by some theorists in the hope of developing a principle (or principles) that simply bypasses public debate and discussion, on the grounds that the moral resources and qualities of the public are not adequate to the task of respecting and protecting the environment. Peter Wenz's book, *Nature's Keeper* (1996), provides a recent illustration. The premise of Wenz's argument is that "our culture" is hopelessly "anthropocentric," and this anthropocentrism is responsible for virtually every evil ever visited upon our society: "Those who consider grass to be at our disposal may be led by degrees to similarly view insects, mice, cattle, and deformed human beings, connecting our culture's anthropocentrism to the Holocaust" (6). The breathtaking sweep of this claim leaves Wenz with a problem, however: If our culture is so uniform and so corrupt that our deepest beliefs lead from agriculture to the Nazis, how can we even begin to reform ourselves? The answer, he suggests, is to turn toward indigenous cultures, for they promote a more harmonious understanding of the proper relationship between humans and the rest of creation. Even though he admits that it is in no way obvious what kind of lessons our society could actually take from small, homogeneous traditional communities, he believes they offer the only alternative to our own perverse culture (13, 157). Regardless of how shockingly and crassly simplistic Wenz's historical, political, and anthropological analysis is in this book, perhaps the most important message is that he simply does not believe that the beliefs and sensibilities of his fellow Americans have any resources internal to themselves with which to address our environmental problems in a satisfactory manner. In his view, his neighbors, after all, hold views that are implicated in the Holocaust.

J. Baird Callicott (1995) faces a similar problem in a recent article. He, like Wenz, believes that only after our conventional worldviews are replaced by a "new nonanthropocentric, holistic environmental ethic" will we be able to articulate a morally defensible understanding of the natural world. In fact, he believes that the problem he faces as a biocentrist is less philosophical than it is political: he holds that a holistic and nonanthropocentric environmental ethic has already been "persuasively articulated by Aldo Leopold, Arne Naess, Holmes Rolston, and Val Plumwood, among others" (24). Thus, the philosophical work is already done, and what remains is the "eventual institutionalization" of this ethic in practice. Callicott suggests that as a matter of practical politics, it may be necessary to speak in more conventional moral language when explaining the reasons for environmental protection to nonphilosophers: "Granted, we may not have the leisure to wait for a majority to come over to a new world view and a new nonanthropocentric, holistic environmental ethic. We environmentalists have to reach people where they are, intellectually speaking, right now. So we might persuade Jews, Christians, and Muslims to support the environmental policy agenda by appeal to such concepts as God, creation, and stewardship; we might persuade humanists by appeal to collective enlightened human self-interest; and so on. But this is no argument for insisting . . . that environmental philosophers should stop exploring the real reasons why we ought to value other forms of life, ecosystems, and the biosphere as a whole" (24).

Unlike Wenz, Callicott's strategy is to appeal to traditional values and try to interpret them in an environmental light. The assumption, however, is that this is nothing more than a rhetorical task, as there is no philosophical merit to these claims. Philosophers have already, as he says, "persuasively articulated" the truth on these matters, but the many simply cannot or will not understand these "real reasons" for valuing nature. Philosophers may be able to fool the many into believing that their religions support environmentalism, but the philosophers understand that these moral traditions are actually a grave threat to the environment. The strategy Callicott proposes appears at most to provide some breathing room for the philosophers while they try to figure out how to pull off the "eventual institutionalization" of "a new holistic, non anthropocentric environmental ethic" (1995, 24). Callicott's strategy is not to engage democratic citizens in public debate about environmental values and policy. On the contrary, it looks as if it is to patronize them while he considers how to gain political power. There appears to be no open toleration for the variety of moral traditions and discourses generated in democratic society. Callicott never attempts to hide his contempt for the moral languages spoken by those around him.

If I am right that the attempt to develop intrinsic value as a trump in moral calculation is motivated by this more general distrust of democracy that drives certain schools of environmental ethics, we are left to consider whether such a distrust is sensible or counterproductive for environmental ethics. My own view is that it is counterproductive, if not insulting. Although my position is not subject to any definitive form of proof, I can point to two pieces of evidence to support my view. First, I believe that our intellectual traditions—religious, political, cultural, and artistic—are rich in respect and concern for nature. Now, this doesn't mean that I believe that these traditions are unambiguously supportive of environmental sensibilities. On the contrary, there is an awful lot in our inheritance that has led to a blindness or a hostility toward and fear of nature that is partially responsible for environmentally irresponsible behavior in the past and present. But I also think it is possible to demonstrate that there are, within these very same traditions, moral resources to help us challenge and confront this environmental irresponsibility.

Let me give one simple example here. John Locke's political theory is not infrequently thought of by environmentalists as an example of the threat liberalism presents to environmental concern. Locke's theory, after all, promotes the view that "the earth has been given to the children of men" (1955 [1689], II.25), and that the private ownership of land and other property is a positive good because it maximizes the degree to which the "industrious and rational" "subdue and cultivate" the earth (II.34–35). For many interpreters, this suggests that we find an early capitalist theory in Locke's work, the defense of a narrow anthropocentrism and a rapacious economic order aimed at the exploitation of nature for no other purpose than the promotion of ever increasing human wealth. But as true as it is that such values can be promoted by an appeal to Locke's theory, it is also true that Locke tells us very early in his discussion of property "Nothing was made by God for man to spoil or destroy" (II.31). My point is not that Locke was an early environmentalist—far from it. It is simply that there are philosophical and moral resources that would suggest to even a Lockean that economic development that seriously damages the natural environment stands in potentially sharp contrast to natural law as understood by Locke himself. I believe that comparable analysis holds true for our religious and artistic and other intellectual traditions as well.

And this leads to the second reason I believe it is wrong to assume that our intellectual inheritances are barren of potential environmental insight and concern, and that environmental ethicists therefore need to develop moral trumps such as intrinsic value: the majority of Americans have, in a very short period of time and to a remarkable degree, adopted the perspective of environmental concern. The data here are well-known: in 1990, a majority of Americans (54 percent) believed that environmental laws did not go "far enough" to protect the environment; in the same year, 73 percent identified themselves as environmentalists, more than 60 percent said they would be willing to pay more taxes for the sake of a cleaner environment, and so forth. All the evidence I know of indicates that the environmental movement has been tremendously successful in encouraging the development of environmental attitudes and values among the citizenry generally. A recent study suggests that there is a "set of widely shared environmental beliefs and values" in American society, that "American environmentalism represents a consensus view" because its "major tenets are held by large majorities, and it is not opposed on its own terms by any alternative coherent belief system" (Kempton, Boster, and Hartly 1995, 5, 215). Far from promoting values deeply at odds with American political culture, environmentalism has expressed concerns and values that have been easily and very quickly recognized and absorbed into everyday and widely used moral language.

In light of this remarkable success, there is reason to be wary of approaches to environmental ethics that in any way promote distrust of or contempt for democratic culture. There is simply no strong reason, it seems to me, for environmental ethicists to be so distrustful of their fellow citizens. Such distrust can only threaten to damage a potentially fruitful relationship between theorists and a wider public. To the degree that intrinsic value is used in the way Hargrove identifies, it is only the product and expression of this more general distrust of democracy and therefore, in my view, not to be encouraged.

It is clear that my real disagreement with theorists such as Hargrove, Callicott, and Wenz is about the relationship between democracy and environmental ethics more than about intrinsic value per se. In fact, if we think of intrinsic value not as a trump, but as a name for a certain kind of moral respect for the natural world that recognizes its (nature's) value beyond human advantage, I not only would not object to it, but would suggest that it is a rich and important part of our common moral traditions. Contrast Mark Sagoff's discussion of the term with Hargrove's: "When we regard an object with appreciation or love, we say it has intrinsic value, by which we mean that we value the object itself rather than just the benefits it confers on us" (1992, 58). It isn't necessarily true that when we love something, we "value it independently of the benefits it confers on us"—Nietzsche's (1967) famous birds of prey love little lambs dearly, but only because they are tasty treats for them. But Sagoff's point is clear enough: there are many times when we experience a respect or appreciation for natural objects and/or systems that is entirely independent of conventional utilitarian considerations. Sagoff charmingly, and appropriately, uses Charlotte's (of *Charlotte's*

Web fame) love of Wilbur as an illustration of such an experience, and suggests that just as we can love humans as bearers of such intrinsic value, so we can and do love nonhuman objects in a comparable or similar way.

What is important to note about Sagoff's discussion is the degree to which theorists such as Norton and Weston would find it conceptually unobjectionable, even if they would not use Sagoff's language. Sagoff's observation is about recognizable moral values, found even in our children's literature, that must be reckoned with in our environmental decision making. Although Weston would not use the phrase intrinsic value to express this point—he reserves that for the kind of knock-down principle or moral trump that Hargrove has in mind—he clearly agrees with Sagoff when he writes, "There is every reason to think that respect for other life forms and concern for natural environments are among" the plurality of values we must consider in our public life (1996, 286). Likewise, Norton writes, "Philosophers, anthropologists, humanists, economists, and citizens must join the search for appropriate public values. In short, all parties must enter the dialogue, rather than creating closed disciplines insulated by jargon and esoteric mathematics" (1992, 38). I see no reason to think that Sagoff's understanding of intrinsic value is guilty of any of the vices Norton is worried about. In fact, our widely recognized moral and aesthetic traditions are full of the kind of love and respect for nature Sagoff refers to, and can therefore be appealed to precisely as the foundation of shared commitments. Rather than shutting off or trumping public debate, intrinsic value in this more conventional sense is entirely compatible with, even critical to, the building of human communities committed to the sustainability that Norton discusses or the kind of mutual moral ground within our complex value systems and traditions Weston hopes to identify (Weston 1996, 322).

So, in the end, I think the battle over intrinsic value is really just a symptom of a bigger issue: the degree to which we believe the public, drawing on our conventional moral traditions, is capable of contending with our environmental concerns and problems. In the end, proponents and opponents of intrinsic value are closer on this particular issue than they first appear. Their real disagreement concerns who they are most inclined to trust: intellectuals and experts of various sorts, or democratic publics. Their difference on this point determines the degree to which they think environmental values, such as theories of intrinsic value, must be developed as philosophical and political trumps in policy debate, or, on the contrary, are to be used in these debates as the grounds for building wide and democratic political constituencies for developing sensible environmental policy.

PART II

Case Studies in Sustainable Environmental Policy and Law

Introduction

In Part 1 of this book we examined the range of normative values that provide a basis for environmental policy discourse. We analyzed the principles of conservation and preservation, we discussed the definition and ramifications of "pragmatism" as a standard for public choice, and we considered the more general role that moral principles play in administrative decision making, given the demands of a democratic political system.

After short arguments by the core participants, we focused on four issues that occupy us throughout the rest of the book. Specifically, we began a debate over the role of science in environmental decision making, the interdependence of social and environmental justice, the adequacy of economic or instrumental evaluation to making "good" environmental policy, and the possibility that only with an acknowledgment of nature's intrinsic value will we properly consider human and natural systems in policy decision making.

The consideration of normative principles and public environmental policy cannot proceed further, however, without a full discussion of the concept of sustainability. Therefore, even though the concept of sustainability has already come up in our discussions, it now becomes the center of concern for this and Part 3 of the book.

The imperative to understand and analyze sustainability comes from our joint acknowledgment that it exists at the center of current environmental philosophy and policy. The discussion of normative or moral principle applied to environmental concerns and the effort to transcend merely economic evaluation of nature, now and in the future, are undertaken in relation to sustainability, which, in terms of environmental policy discourse, has become the rallying cry for everyone from academics to grassroots environmentalists to business leaders.

Both those who decry too much regulation and those who clamor for more have adopted the goal of sustainable development. In this nation, abroad, and in the world of international environmental law, sustainability has become the standard for judging "better" environmental policy. Those who worry about intergenerational justice and those who would apply today's preferences as the moral imperative for the law of the future agree that sustainability is a "reasonable" goal. Conservationists, preservationists, pragmatists, and even bureaucrats and economists promote the concept. With agreement like this, sustainability demands attention.

Therefore, we now proceed to the consideration of seven case studies that concretely address the concept of sustainability as an adequate standard for environmental policy and law. The following case studies, originally given as MacArthur Lectures at Bucknell University, apply sustainability to a cross-section of environmental cases toward the end of understanding what using sustainability as the core principle of modern environmental argument means, what a sustainable public policy is, and how it is different from the status quo.

Specifically, taking the concept of sustainability as a common point of departure, Barry Rabe examines the role of decentralization in fostering sustainable subnational environmental policy in Canada and the United States, and Anna Schwab and David Brower argue for the use of sustainable development to control human growth patterns and mitigate natural hazards. Celia Campbell-Mohn then takes up the legal ramifications of the concept of sustainability for private property, Jonathan Wiener argues that sustainable levels of environmental risk require the constraint of government action, and Robert Percival takes an international viewpoint in his argument for a more active legal and administrative structure to create and implement sustainable environmental law. Finally, Jan Laitos argues that sustainable resource law on public lands has come to mean the ascendance of preservation and recreation over multiple-use policy, and William Lowry investigates the role of Canadian and American political institutions in the sustainable preservation of national parks for future generations. Overall, these essays provide case studies in sustainability for a cross-section of pollution abatement and natural resource issues, applying this normative standard to national, comparative, and international law.

In Part 3, the discussants again address the issues of science, social justice, and the instrumental or intrinsic value of nature but with special attention to how the following case studies define and utilize the concept of sustainability as a normative standard for environmental policy and law. However, before this analysis, we need to see how sustainability is being used currently to analyze environmental policy and legal issues. This is the role of the case studies.

The Subnational Role in Sustainable Development: Lessons from American States and Canadian Provinces

Barry G. Rabe

It is difficult to quibble with the basic tenets of sustainable development, as set forth in the 1987 Brundtland Commission report (World Commission on Environment and Development [WCED] 1987) and in related documents. The ideas of reconciling economic development with long-term resource capacity and the protection of future generations have been enthusiastically received by most national and subnational governments. This has been especially evident in the North American nations of the United States and Canada, where both federal and virtually every state and provincial government have gone on record in support of sustainable development. Many have begun to use the language of sustainability extensively in describing new policy initiatives relevant to environmental protection, energy distribution, transportation, and economic development, as well as other areas. Indeed, it has become increasingly fashionable, from Tallahassee to Victoria, to refer to all sorts of governmental activities as compatible with sustainability.

Support for decentralized approaches to sustainable development is evident in a variety of international agreements and scholarly analyses. A core premise of the Brundtland Commission report is unleashing localized creativity in devising new policies compatible with the broad goals of sustainability. The Agenda 21 principles, established to guide the U.N. Commission on Sustainable Development, strongly endorse a "bottom-up" approach that calls for subnational experimentation and mechanisms for providing "decentralized feedback to national policies." Even the 1992 Rio Declaration on Environment and Development declares that pressing environmental problems such as global warming are most likely to be addressed effectively through broad political participation "at the lowest, most accessible, and policy-relevant level." As policy analysts David Feldman and Catherine Wilt have noted, this approach "clearly rejects centralized, bureaucratic approaches" (1998, 140).

This emphasis coincides with an increasingly common theme in much environmental policy analysis that endorses far-reaching decentralization. The conventional wisdom of the 1960s and 1970s anticipated severe collective action disincentives to decentralized environmental protection, given the inherent cross-boundary nature of many environmental problems and the capacity of localized units to shirk responsibility. However, the overwhelming emphasis of more recent scholarship has embraced state, provincial, local, and regional units as highly committed to innovative approaches to environmental protection and sustainability. In such analyses, these subnational authorities are increasingly depicted as more capable and innovative than their central-level counterparts. Indeed, in some areas of environmental policy analysis, such as the growing body of scholarship on "common-pool resources," subnational units are regularly viewed as capable of doing little wrong, whereas their national counterparts are perceived as capable of doing little that is effective.

Such an emphasis on decentralization as a stepping stone to sustainability would appear to provide Canada with a tremendous advantage over the United States in devising an innovative, meaningful response to the challenge of sustainable development. Canada simply has none of the centrally directed, medium-based, command-and-control legislation that has long been a hallmark of American environmental policy. In air pollution policy, for example, Canada has never enacted a law that either establishes national air quality standards, imposes a uniform framework that mandates specific pollution-reduction technologies, or creates a national mechanism for emissions trading across jurisdictional boundaries. Instead, Ottawa has set national "objectives" for air quality, but allows individual provinces to "set air quality standards through their own environmental protection acts or specific air management legislation" (Davies and Mazurek 1997, 205). This pattern of provincial deference is also evident in other areas of environmental policy, including water pollution and hazardous waste management, giving Canada one of the most decentralized environmental policy systems of any Western government (Harrison 1996). Canadian delegation of authority to individual provinces has been further expanded through the Canada-Wide Accord on Environmental Harmonization reached between Ottawa and the provinces. Under this agreement, which took effect in January 1998, the federal government formally confines its environmental jurisdiction to "federal lands," whereas "the provinces will regulate everywhere else (although exceptions may be made in certain circumstances)" (Elgie 1998, 10).

In contrast, the U.S. environmental regulatory system is far more oriented toward central government influence, most notably through rigorous sets of standards and rules imposed through medium-based federal programs. In virtually every area of environmental policy, U.S. federal legislation is more detailed than its Canadian counterpart and far more oriented to impose formal requirements on state environmental agencies and the private sector. The degree of federal influence varies by program and is particularly strong in such areas as air and water pollution control, hazardous waste management, and protection of endangered species. During the Clinton administration, several efforts were undertaken to allow greater state flexibility, particularly among those states demonstrating a particular commitment to many of the sorts of goals central to sustainable development (Kraft and Scheberle 1998). In addition, the U.S. federal government has a much stronger tradition of providing grant funding and technical support for states. States on average receive approximately 20 to 25 percent of their total environmental funding from federal grants. Nonetheless, many analysts of U.S. environmental regulatory federalism view the federal role as excessive, squelching state and local innovation by imposing rigid national standards.

Given these distinctions, most observers would anticipate that the more independent Canadian provinces would have made far more progress than American states in responding to the challenge of sustainability. But is that expectation reflected in actual policy formation and implementation? Almost fifteen years after the issuance of *Our Common Future* (WCED 1987), it is appropriate to examine the translation of the concept of sustainability into actual policy.

The review of recent experience, based on comparative case study analysis of twenty subnational units (the states of Arizona, Colorado, Georgia, Michigan, Minnesota, New Jersey, North Dakota, Oklahoma, Oregon, and Pennsylvania and all ten Canadian provinces) conducted between 1996 and 1998, indicates a very limited, largely symbolic response to these challenges in most provinces and states. But contrary to the conventional wisdom on decentralization, American states have on the whole made more progress than their Canadian provincial counterparts, particularly in the areas of pollution prevention, regulatory integration, environmental indicator development, and active exploration of greenhouse gas reduction options. A more active federal government role in fostering innovation and greater policy entrepreneurship within state (versus provincial) agencies appears to have directly contributed to this comparative difference (Rabe 1999). At the same time, there remains tremendous variation in the response to sustainability across the respective state and provincial cases. The state of New Jersey, for example, has advanced particularly far in regulatory integration and pollution prevention planning, moving toward a system of facilitywide permits that dramatically reduces overall emission levels. New Jersey has also taken some pioneering steps in attempting to combat global warming. The province of Prince Edward Island has made considerable progress on land use protection and addressing nonpoint sources of environmental contamination. The province has also established seven core sustainability goals that promote integration across agency boundaries, pollution prevention, and waste reduction. But these cases generally remain exceptions to the prevailing pattern, with far less overall innovation than would be anticipated, especially in Canada. Instead, observers in most jurisdictions submit a fairly uniform set of perceived impediments, including resource constraints, paucity of political leadership and support, resistance to change within agencies and regulated firms, and the absence of immediate and visible threats to economic sustainability, especially in the relatively prosperous late 1990s. In short, many provincial and state officials suggest that sustainability is a noble goal but one that is exceedingly difficult to translate into sustainable policy.

One subnational case stands out, however, as exceptional in the extent to which it has mounted a significant response in each of the dimensions of sustainability noted above. The state of Minnesota, far more than any other state or province examined, has taken seriously the challenge of sustainability and responded with an array of significant policy initiatives throughout the 1990s. Consistent with federal and state law, Minnesota continues to operate all standard environmental protection programs. But it has closely examined these for creative ways to make them more compatible with sustainability goals and has also explored actively a host of new policy initiatives. Consequently, Minnesota comes far closer to meeting the expectations of decentralization advocates, giving some glimpse into what a "sustainable state" or "sustainable province" might look like. But its exceptional features also underscore the apparent limits of imposing excessive expectations on decentralized units in responding to the challenge of sustainability.

This distinct case is the primary focus for the balance of this case study. First, we consider some of the factors underlying this unusually rigorous policy response. In short, what has happened in Minnesota and what has been missing in other American states and Canadian provinces? Second, we examine a number of policy innovations being undertaken in Minnesota and review their performance to date. What might the next generation of environmental and related policies look like, to the extent that the Minnesota experience can be deemed to provide a model worthy of emulation by other governments? Finally, we explore the prospects for long-term commitment to these initiatives in Minnesota and for their replication by other state, provincial, or federal governments.

The Political Context for Innovation

Traditional analyses of state and provincial policy innovation or rigor have tended to focus on variables such as fiscal resources, institutional capacity, and problem severity. These variables may provide some of the impetus behind the unique Minnesota response but must be supplemented to provide a fuller explanation of its deviance from other cases. Minnesota traditionally has been ranked among a group of about a dozen states, including Connecticut, Iowa, and Kentucky, that are designated as "strugglers." According to political scientist James Lester, such states generally "have the will but not the resources (fiscally and institutionally) to pursue aggressive environmental protection policies" (1994). Minnesota has tended to rank somewhat above average in per capita expenditures on environmental protection in the 1980s and 1990s, but well behind a number of other states (Council of State Governments 1999). Its main environmental agency, the Minnesota Pollution Control Agency (MPCA), has historically retained a fairly traditional, medium-based, pollution control focus. In turn, the state has not had the sorts of riveting environmental disasters that tend to trigger major policy initiatives (Birkland 1997) or particularly high measures of environmental risk or contamination. Consequently, it would not initially appear likely to be such an outlying case.

However, Minnesota is distinctive in having a political culture deemed highly receptive to public sector innovation. Consistent with this culture, it has an unusually large and diverse collection of leaders with a long-standing interest in finding new approaches to environmental and related challenges. This has provided a rich network of "policy entrepreneurs" increasingly thought to be central to policy innovation (Mintrom and Vergari 1998). "There's almost a natural psychological progression for people in the field, from conventional pollution control to thinking about sustainability," explained one state agency official. "For a lot of people in this state, that light has gone on, that you can't separate these things in separate boxes. From the start, we have just had a lot of stakeholders involved and interested."

Among elected officials, Republican Governor Arne Carlson served from 1991 to early 1999 and is widely seen as a champion of the shift toward sustainability. In turn, a series of key state legislators from both political parties in both the House and Senate have been active and consistent supporters of new approaches. Within state agencies, a number of officials in the MPCA and other related units of state government, including the Office of Environmental Assistance, Department of Natural Resources, Environmental Quality Board, and Public Utilities Commission, are commonly noted as sources of new ideas and support for implementation. Fragmentation problems endure, as they do in all states and provinces, but Minnesota's elected and agency officials have demonstrated unusual willingness to consider—and undertake—far-reaching innovations. Furthermore, alternatives to conventional approaches to environmental regulation have a long-standing tradition in the private sector in Minnesota, most notable in the pioneering work of the 3M Corporation on pollution prevention. Indeed, many of the major policy initiatives launched in Minnesota have had extensive corporate stakeholder input and support, even in instances when these have required significant new investment of resources. These efforts have also tended to draw support from environmental advocacy groups, even in those types of initiatives that afford compliance flexibility for regulated parties and have been known to trigger opposition in other settings.

This broad base of entrepreneurial talent has led to the creation of a diverse network of supporters within the state, many of whom are well connected to regional and national institutions. In particular, the state has been extremely successful in leveraging federal government grant funding as well as securing U.S. Environmental Protection Agency authorization to allow innovation within the vast framework of federal regulatory programs. Such a base has not only put sustainability and environmental policy innovation on the agenda in Minnesota, but has served to support and literally sustain new efforts in a way not evident in the other states and provinces examined. In turn, a series of national awards for various state policy initiatives in recent years has served to further elevate these efforts and provide supporters with opportunities to claim credit and seek to solidify their standing.

Minnesota Policy Initiatives

The shift toward sustainability in Minnesota has been multifaceted, with active exploration of alternatives in numerous policy areas. The breadth and rigor of these initiatives are among the factors that distinguish Minnesota from other states and provinces. These have ranged from preventive initiatives designed to minimize the generation of pollutants and toxic substances to local experiments in developing "sustainable communities." Each of the major areas of experimentation is discussed below, beginning with the series of state initiatives in the area of pollution prevention.

Pollution Prevention

Minnesota has probably received the most national notoriety for its early and extensive efforts in the area of pollution prevention. All fifty states and ten provinces now have some form of pollution prevention legislation. But most are confined to relatively narrow programs that provide technical or financial assistance to private firms willing to explore methods to eliminate or reduce pollutant generation and release. Minnesota, in contrast, moved into this area far earlier—and more aggressively—than most state or provincial counterparts. Building on its 1980s work with technical assistance programs and a series of legislative hearings in 1989 and 1990 to consider the limitations of the existing pollution control system and possible alternatives, the Minnesota Toxic Pollution Prevention Act (Minnesota Statute Chapter 115D) was signed into law on May 3, 1990. Enacted in concert with the 1990 U.S. Pollution Prevention Act, the Minnesota legislation established the goal of "eliminating or reducing at the source the use, generation, or release of toxic pollutants, hazardous substances and hazardous waste." The legislation had broad, bipartisan support and has been a cornerstone in all subsequent pollution prevention initiatives. It was amended in 1993 to further expand the scope of its activities.

The Toxic Pollution Prevention Act required all Minnesota facilities reporting toxic chemical releases under the U.S. Toxic Release Inventory (TRI) to establish comprehensive plans for pollution prevention and submit progress reports to the state each year. In turn, the state attempted to assist facilities through these processes by expanding technical assistance and grant programs. Minnesota also established a process for selecting annual Governor's Awards for Excellence in Pollution Prevention and disseminating information about pollution prevention through conferences, workshops, and continuing education programs.

The plans and progress reports remain, however, the central focus of the legislation. They have enabled Minnesota to make extensive use of the unique data on chemical releases provided each year through the TRI. Facility pollution prevention plans must include a description of the "types, sources and quantities" of TRI chemicals at the site and a "description of processes that generate or release" them (Minnesota Office of Environmental Assistance [MOEA] 1998, 13). They must also specify prior and current pollution prevention initiatives undertaken at the facility, along with an evaluation of their effectiveness.

This sets the stage for perhaps the most important aspect of the planning process, an assessment of pollution prevention alternatives as well as a "statement of pollution prevention objectives and a schedule for achieving these objectives" (MOEA 1998, 13). Through the annual progress reports, each pollution prevention objective is reviewed, involving an assessment of the methods used, actual progress achieved, and any impediments faced in attempting to attain the objective. These progress reports, unlike the plans, are public documents. In conjunction with the overall TRI scores released each year, this process both provides considerable incentive for facilities to pursue pollution prevention opportunities in an active and systematic way and a method to measure progress. Legislative amendments approved in 1993 served to expand this program by extending the planning process requirement to additional firms, particularly nonmanufacturing facilities that use large amounts of chemicals covered by the TRI. Ironically, as the scope of the program has expanded, the number of Minnesota facilities required to submit plans and progress reports has declined, from 562 in 1992 to 418 in 1996. This may reflect the impact of implementing pollution prevention provisions, with firms achieving significantly lower levels of chemical releases that place them below the threshold that requires plan development.

Fees, Taxes, and Economic Incentives. Minnesota has also attempted to increase involvement in pollution prevention through creation of economic incentives. Extensive use of a series of fee and tax programs is intended to encourage prevention and related waste reduction activities. The Toxic Pollution Prevention Act has maintained a two-part fee structure throughout its existence. All Minnesota facilities required to participate in the planning process must pay $150 annually for each of the nearly six hundred chemicals listed in the TRI that they release. In turn, facilities that release less than 25,000 pounds of TRI chemicals each year also must pay a fee of $500. But those facilities that release more than 25,000 pounds of these chemicals annually must pay a fee of $.02 per pound of toxic chemicals released. This program generates approximately $1 million per year, all of which is allocated to pollution prevention assistance programs established by the legislation.

This type of initiative, long advocated by environmental policy analysts as a way to better link environmental and economic impacts, is also evident in other environmental programs in Minnesota. For example, the state has experimented with municipal solid waste taxes for about a decade. These have been designed to create an incentive for waste generators to pursue waste reduction options and provide a source of funding for related programs such as cleanup of closed solid waste landfills. In the latest version of this initiative, Minnesota enacted the Solid Waste Management Tax in January 1998. This legislation establishes a 9.75 percent tax on the solid waste bill of each homeowner or renter and a 17 percent tax on commercial waste bills. In turn, nonindustrial solid waste is taxed at a rate of $.60 per cubic yard. Many observers contend that this decade of experience with such taxes has contributed to the dramatic reduction in solid waste generation and increase in recycling rates in Minnesota during the 1990s. By 1998, Minnesota recycled 42 percent of its solid waste. Mindful of the potential incentive such a tax might create for noncompliance, particularly given Minnesota's large decentralized areas, the state has begun to consider ways to assure full participation. For example, the state has sponsored a pilot program in a rural county whereby citizens who relinquish their solid waste "burn barrels" and agree to use authorized solid waste services will receive six months of waste collection at half price.

Substance and Product Bans. Alongside planning and economic efforts, Minnesota has also increasingly turned to the option of preventing environmental contamination through outright prohibitions on the use of various chemicals or substances in any commercial activity. Legislation that went into effect in July 1998 made Minnesota the first state or province to prohibit the "deliberate introduction" of lead, cadmium, mercury, and hexavelent chromium into a series of products such as inks, dyes, paints, pigments, and fungicides (U.S. Environmental Protection Agency [EPA] 1997b, 5). The state had previously banned the use of these heavy metals in packaging materials and has targeted mercury throughout the 1990s in an all-out effort to minimize its continued introduction in any form to the Minnesota environment (MPCA 1996, 14). Minnesota legislation has also extended this strategy to municipal solid waste, with a lengthy list of items and materials that are banned from disposal, including nickel-cadmium and sealed lead acid batteries, motor vehicle fluids and filters, all mercury-containing products, and major appliances (MOEA 1997, 21).

In conjunction with substance and product bans, Minnesota has also begun to explore so-called product stewardship requirements. Far more common in Europe than in North America, such programs require manufacturers and distributors of designated goods to establish collection systems to "take back" and safely manage them once they have been used. Minnesota statutes already have established such stewardship requirements for vehicle batteries, some rechargeable batteries, fluorescent tubes, mercury-containing manometers and relays, and used oil and oil filters. Consideration is being given to add various other commercial items to such lists in future years.

Intrastate and Intergovernmental Partnerships

Pollution prevention and related sustainability initiatives have required not only the passage of legislation but also implementing institutions with commitment and capacity to move the state in new directions. Although Minnesota environmental and related agencies maintain many traditional dividing lines between respective media and function, the state has established a series of supplemental organizations that play essential roles in fostering new approaches. Such organizations have also, in many instances, helped forge intrastate partnerships necessary for implementation. In turn, the state has worked aggressively to take full advantage of opportunities presented by the federal government to acquire added financial resources to help underwrite many of the costs of new programs as well as receive unprecedented flexibility to pursue new initiatives while remaining in full compliance with applicable federal law.

Office of Environmental Assistance. In many states and provinces, pollution prevention and regulatory alternatives get shoved into corners of existing pollution control agencies. They quickly become overshadowed by well-entrenched programs and agency units. In Minnesota, many of these functions remain under the jurisdiction of the Pollution Control Agency. But a separate organization, the MOEA, maintains a largely independent set of environmental programs and has put considerable pressure on the MPCA and other state government units to maintain a major focus on sustainability objectives.

The MOEA was created in the mid-1990s, an offshoot of the former Office of Waste Management (OWM). As the prior OWM emphasis on waste management and facility siting began to shift toward waste reduction, it was revised to incorporate a series of broader responsibilities. The MOEA director reports to the MPCA administrator and both organizations are lodged in the same building in St. Paul. However, MOEA's somewhat autonomous status and broad charge have given it a unique opportunity to pursue a series of initiatives that might easily be quashed elsewhere. It lacks formal regulatory authority but exercises considerable influence through other mechanisms. "The MOEA has been somewhat entrepreneurial, able to take advantage of different opportunities to educate and excite people about real options," explained one state official.

First, the MOEA is responsible for administering a good deal of the state's grant funding, particularly in pollution prevention and related areas where it operates the Minnesota Waste and Pollution Prevention Assistance Grant Program. In fiscal year 1997, for example, the office distributed more than $750,000 for pollution prevention and related sustainable development initiatives across

Minnesota. In recent years, it has targeted local, community-based initiatives, discussed further below, for a large portion of this assistance. Second, the MOEA handles considerable program evaluation duties. State legislation authorizes MOEA to complete a comprehensive report on Minnesota's progress in pollution prevention and related areas every other year. These reports, the most recent of which was released in summer 2000, include detailed analyses of all essential dimensions of pollution prevention in Minnesota. They have been well publicized as "progress reports" on state efforts and also provide a forum for MOEA to make recommendations for future policy. In fact, each of the reports issued during the 1990s has featured numerous recommendations, many of which have either influenced policy deliberations in the legislature and agencies or actually been enacted into law. Third, the MOEA has worked closely with diverse stakeholders and attempted to build bridges among differing constituencies as policy alternatives have been considered. These efforts have ranged from consultant-like assistance to local governments to an electronic newsletter on a wide range of statewide activities pertinent to sustainable development.

Interagency Integration. The role of the MOEA is complemented by a set of initiatives designed to compel individual state agencies to work toward the common goal of environmental protection, pollution prevention, and the broader goal of sustainability. Foremost among these is the Minnesota Environmental Quality Board (EQB), which serves to further support policy innovation as well as foster integration of perspectives across traditional sectoral and agency divides. The EQB membership consists of the heads of ten state agencies with direct relevance to environmental quality, including the MPCA and such departments as Natural Resources, Agriculture, Energy, Transportation, Health, and Economic Development. The Board also includes five representatives from the general citizenry and its chair is appointed by the governor. During the 1990s, the EQB played an active role in fostering interagency collaboration and advancing possible initiatives in pollution prevention and sustainability in diverse policy areas.

In 1993, the EQB developed the Minnesota Sustainable Development Initiative, which was intended to allow its diverse membership to examine the long-term ramifications of linking economic development with environmental sustainability. This led to the publication of a report, *Challenges for a Sustainable Minnesota,* which outlined a future agenda for environmental policy in the state. In response, Governor Carlson established a process for roundtable deliberations, designed to bring diverse stakeholders together to consider options for sustainability in such sectors as agriculture, energy, forestry, minerals, and manufacturing. These operated on a consensus basis and resulted in a wide range of policy recommendations, including possible directions for environmental regulatory reform. Sustainability was a central focus for these respective groups, and a 1998 final report was intended to help set the agenda for the next state administration and generation of state environmental policy efforts.

An additional experiment in multiagency integration is the Interagency Pollution Prevention Advisory Team (IPPAT), which formally involves more than twenty Minnesota departments and agencies in pursuing pollution prevention within their own areas of concentration. The IPPAT was formed in response to a 1991 gubernatorial executive order that required all state departments and agencies to develop pollution prevention policy statements, undertake prevention efforts in their operations, and prepare plans and progress reports. It convenes regularly to facilitate communication on prevention opportunities among otherwise isolated agencies. A major focus for IPPAT has been examining processes of state purchasing decisions and specification requirements, with the intent of modifying these wherever possible to maximize the prospects for pollution prevention.

Regulatory Integration and Federal Support. Even before Clinton administration initiatives to encourage states to experiment with conventional environmental and grant programs in pursuit of greater cross-media integration and pollution prevention, Minnesota programs and the MPCA were actively exploring such alternatives through their federal government relationship. In particular, they made extensive use of existing federal grant programs and negotiated actively with U.S. EPA officials to explore ways to redesign traditionally fragmented regulatory tools to promote prevention and integration as well as greater efficiency. This resulted in a series of experiments in integrated inspection and monitoring for a number of facilities located along the Minnesota shore of Lake Superior. In addition, MPCA officials began to experiment with a so-called flexible permit, whereby individual facilities would receive streamlined review of multiple permits in exchange for substantial emission reductions well below existing standards. One such permit was granted to a major 3M facility in 1993, with particular attention to a variety of air contaminants, and other firms have expressed willingness to pursue such a project (Rabe 1995a). Both of these experiments took place with the active collaboration of the federal government.

This relatively early exploration of regulatory integration effectively positioned Minnesota to take full advantage of Clinton-era programs to devolve regulatory functions to states and encourage state innovation. In fact, Minnesota has been an unusually active participant in new federal programs intended to foster greater environmental policy integration and efficiency, such as Project XL and the National Environmental Performance Partnership System. The Minnesota legislature, drawing in part on some of the earlier state experiments in regulatory integration, further paved the way for this participation through passage of the Minnesota Environmental Regulatory Innovations Act of 1996 (Minnesota Statutes Chapter 114c). Passed unanimously by both legislative chambers and signed into law by Governor Carlson in April 1996, this legislation concluded that "environmental protection could be further enhanced by authorizing innovative advances in environmental regulatory methods." In response, the state authorized and encouraged MPCA involvement in new federal initiatives intended to allow such innovation. Moreover, the legislation also gave the MPCA the authority, according to one analyst, to "vary requirements" in other pollution control programs "if the agency finds that doing so would promote pollution prevention or reduce the administrative burden associated with regulatory requirements" (Buelow 1997, 13).

Project XL was a direct output of the Clinton administration's National Performance Review, the centerpiece of its "reinventing government" initiative. Under this program, according to an MOEA report, participants "must commit to capping each of their facility's regulated emissions or discharges at levels significantly lower than currently allowed. In return the agency offers a single, multimedia operating document" that allows far greater operational flexibility than under conventional methods (MOEA 1996, 69). Minnesota was one of a handful of states approved for initial participation in the program, the first case involving a 3M plant in Hutchinson. Observers concur that the Hutchinson case was making considerable progress toward identifying pollution prevention and regulatory integration opportunities. However, differences between 3M and EPA Region V officials became so severe that 3M decided to withdraw from the program in December 1996. Still, the MPCA remains one of the few state environmental agencies with EPA approval to participate in XL, although the future direction of the program remains unclear. At the same time, MPCA authorities are weighing options for pursuing their own regulatory integration initiatives in response to the latitude provided by the 1996 Regulatory Innovations Act.

Minnesota was an early and active participant in another major Clinton devolution initiative, the National Environmental Performance Partnership System (NEPPS). This effort, intended to realign intergovernmental management in environmental policy, stemmed from an agreement signed by EPA and state environmental agency heads in May 1995. Under NEPPS, states are encouraged to concentrate their energies on their most serious environmental problems. In turn, the federal government agrees not only to provide that added latitude but also to "give states with strong environmental programs more flexibility, both in program operations and in spending authority." At the same time, this allows the federal government "to concentrate its oversight and technical assistance efforts on weaker state programs" (Kraft and Scheberle 1998).

Agreements under NEPPS must be formally negotiated between the U.S. EPA and the counterpart state agency. Minnesota quickly began this process and was among the first states to be approved for participation in all key aspects of NEPPS. Its Environmental Performance Partnership Agreement (EPPA) was approved in October 1996, then revised in February 1997 (U.S. EPA 1997a). This agreement features a series of medium-specific and multimedia goals, each area followed by specific subgoals, objectives, and outcome measures to determine success. Many of these goal descriptions and related sections are quite detailed, reflecting the considerable advance work completed in Minnesota on priority setting and serious exploration of regulatory alternatives. In fact, this process appears to have been far easier to complete in Minnesota than in other participating states in which consideration of such innovative activity remains in very early stages. Among multimedia goals, the agreement specified three priority areas: brownfields reclamation, mercury pollution reduction, and development of sustainable ecosystems. In the case of brownfields, the Partnership Agreement outlines an environmental objective of removing, by 2006, "environmental barriers that deter productive use at 90 percent of the estimated 4,000 acres of commercial and industrial properties in Minnesota which are currently under-utilized due to contamination" (U.S. EPA 1997a, 34). For mercury pollution, the agreement features a series of specific objectives to reduce substantially the future use and release of mercury into Minnesota's environment.

Minnesota is also heavily involved in the other key component of NEPPS, the Performance Partnership Grants (PPG) program. The program involves a diverse series of federal environmental protection grants. Whereas such grants have historically been distributed on a highly fragmented, medium-specific basis, PPG is designed to allow recipients to combine funds from two or more categorical grants and address state-determined environmental priorities (Kraft and Scheberle 1998, 6). According to the Partnership Agreement established by EPA and Minnesota, the MPCA intends to use this unprecedented flexibility to "maintain consistency with national program guidance, more effectively link program activities with environmental goals and program outcomes, develop innovative strategies for pollution prevention, multimedia permitting and enforcement, ecosystem management and community-based management, and provide savings by streamlining administrative requirements" (U.S. EPA 1997a,

3). Minnesota has long been unusually adept at securing substantial amounts of federal grants-in-aid assistance and using those funds for creative purposes, most notably in the area of pollution prevention. However, the PPG program offers unprecedented opportunity for the state to pursue innovative objectives with external funding.

Emphasizing Environmental Outcomes

Environmental policy in most Western democracies, including the United States and Canada, has long been criticized for its emphasis on *output* measures that attempt to quantify the amounts of work completed by an agency. Instead, according to a wide range of policy analysts, environmental agencies should strive to develop *outcome* measures that take into account environmental performance and the actual impact of policy instruments on those measures (Davies and Mazurek 1997; U.S. Office of Technology Assessment [OTA] 1995). Both nations have begun to make this transition, although the preponderance of environmental agency reporting and evaluation remains focused on outputs. Indeed, state and provincial agencies in general remain most adept at reporting activities such as numbers of permits issued, inspections completed, complaints handled, and memoranda drafted. They tend, by contrast, to offer only a very limited understanding of environmental quality trends or the impact of various programs on environmental quality (Rabe 1999).

Minnesota, far more than any other state or province examined, has taken the challenge of developing outcomes measures and attempting to use them to guide policy. Indeed, Minnesota environmental agency documents (including those from the MPCA and MOEA) read very differently from those in most other states and provinces, given their substantial emphases on presenting and interpreting outcomes. Moreover, the state increasingly uses such information in weighing alternative policy priorities and in establishing publicly accessible measures of progress in key areas.

The state has made particularly extensive use of the data generated annually through the TRI, as have a few other states such as New Jersey and Oregon. However, a number of other states do not exceed minimal reporting requirements and conduct only superficial analyses of trends. Provinces have just begun to think about, much less utilize, such a data source, given the very recent initiation of a roughly comparable reporting system, the National Pollutant Release Inventory, by the Canadian federal government. In contrast, Minnesota has regularly and systematically evaluated TRI data and sought to use its findings for a number of key measures of trends and priorities. In particular, the state has discovered that TRI data offer abundant opportunity to concentrate agency regulatory efforts on regions or facilities with unexpectedly high levels of pollutant releases.

Overall findings from the Minnesota TRI analyses during the 1990s suggest a decline in total chemical releases and a general shift from direct releases of chemicals to the environment to other options such as on-site recycling and recovery and off-site treatment (MOEA 1996, 1998). Total environmental releases in Minnesota have dropped steadily during the decade, from 46.4 million pounds in 1991 to 22.1 million pounds in 1996. But the findings also indicate that the bulk of pollution prevention gains among TRI chemicals have tended to be concentrated among a relatively small number of large firms and that overall reductions in releases are not attributable to as high a level of source reduction as the state had hoped.

At the same time that reporting and analysis of TRI data have become central components of environmental planning in Minnesota, the state has continued to refine its methods for gathering and interpreting this important information source. The state has worked to normalize TRI trend data to assure comparability from year to year. It has also attempted to refine key measures such as "production ratios" to "determine if there is progress towards pollution prevention for individual chemicals, and for a facility as a whole" (MOEA 1998, 28). Such definitions are particularly crucial in implementing the pollution prevention planning and progress report aspects of the 1990 state prevention legislation.

Minnesota's efforts to increase consideration of environmental outcomes are not confined to the TRI. Indeed, various initiatives have attempted to make fuller use of data from medium-specific and waste management programs. In recent years, Minnesota has made unusually extensive efforts to publicize a range of outcomes findings, from quantitative measures of agricultural feedlot waste control improvements to detailed trends in the volumes of solid waste generated and the extent to which respective management strategies are used. The latest state effort in this area involves greater data integration and accessibility. The MPCA, supported through a federal grant, is attempting to integrate all of its environmental data into a single, multiprogram database that is also designed to include region-specific information. These data, and related initiatives involving MOEA and the Minnesota State Planning Agency, are intended not just to dump more information into the public domain but to provide it in highly usable formats and venues.

Long-term plans include active consideration of possible core measures of environmental quality and trends to guide the future direction of state environmental policy, as well as specific environmental goals for which quantifiable targets can be established. This has been a central consideration in recent strategic planning by the MPCA and an ongoing concern of the MOEA. The 1997 NEPPS agreement between the state and EPA emphasizes prior Minnesota experimentation in this area and establishes it as a key challenge for future policy planning. In fact, the formal agreement notes, "In the future, environmental indicators will achieve as much attention, if not more, than the administrative measures that have been previously tracked . . . we should be guided more by the outcomes of our actions, such as improving the environment, rather than focus on the administrative 'beans' that we have traditionally measured" (U.S. EPA 1997a, 6–7).

Locally Based Sustainability Initiatives

Sustainability and related goals of regulatory innovation have also become more salient issues at the local level throughout many areas of Minnesota. In fact, the state has made a concerted effort in recent years to promote active exploration of *community-based* innovation through a series of initiatives. These have ranged from financial and technical support for sustainability experiments to a major new piece of legislation passed in 1997 that was intended to incorporate a number of sustainability concerns into land use management.

The MOEA has played a major role in many of these developments. The focus of its Community Assistance Program was expanded substantially in 1996 to emphasize "community sustainability." This has resulted in a series of MOEA sustainability grants to local communities, as well as conferences and a Sustainable Communities Partnership program designed to provide a wide range of support from both the private and public sectors (MOEA 1997). Such support is intended to assist communities to "leverage resources" to promote sustainability goals and may include community assessments, comprehensive planning and design, business and technical assistance, and project development. One goal of these efforts is to enable local communities to develop their own capacity to forge a sustainable future. Steele County, an agricultural and manufacturing county with 31,000 residents in southeastern Minnesota, has exemplified these goals through its multifaceted effort to become a "model" sustainable community.

Locally based sustainability initiatives also have been bolstered through a set of laws enacted in the mid-1990s. For example, the Livable Communities Act of 1996 shifts funding from programs such as mosquito abatement to brownfields reclamation and the renewal of housing in inner-city areas. Perhaps the most important of these new steps is the Community-Based Planning Act of 1997. This legislation integrates eleven specific sustainability goals into a framework for land use planning at the local level. It is designed to formally bring together analyses from such diverse sectors as transportation, housing, economic development, and community design within a unified planning framework. The implementation of this new program will be overseen by a team established within the state planning agency. "This legislation has a very strong sustainable development twist and offers a nice interconnection among these areas," noted one state official.

Addressing Global Climate Change

Perhaps the greatest single threat to long-term sustainability involves global warming due to growing accumulation of greenhouse gases in the atmosphere. Much like the international agreements on sustainability more generally, the Kyoto protocols and related statements emphasize the desirability of allowing decentralized governmental units to devise innovative approaches to greenhouse gas reduction. In virtually all of the provinces and many of the states examined, little if any activity in this area is evident. However, Minnesota once again has begun to explore promising policy innovations and, along with New Jersey and Oregon, has taken a lead role.

One obvious area where provinces and states could address greenhouse gas emissions is through their considerable oversight of electricity generation and distribution. Many of these governments are currently exploring alternatives to traditional utility regulation, although most of these are not formally considering environmental ramifications of greenhouse gas emissions in the process. Ironically, Minnesota is not currently planning to deregulate its electric utilities or dismantle its traditional system of utility oversight under the auspices of the Minnesota Public Utilities Commission (MPUC). Electricity costs are comparatively low in Minnesota and the state has decided to delay any decision on deregulation until after it can observe the experience of other states. But the MPUC has a long-standing tradition of innovation in environmental and energy policy. This includes a "renewable default" policy that requires electricity generators to demonstrate that renewable energy options are not feasible before using nonrenewable sources and aggressive "minimum efficiency standards" for utilities. In addition, the Commission has long supported one of the nation's most active programs of energy conservation, through so-called demand-side management initiatives (Smeloff and Asmus 1997, 139).

During the early 1990s, the MPUC joined other state environmental agencies in pressuring the Minnesota legislature to pursue a range of environmental policy innovations, many of which called for expanded use of pricing mechanisms in various policy areas. In the area of electricity regulation, legislation passed in 1993 called on the MPUC to "quantify and establish a range of environmental costs associated with each method of electricity generation. A utility shall use the values established by the Commission in conjunction with other external factors, including socioeconomic costs, when evaluating and selecting resource options in all proceedings before the Commission, including resource plan and certificate of need proceedings" (Minnesota Statute 216B.2422 1993).

Staff from the MPUC and the MPCA worked to develop a range of interim values for six types of emissions: sulfur dioxide, nitrogen oxides, lead, volatile organic compounds, particulates, and carbon dioxide. The Commission adopted specific interim values in March 1994, which launched a formal set of prehearing conferences and public hearings. During 1994 and 1995, an extraordinary range of individuals and organizations participated in this process. During a nine-day period in April 1995, six public hearings were held throughout Minnesota, including a three-day videoconference, and more than 160 individuals presented testimony (State of Minnesota 1996, 6).

Unlike climate change policy formation in other states, this process clearly indicated that the proposed policy initiative had attained a high level of saliency, particularly with organized interests opposed to the proposal. Opposition came from a variety of Minnesota electricity generators and industrial consumers. Coal mining interests and state government officials from North Dakota, which is a major supplier of energy in the coal-dominated Minnesota utility field, also took an active role in opposition. In response, a key staff member of the MPCA defended the "damage cost" estimation process through a detailed cross-examination. In 1996, Minnesota Administrative Law Judge Allan Klein issued a decision that endorsed the proposed values in principle, although he made some adjustments in exact cost ranges. His decision included an extensive review of the MPCA analysis as well as the credentials of the staff member who took the lead in defending the proposal. Judge Klein established a dollars-per-ton range of environmental costs for the pollutants specified in the 1993 legislation, including a range of $.28 to $2.92 per ton as the "environmental cost" of carbon dioxide. These calculations were based on 1993 dollars, and Klein called for appropriate adjustments to current dollar levels. Four of the other five pollutants included in the valuation process received a much higher cost calculation per ton. However, the far greater volume of carbon dioxide emissions for most electricity-generating units, particularly those burning coal, reflects a potentially enormous estimation of environmental costs. "This process demonstrated that you could begin to estimate the costs of a variety of pollutants, including carbon dioxide, in a large sector such as electricity," noted one state official who played an active role in the policy formation process. "At the very least, it offered a range of cost numbers that could be used as a planning tool."

Impediments to Sustaining and Expanding Sustainable Approaches

This combination of developments in various areas of policy in Minnesota during the 1990s suggests that it is indeed possible to begin to respond to the sustainable development challenges set forth by the Brundtland Commission in 1987. Many aspects of sustainability remain difficult to define, much less translate into concrete policy initiatives. Nonetheless, the range of Minnesota innovations in such areas as pollution prevention, regulatory and intergovernmental integration, outcomes measure development, community-based sustainability experiments, and greenhouse gas reductions offers considerable promise. Indeed, the mere creation of these innovations runs counter to the extremely pessimistic tone of much of the conventional wisdom on U.S. and Canadian environmental policy, which is highly critical of prevailing approaches and skeptical about the political feasibility of most serious reform options. The Minnesota case thus takes on added significance, as it challenges all of these common assumptions, particularly in considering the broad range of stakeholders who have participated in and supported the development of these new approaches. It is in many respects a model for deliberation on the future of environmental policy in North America.

At the same time, the Minnesota experience also serves to further underscore the enduring impediments to more complete realization of sustainability goals. Such limitations reflect continuing problems for other states and provinces to even begin to approach the degree of innovation being undertaken in Minnesota. In addition, there are significant questions concerning the long-term political and managerial viability of Minnesota's new wave of initiatives.

First, the Minnesota case illustrates all of the inevitable limitations to "unilateral sustainability." No other state or province examined begins to approach Minnesota's degree of commitment to transforming its basic approaches to environmental policy. Lack of innovation is particularly notable in the majority of provinces. Moreover, major environmental funding cuts in several provinces in the late 1990s have served as a further disincentive to innovative policy development related to sustainability. Even if Minnesota's efforts continue to demonstrate substantial reduction in chemical releases and other environmental improvements, the inherently cross-boundary nature of most environmental problems indicates that the state will continue to unwillingly "import" other problems. For example, many of the most serious aspects of toxic pollution off the Minnesota shore of Lake Superior represent chemical releases from other states and environmental media over which Minnesota has virtually no influence.

In turn, Minnesota's rigorous efforts in the late 1980s and early 1990s to encourage all of its counties to actively pursue solid waste reduction and recycling options and assume responsibility for the bulk of their own wastes backfired. A series of federal court decisions in the mid-1990s ruled that so-called flow control programs, which attempted to discourage waste exports and build local waste management capacity, violated the commerce clause of the U.S. Constitution. Consequently, many Minnesota communities that invested heavily in recycling and management alternatives to traditional landfills now face significant debt as commercial landfilling in and outside of Minnesota is proving very cost-competitive. Hence, the realities of U.S. judicial federalism and the unwillingness of other states to promote environmentally preferable waste management alternatives as aggressively as Minnesota illustrate the potentially sharp limitations that can be imposed on how far a single state can move unilaterally to protect its environment.

Even Minnesota's pathbreaking efforts in conjunction with reducing greenhouse gas releases may be jeopardized by opposition from a neighboring state. North Dakota mines large quantities of coal and is a major supplier for Minnesota-based electricity plants. North Dakota officials actively opposed the Minnesota value assessment process for carbon dioxide and other pollutants and its attorney general has brought a formal challenge of its legality through litigation that is now pending in the courts.

Second, although many states and provinces have made some steps along the lines of Minnesota, many of these are extremely limited and often purely symbolic in nature. In short, virtually all state and provincial environmental agencies and stakeholders talk the language of sustainability but most give little indication of translating these rhetorical flourishes into concrete policy. In many instances, initiatives with a sustainability label give scant indication in practice of moving beyond mere regulatory streamlining. Efforts to reduce paperwork burden and move toward a more voluntary system of compliance offer many advantages, but few of these state or provincial steps give any indication of being designed to refocus resources on environmental quality. Instead, the overriding emphasis in the vast majority of these initiatives appears to be minimizing regulatory pressures or environmental considerations to maximize the likelihood of economic development.

Even among those states and provinces that appear to come closest to emulating the pattern of Minnesota, the resiliency of political support for sustainability initiatives appears suspect. In recent years, New Jersey has more closely paralleled the Minnesota path than any other state or province examined. It has launched a range of innovative programs, including an effort that goes well beyond Minnesota's in attempting to integrate pollution prevention with core regulatory functions such as permitting. However, a series of political shifts in the past few years and the implementation of major reductions in budget and staff in the New Jersey Department of Environmental Protection have left the future of a number of these innovative steps highly uncertain. In fact, these efforts may never move beyond the experimental stage, and are likely to be replaced by a combination of traditional regulation and efforts to ease regulatory burdens to foster economic development. Similarly, on Prince Edward Island, a change in government has led to substantially reduced emphasis on establishing key sustainable goals and measuring progress in attaining them during the late 1990s.

Third, Minnesota is not exempt from the possibilities of major shifts in political currents or of philosophy that has undergirded its recent steps toward sustainability. The vast majority of MPCA efforts remain concentrated on conventional pollution control-by-medium approaches, much of it dictated by federal and state legislation. In fact, many MPCA officials and other key stakeholders may either be resistant to such changes or worn down by other forces over time. As one state official noted, "Everyone will talk the happy talk of sustainability but may not be willing to weave it fully into the fabric of environmental policy. We still see new people come into the agencies with enthusiasm for new approaches, but they get assigned to a specific medium or program, and if they try to venture out and do something creative, they get whacked."

Other observers note that the past decade of innovation in state environmental policy has benefited from extensive entrepreneurial support and a constructive partnership between former Republican Governor Carlson and a legislature that has fluctuated between Democratic and Republican Party control. Carlson was prevented from seeking a third term in November 1998 due to term limits and retired in January 1999. "It's not fair to call Carlson an active champion, but he was supportive and visible, and allowed us to do a lot of things," explained one state official with considerable expertise in pollution prevention and other areas of innovation. "It's not clear what is going to happen next." Carlson's successor, former professional wrestler Jesse Ventura, is the first Reform Party candidate ever elected governor and did not make environmental policy a major part of his campaign. Some state environmental policy officials who were pivotal figures in building support for sustainability-oriented innovation have left office and there has been controversy surrounding appointees to senior environmental posts. Most observers concur that the future direction of the governor on environmental issues remains highly uncertain.

Minnesota's legislature is also uncertain in its future commitment to environmental policy. Since the 1998 elections, Republicans control the House and Democrats control the Senate, leaving three separate parties sharing governance duties. Although the state has not formally imposed term limits on legislators, Minnesota has joined many other states in experiencing unusually high rates of legislative turnover in the past decade. Between 1987 and 1997, 74 percent of House seats and 61 percent of Senate seats changed hands in Minnesota (Rosenthal 1998, 73–74). Given such rates of transition, many of those legislators most conversant with and supportive of 1990s sustainability initiatives are likely to have already left office or soon be replaced by newcomers with uncertain commitments to sustainability in the near future. This may be particularly important as so many of the programs discussed above are less than a decade old and not yet well established.

Amid all of this likely change among elected officeholders, the MPCA is launching a massive reorganization that will ultimately move about 80 percent of its staff from their current office or laboratory settings. A central goal of this change is decentralization, shifting authority from agency headquarters in St. Paul to regional offices in the northern, central, and southern portions of Minnesota. Such a shift could conceivably benefit sustainability initiatives, including those with a community-based focus. However, observers concur that the long-term impact of the reorganization is uncertain and that intensified pursuit of sustainability does not appear to be one of its central objectives. In fact, some fear that so many resources will be devoted to the administrative changes that many program initiatives could get stalled during the transition.

These three factors converge to underscore the enduring impediments to sustainability-oriented initiatives—in Minnesota or any other jurisdiction in North America. Nonetheless, the most important lesson to draw from the Minnesota experience of the 1990s is that sustainable development—and related concepts such as pollution prevention and regulatory integration—need not be dismissed as ethereal. To paraphrase one state agency official, "happy talk" may prevail in many aspects of deliberation over sustainable development, but the sorts of departures underway in Minnesota indicate that it is possible for sustainability to involve far more than rhetoric and symbolism. At the same time, the experience of other states and provinces offers caution against excessive expectations that political decentralization is most likely to facilitate the goals of sustainable development.

Sustainable Development and Natural Hazards Mitigation

Anna K. Schwab and David J. Brower

Living Beyond Our Means

The United States of America is fast growing beyond its limits. Our growth-acclimated society has exceeded the Earth's capacity to sustain us, while we continue to follow a growth-at-all costs lifestyle. We are consuming natural resources and producing wastes at ever increasing rates. Nonrenewable resources are being used faster than renewable substitutes are being found, and pollutants and toxins may soon reach the assimilative limit of the oceans. Clearly, these patterns of consumption and overuse are unsustainable; many believe we are approaching a "point of no return" in our relationship with the Earth. There are also those who believe that our current patterns of behavior regarding natural resources are a dereliction of duty, a duty that is imposed on all living things to live within the carrying capacity of their environment.

These unsustainable patterns permeate nearly every aspect of our collective lives. We are even transgressing the limits placed on us by the natural world in our choices of habitat, and have pushed development into areas that are inappropriate or dangerous for human settlement. Human activity is routinely located so that it creates a serious threat to ourselves as well as to a wide variety of natural resources and functions, many of which are beneficial to people as well as valuable in themselves as part of an interrelated living ecosystem.

Level, dry, and stable construction sites are long gone in many communities, yet the pressure to build more commercial venues, production facilities, employment centers, and residential units continues to be steady and strong in much of the country. Left to develop are only previously unused lands: wetlands, mangroves, oceanfront beaches and dunes, floodplains, steep slopes, fault zones, fire-prone areas, and other wild spaces. Many are altered to suit the builders' needs—wetlands are drained, dunes are leveled, vegetation is planted in fire-break zones—and the natural integrity of the area is forever impugned. In choosing these building sites and changing the landscape, we not only lose the inherent value of these areas, but we also expose ourselves to forces beyond our control.

Natural Hazards and Disasters

The forces to which our development decisions expose us often present themselves as "natural hazards," including such recurring extreme events as floods, earthquakes, hurricanes, erosion, wildfires, tornadoes, and volcanic eruptions. These occurrences are indeed tragic for the people living in the area who lose homes, farms, businesses, family and loved ones. However, despite the magnitude of a particular disaster from the human perspective, most such incidents themselves, though perhaps unusual, are not an aberration of nature or "freak" occurrence.

True, these events can result in massive damage to the ecological environment: fire can destroy grasslands and forests, coastal storms can move barrier islands, tornadoes can uproot trees, earthquakes can alter the landscape. Yet these occurrences, as well as their destructiveness, are part of the natural system. Mother Nature is amazingly recuperative from the forces of wind, rain, fire, and earth, and the natural environment can regenerate with remarkable resiliency, often restoring habitats and ecosystems in time for the next generation of plant and animal life to begin anew in a continuous cycle of destruction and renewal.

For instance, a flood, even one as monumental as that which occurred in the Midwest in 1993, is a naturally occurring, inevitable, largely unstoppable geophysical phenomenon. Since the dawn of time, riverine systems of the world have dealt with water flows that exceed the capacity of their channels by allowing the excess to spread out over the adjacent floodplain. Such events are an integral aspect of the life of a river, and occur as a result of rainfall, snowmelt, and other intrinsic components of the Earth's hydrological cycle. The floodplain is designed to absorb the overflow of its river, dissipating the impact of flooding over a wider area. This process has resulted in riparian soils that are rich in alluvial deposits, highly beneficial for the growing of crops, and attractive for the establishment of settlements on the river's banks.

It is not until the crops are planted and the communities built that flooding can be characterized as *hazardous.* It is only when the manufactured environment intersects with the extreme events of nature that *disasters* result. Disasters occur when human activity, such as construction and agriculture, take place in the path of the forces of nature. The human environment, particularly the built environment, is not nearly as resilient or recuperative as the natural environment, and the occurrence of a natural hazard can result in the debilitation or destruction of an entire community for many years following the event. In typically anthropocentric posture, then, we consider the naturally occurring geophysical processes of the Earth to be hazardous when they prove detrimental to *human* lives and property, rather than examine our own behavior.

Human Geohazards

To a large degree, the level of vulnerability of a particular community to hazard events can be attributed to the recognition and acceptance (or lack thereof) of the dichotomous relationship between natural geophysical events, such as floods, and human activity, such as development. In addition to a tendency to underestimate the forces of nature, the relationship between these geophysical events and human activity is often more complex than humans merely "getting in the way of" naturally occurring phenomena. In some instances, human activities can themselves exacerbate or even cause hazards and create disasters. "Human geohazards" is a term that describes human enterprise that accelerates or interferes with an otherwise innocuous natural process.

For instance, not only does inadequately planned and designed development often place people and property in harm's way, such development can also negatively impact the natural environment within which structures are built. The natural functions of the ecosystem can thus become greatly impaired, reducing the ability of the environment to absorb the impact of future hazard events. A cycle of lowered protection and higher levels of loss ensues in these communities. Consider buildings located on the site of leveled or reduced oceanfront dunes. Although providing a lovely view and easy beach access for the occupants, such ill-advised structures are subject to the full impact of coastal storms as well as the ravages of normal rates of shoreline erosion. Not only has this type of development placed people and property in harm's way, it may have ramifications beyond the immediate and obvious dangers: pollution runoff and exacerbated sedimentation rates can damage nearby wetlands and coral reefs, reducing their ability to deflect some of the stresses associated with wave action, flooding, and hurricane impact. Furthermore, the structures themselves can become floating battering rams or wind-borne missiles during violent storm events, creating an increased risk of damage to neighboring buildings and imperiling human lives.

Structural Solutions

Traditional attempts to manage the intersection of geophysical events and human behavior have focused on physical manipulation of the natural environment. In some instances, structural activities are the most practicable and provide the greatest degree of protection, yet ironically, these engineering methods sometimes worsen the very problems they were designed to solve (if not at home, then downstream), creating their own geohazard. For years, property owners along the Atlantic coast have attempted to slow down the rate of erosion, which in some areas steadily eats away at their lots and threatens their cottages, condominiums, and hotels, by setting up hardened structures perpendicular to the shoreline. These groins and jetties are designed to "capture" the downshore drift of sand that occurs naturally along the coastline. Happy owners find "their" beach accreting nicely with the captured sand, protecting their investment for a few more years (or at least until the next big storm). However, just a short walk down the beach will reveal that lots on the downshore side of the groin or jetty are experiencing accelerated erosion. These lots have been deprived of the sand that would have been deposited naturally on the beach by the longshore current, sand that is now lying on the beach of upshore neighbors. These downshore property owners are thus put at increased risk from storm damage and rapid erosion. Clearly, this irresponsible method of preventing a natural hazard has served only to make the condition more hazardous for others.

A similar problem has occurred in some river communities that have built extensive levees to prohibit riverine floodwaters from encroaching on their settlements. By doing so, however, discharge in excess of the river channel's normal capacity is prevented from reaching its floodplain. In such a case, the flood waters have not been eliminated, they have merely been transferred elsewhere; the flow of water has no alternative but to continue until it reaches a place where it can dissipate, often into the nearest community without containment works, or to breach the levee and flood the community it was designed to protect.

Not only do structural engineering methods have the potential to create a human geohazard, but the level of damage that results is often greatly exaggerated as well. By encouraging intensive land uses in the floodplain or on oceanfront beaches, for example, protection via engineering devices may create a false sense of security, leading to further development in these hazardous areas and contributing to an increase in the community's vulnerability to future hazard damage.

A New Approach to Development Decision Making

It is clear, then, that our current modus operandi cannot be continued indefinitely. Human use (overuse) of the world's resources, including our methods of staking out new territory, are unsustainable. That is, we cannot expect to carry on at this rate with the same returns forever and without irreparable damage to ourselves and future generations. In fact, children alive today in all likelihood will face a very different world when they are adults and making development decisions of their own. We must do what is in our power now to see that we do not limit their choices through thoughtless decisions we make.

Fortunately, *sustainable development* has emerged as a paradigm with the potential to give human beings the perspective and the power they need to rediscover our proper niche in the Earth's panoply. Indeed, sustainable development may be seen as a moral imperative that we must pursue more thoughtful ways, must change our values and assumptions, must consider beyond the here-and-now.

The concept of *natural hazard mitigation* falls neatly under the broader umbrella of sustainable development as one of the ways by which we can change our current self- and eco-destructive habits. With hazard mitigation as one of the pillars of sustainable development, we can make our development decisions in such a way as to make the built environment more resilient to the impacts of natural hazards, thereby decreasing the future vulnerability of human life and property while bolstering the long-term viability of natural ecosystems and human communities.

Defining Sustainable Development and Hazard Mitigation

The literature is replete with definitions of sustainable development, but the one that is nearly universally accepted today emanates from the report published in 1987 by the United Nations World Commission on Environment and Development entitled *Our Common Future,* commonly referred to as the Brundtland Report. Sustainable development is "*development that meets the needs of the present without compromising the ability of future generations to meet their own needs.*"

There are also myriad interpretations of what constitutes hazard mitigation, but one standard definition used by the Federal Emergency Management Agency (FEMA) describes natural hazard mitigation as "*any action taken to reduce or eliminate the long-term risk to human life and property from natural hazards.*" This can involve activities ranging from minor structural changes to an existing building that make it more resistant to the impacts of natural hazards (such as extra nails to hold roofing material in place during high winds) to major avoidance policies that permanently remove particularly hazardous areas from the development marketplace (such as public acquisition of hazardous sites).

The Shared Principles of Sustainable Development and Hazard Mitigation

Neither sustainable development nor hazard mitigation are new ideas. Yet it is not until recently that these concepts have become widely recognized as legitimate, "doable" principles to be incorporated into decision making. And it is not until even more recently that the two concepts have been coupled as complementary methods for reaching the same broad goals. Though the concept of sustainable development may be wider in scope, both concepts clearly have many salient aspects in common.

The first such important common element is the recognition that these are *qualitative* concepts and do not necessarily involve quantitative measures. Sustainable development communicates a concern with *what kind* of development, rather than how much, and hazard mitigation encourages development that is built to standards designed to withstand likely hazard impacts and is located in areas that minimize those impacts. Neither principle necessarily proposes a "no growth" policy for communities to become less vulnerable and more sustainable. Rather, these concepts advocate for the safe accommodation of future population rise through conscientiously controlled growth and development.

The second common bond between the concepts of sustainable development and hazard mitigation involves an ethic of conservation and preservation. Natural hazard mitigation calls for conservation of natural and ecologically sensitive areas, such as wetlands, floodplains, and dunes, features that enable the environment to efficiently and cost-effectively absorb some of the impact of hazardous events. These ecosystems also serve as important pollution filters as well as provide habitat for a number of species of fish and wildlife. In this way, preservation and protection for mitigation follow one of the fundamental premises of sustainable development: that we respect our natural heritage and allow its systems to operate as designed, without alteration or interference. By allowing the environment to perform its functions unimpaired, mitigation-through-preservation programs can help communities attain a level of sustainability, ensuring public and environmental health for the community as a whole.

Third, proponents of sustainable development theory recognize that our economic structure and the natural environment are not necessarily in conflict, but instead are irrevocably interconnected and interdependent. Despite our seeming "dominance" over the natural world, humans are still dependent on the bounty of Mother Nature for our own viability. This is clearly evident in our reliance on the Earth's natural resources for survival. However, natural ecosystems that are not operating at optimum levels due to pollution or other human-induced trauma do not produce the staples of a firm economy. In turn, economies that are faltering do not allow people the "luxury" to invest wisely and consider the long term, which can often put natural resources in peril as they are exploited for immediate gain. A vicious cycle of environmental degradation and economic decline may then be established, producing a severely lowered quality of life for people and an uncertain future for the vitality of the area's ecosystems.

Hazard mitigation can play a vital role in maintaining a balance between a community's economic condition and its natural setting. A core assumption of mitigation strategy is that current dollars invested in mitigation will significantly reduce the demand for future dollars by lessening the amount needed for emergency recovery, repair, and reconstruction following a hazard event. A greater degree of community resiliency to the impacts of natural hazards enables local businesses and industries to reestablish themselves in the wake of a disaster, getting the economy back on track sooner and with fewer interruptions in the flow of goods and services.

Mitigation can also provide a degree of socioeconomic continuity in the community by reducing the social upheaval that often accompanies a hazardous event. Damage to transportation and communication systems, dislocation of people, loss or interruption of jobs, and closing or disabling of businesses, schools, and social centers often create personal and family stress for disaster victims in addition to financial hardship. By minimizing the causes of these stress factors, untold repercussions of disasters may be avoided, including such human tragedies as domestic violence, child abuse, depression and anxiety, and even suicide, all of which have been shown to increase in the aftermath of severe disasters.

Fourth, sustainable development implies a change in values and speaks in terms of *needs,* not desires. Adherence to the principles of sustainable development does not guarantee a life of luxury for all, but neither does it demand major sacrifice. We must be willing to give up the oceanfront homesite when a structure there is clearly in a fragile and hazardous location. We must cease to view ourselves as merely consumers of the world's goods; instead, we must recognize our role as stewards of the planet.

Fifth, to bring these measures of sustainability to fruition, sustainable development theory requires that we focus on intergenerational equity: we must meet the needs of the present generation, but not at the expense of what future generations may need. In similar forward-looking fashion, hazard mitigation requires that we build, rebuild, and plan for today's development while considering the impact of hazards yet to come on inhabitants in the years ahead. A community's future vulnerability can be determined by projecting various development scenarios and assessing the number of people that would experience harm and the amount of property that would be damaged were a hazard event to occur. Armed with such knowledge, proactive communities can take action to reduce this level of vulnerability, strengthening the community as a whole for today and tomorrow.

Implementation at the Local Level

Although these shared principles of sustainable development and natural hazard mitigation intimate that the focus is solely directed toward global concerns, the very nature of the concepts makes them decidedly local in nature. It is at the local level that most land use patterns are determined, infrastructure is designed and provided, and many other development issues are decided. It is also at the local level that hazards are experienced and losses are suffered most directly. Because of our decentralized approach to many of these types of issues, there is much within the power of a typical unit of local government to bring about a sustainable and mitigative approach to growth and development within its jurisdiction.

Many American communities have already instituted programs designed to influence various characteristics of growth within their planning jurisdictions, such as the type, amount, density, and timing of new development that will be permitted. Local measures are also routinely employed to control the overall mix of land uses to ensure that incompatibility and inefficient use of resources are minimized. Other tools available to local governments are used to reduce or distribute the costs of growth. These include reducing economic costs, referring to avoidable financial outlays associated with new development, as well as distributional costs, which entail distributing the economic costs of growth among current and incoming residents fairly and equitably. Environmental costs are also influenced by local action to reduce the damage to natural ecosystems that can result from ungoverned development. These local growth management activities can be used to promote a higher quality of life, a safer built environment, and sustainable patterns of growth and development in the communities that choose to engage in them.

Controlling Growth and Development: The Tools of Local Government

There are many opportunities open to communities that wish to guide their growth and development in a principled and responsible fashion. The police power, which is bestowed on local governments by the state in which they are located, authorizes local government actions that protect the public health, safety, and general welfare. In fact, it can be argued that local governments are under an *affirmative duty* to promote the health, safety, and general welfare of their citizens; it can be further argued that refraining from managing growth and development in a responsible manner is a dereliction of that duty.

As a general rule, local governments have four major areas of authority, all of which can be infused with the ethics of sustainable development as that community expresses and implements them, and carried out so that the principles of mitigation are followed. These areas are regulation, taxation, acquisition, and spending.

Regulation. The first category of government activity in controlling growth and development is *regulation,* which includes regulation over land uses as well as other human activities. Regulation of land uses can take many forms, including the enactment of zoning ordinances, subdivision regulations, planned unit or cluster development provisions, floodplain management ordinances, critical area management laws, and numerous other regulatory activities.

Of all the local regulatory options, zoning is perhaps the most ubiquitous. A zoning ordinance authorizes the government to divide its jurisdiction into various zones and designate which types of land uses will be permitted in each zone. Zoning can be used as an effective tool for achieving a wide variety of mitigative goals, such as easing congestion on public roadways, thereby augmenting evacuation capacity; reducing undue concentrations of the local population who may be at risk from natural hazards; limiting the density and/or increasing the minimum lot size of parcels located in designated hazard areas; restricting development in areas with inadequate access to protection services, such as fire or emergency medical services; and zoning to preserve natural areas that mitigate against hazards, such as wetlands and dunes.

In addition to zoning, many local communities enact subdivision ordinances as a method to control development. Subdivision regulations are activated on the division of land for development or sale, and though they do not control the type and quality of development as directly as does zoning, they can serve several mitigation objectives. Subdivision ordinances can be quite effective in controlling flooding by prohibiting the subdividing of land that lies within mapped floodplains. When other types of hazard areas are officially mapped by the local government, subdivision ordinances can prohibit subdivision of parcels located in these areas as well. Many communities also include standards for infrastructure and facilities provided by developers in their subdivision regulations, as well as performance standards for the buildings. These provisions can be quite effective in ensuring that infrastructure and facilities are adequate for the hazard risk posed in that area, including such fundamentals as sufficient drainage and stormwater management facilities. Subdivision ordinances can also require that buildings be sited away from hazard-prone sections of land parcels, such as shoreline erosion points, or that developers undertake specific mitigative activities, such as protecting or creating wetlands, augmenting dune systems, or planting vegetative buffers.

Local ordinances that regulate planned unit developments or cluster development can be another useful legal tool when enacted with sustainable development and hazard mitigation as a guide. These ordinances typically allow density of new development to be concentrated at higher than normal levels in certain parts of the parcel being developed. The remaining land is then designated for less intensive uses. This approach can effectively preserve open space and protect sensitive natural areas or high-hazard areas.

Environmental quality and hazard mitigation through regulation can be implemented by local communities through management of critical or sensitive areas within their jurisdictions. For instance, many governments have realized the value of wetland preservation and estuarine or riparian habitat protection, and have enacted ordinances that prohibit development in or around areas that serve as essential habitat, pollutant filters, and storage areas for flood waters. Other communities restrict the grading of hillsides and limit development on slopes prone to landslides through soil conservation and steep slope preservation programs. Such measures can control erosion and stabilize slopes by prohibiting inappropriate land disturbance processes or by requiring terracing or planting of vegetation. Shoreline communities may choose to call for preservation of beaches and dunes through shoreline setbacks that establish a minimum distance between the shoreline and where buildings will be permitted. These measures allow the beach and dune systems to act as a first line of defense against the impacts of wind and waves, as well as keep intact these ecologically critical areas.

By enacting such regulatory provisions, the community promotes sustainable patterns of development that decrease the level of vulnerability to natural hazards, at the same time enhancing water quality, preserving wildlife habitat, conserving the natural environment, and enhancing the quality of life for citizens.

Taxation. The second major power of local government is *taxation,* which extends beyond the mere collection of revenue and can have a profound impact on the pattern of development in the community. For instance, some communities use a preferential or use value assessment system for taxing certain types of property that are environmentally sensitive or otherwise valuable to the community, including farmland, forestland, historic properties, open space, wetlands, and riparian areas. Under this type of program, certain parcels of land are assessed according to their current income-producing capacity, rather than their value on the open market. This results in a reduced tax burden on lands that are under development pressure, but that are valuable in their current low-intensity state or would be hazardous to develop.

Land gains taxation involves assessing property in inverse proportion to the amount of time the land is held, thereby discouraging speculation. Similar taxation devices include transfer taxes, assessed against the sellers of land of certain types of uses, and development taxes, which are charged against developers upon the conversion of land to uses of greater intensity. Although these taxation methods may not provide for long-term protection, they can provide a disincentive to convert land to a higher density, thereby slowing rapid growth in a community, particularly in high-hazard areas such as floodplains. Land transfer taxes have been used to fund *land banks,* allowing the community to purchase open space or other types of conservation easements, protecting the land from inappropriate development.

Some communities have instituted impact fees or system development charges that require developers to contribute to the financial outlays imposed on the local jurisdiction to support the new development. The amount charged is proportional to the cost of the impact that that development will cause. Such fees are typically used to provide for improvements such as roads, water, sewer, and schools, but can also be employed to provide hazard mitigation features, such as flood storage facilities in areas where new development will contribute to an increase in flood heights.

Impact fees are a use-based charge against new development; exactions require developers to directly invest in the facility and service needs created by their projects. Cash exactions can be put toward on- or off-site improvements, including such mitigative uses as poststorm reconstruction funds, and exactions in the form of a land dedication or grant to the community can be used to acquire open space in floodplains or other hazard-prone areas.

Acquisition. Acquisition is a third useful government tool for pursuing sustainable development goals as well as hazard mitigation in a community. This includes the power of eminent domain as well as the authority to purchase property on the free market. The community can thus proactively acquire land to absolutely control its use, thereby removing the property from development potential. The community may also acquire easements in sensitive or hazardous lands, including negative easements, which prevent the owner of the property from building or engaging in other specified uses, and affirmative easements, whereby the public is granted a right to use the property in a specified manner, such as for beach access. In many areas of the country nonprofit land trusts and conservancies have played a vital role in the acquisition and management of environmentally sensitive, hazardous, or other important lands, often in cooperative ventures with local governments. Land trusts can be used to promote farmland conservation, provision of sites for low-income housing, public recreation, nature areas, and other sustainable uses.

Spending. Spending is the fourth major power, whereby local governments make expenditures of public funds that impact the community. Government expenditures include payment for public infrastructure such as parks, roads, water and sewer lines, and public buildings, as well as payment for public services such as police and fire protection. The decision of when and where to provide infrastructure and services in large part determines where, when, and how intensely development will take place in the jurisdiction. The spending decisions made by local governments can be effective in directing new development away from hazardous and otherwise inappropriate areas and maintaining a level of growth that is conducive to the long-term viability of the community.

Capital improvements programming is one method local governments can use to define when, where, and what level of municipal services will be supplied. Setting up a capital improvements spending timetable can be very effective at managing growth, because few developers can afford to provide all the facilities and services that their projects will require without some public investment. Development can therefore be effectively limited in hazard or otherwise sensitive areas if the community does not extend infrastructure to these places.

Some communities have imposed concurrency or adequate public facilities requirements on new development as part of their spending program. These provisions are implemented to ensure that public services are provided simultaneously with the demand for those services (concurrency) or that a certain level of services be made available upon completion of the development project or within a designated time period following completion (adequate public facilities). These practices can be used to direct development into areas that are less hazard-prone, although if not implemented carefully, they can also produce the opposite effect. Creation of urban service districts in conjunction with concurrency and adequate public facilities requirements can help shift the direction of growth into appropriate locations by defining where certain services will be provided within the community.

Windows of Opportunity

Despite the wide array of tools and techniques that are available to governments to operationalize the principles of sustainability and mitigation, one of the roadblocks to implementation is the fact that much of the land within local jurisdictions has already been developed according to practices and traditions that are far from sustainable. Ironically, the time immediately following a natural disaster provides a community with a unique window of opportunity for inserting an ethic of sustainability in guiding development and redevelopment in high-risk areas. With forethought and planning, communities that are rebuilt in the aftermath of a natural hazard can be built back so that they are more resilient to future hazards, breaking the pattern of repeated hazard-destruction-rebuilding. At the same time, the community is given the opportunity to incorporate other attributes of sustainability into its "second chance" development, such as energy efficiency, affordable housing, use of recycled building materials, reduction of water use, and environmental protection.

A Moral Obligation

The holistic approach of infusing hazard mitigation into the major tenets of sustainable development and using the principles to guide future decision making is considered by some to be our ethical obligation to future generations: our children and grandchildren. At the very least, we have some duty to refrain from detracting from the long-term viability of our living places. Many would say we have an affirmative duty to increase their safety over time, a duty that can be fulfilled only through hazard mitigation. We must do what is within our power to make the future safer. We have control in terms of land use, the direction and nature of economic development, capital facilities and societal infrastructure—and all these will impact the vulnerability of our descendants. It is therefore incumbent upon us as thinking, reasoning beings to minimize those impacts.

We do not imply that answering this ethical call to duty is as simple as doing what's "right." Once the moral issue is raised, often even more questions arise regarding the duties and obligations that are owed, by whom, and to whom. Indeed, there are many and disparate players in the movement toward sustainability, each with a valid role to play, but often at odds with one another.

Competing Values

In the context of hazard mitigation we can see the tensions among ethical responsibilities. Government agencies and regulators; the private sector, including building owners, corporations, and merchants; professionals in architecture, construction, engineering, and related fields; as well as individuals such as homeowners, consumers, farmers, residents, and tax payers all play a role in affecting our present and future vulnerability. Although all these players may act with the best intentions, each may have a view of what sustainability or mitigation should involve limited to his or her own interests.

Sometimes judgments must be made as to relative priorities among values, such as protection of public health and safety versus protection of property. Most people would probably choose human lives over financial concerns in an emergency situation; however, arguments for property protection where the danger to life and limb is not immediate can (and have been) made.

The precedence of protection of the natural environment over protection of development may also cause a series of "trade-offs" among values that appear to be incompatible. We are duty-bound to pass on a natural legacy to future generations. This we cannot do if the physical manifestations of such a legacy are allowed to be destroyed by impacts that we can take steps to lessen. Yet many feel these steps are taken at the expense of private property rights, which, in this country, are deemed by some to be nearly inviolate. Setback lines, for instance, commonly prohibit construction in oceanfront erodible areas, areas that are very vulnerable to coastal storm and erosion hazards. Although an effective mitigation technique, such regulations prevent developers from creating jobs for local residents, increasing the local tax base, realizing their own economic gain, and providing housing consumers the opportunity to purchase property there.

A moral quandary may also arise when choices appear to pit environmental protection against public health and safety. For instance, when a community expands its road and bridge system to increase the efficiency and capacity of evacuation routes in the event of a natural hazard, the construction may prove detrimental to fragile ecosystems. It could also lead to unanticipated impacts if such road work encourages denser development in areas that hitherto had been of limited accessibility.

Multiobjective Solutions

Despite the many differing ethical priorities involved, there are often alternative solutions that can protect people, property, and the economy as well as advance our role as stewards of the natural environment. In fact, the most effective mitigation strategies involve protecting and restoring the natural functions of the ecosystem, with the dual purpose of environmental protection *and* life and property protection. For example, conservation of wetlands promotes flood control, and preservation of the coastal dune system provides a natural seawall. In turn, these protected areas will withstand the impacts of natural hazards much more steadfastly, thereby safeguarding both the public and private investments made in improvements in proximity to these natural features, and preventing the economic calamity that can occur when development takes place on sites where natural mitigation measures have been removed.

Some scholars have expanded the ethic of conservation and protection to include a "restorative" value. According to this view, it is not enough to maintain the existing natural environment. Because the destruction and degradation of the environment that has occurred already in many places has led to many of our worst disasters, we are not carrying out our duty adequately by merely continuing the status quo. We must instead act on the moral duty to rectify the damage that has been done, and enhance, not just preserve, our natural defense mechanisms.

Defining Our Ethical Obligation: To Whom Is Our Duty Owed?

Solutions such as these are more likely to emanate from a milieu of expansive thinking about the ethics of our development decisions. In particular, we must open wide the definition of who is owed our moral duty. Not only must we consider our temporal wards—those future generations whose well-being is entrusted to us—we must also expand our sense of spatial responsibility. We can no longer take a parochial attitude with regard to our moral obligations; the geographic scope of hazard mitigation and sustainable development theory requires that we consider much more than the immediate vicinity in which we live. Because the consequences of pollution and misuse of natural resources transgress jurisdictional lines, a wide-angle view is critical to overcome the artificial limits that are imposed by the politically delineated boundaries that separate our communities. Actions taken and decisions made in one jurisdiction can have profound ramifications for neighboring towns, states, and regions. In a ripple-like effect, even the world at large is impacted by individual and collective behavior. Our "neighbors" are citizens of the world, some of whom are not yet born. When we consider our alternatives, development decisions should always be made for the wider moral community.

A Dynamic Duo

It is clear that there is much that can be done at the local level to promote development that is sustainable, including land uses that help mitigate the impacts of natural hazards. However, the local implementation of various tools and techniques does not imply a quest to reach nirvana. Sustainable development is process-oriented and does not focus on a static world order; instead, it involves a dynamic, evolutionary continuum of action that will forever need readjusting to fulfill its mission. As a part of this movement, hazard mitigation must also be seen as more than an end-state. We know we must do more than merely nail shutters over the windows when gale-force winds are predicted. Likewise, hazard mitigation involves a constant search for ways to incorporate mitigative concepts into development decisions to reduce our vulnerability to natural hazards for today and tomorrow.

Sustainable Governance

Jonathan Baert Wiener

It seems to be obligatory to begin any case study on sustainable development by noting that the term eludes definition (see Hodas 1998; Howarth 1997; Norton and Toman 1997; Pezzey 1997; Stone 1995). Indeed, the ambiguity or even meaninglessness of "sustainable development" is by now such a well-worn cliché that it is even a cliché to say so. I will not attempt here to define the term. I want to take a very different tack from most of the literature on sustainable development, to the extent that the aim of this literature has been to define the term in its totality, top-down. That is, it has been an effort to define the upper limit of aggregate development (or per capita aggregate resources) that would satisfy a criterion of sustainability across generations (see, e.g., Howarth 1997; Pezzey 1997; Solow 1992; World Commission on Environment and Development [WCED] 1987). Instead, I want to come at the question from the other side, bottom-up or inside-out. I want to ask: What criteria for microlevel institutional activity are needed for some version of macrolevel sustainable development to succeed?

In particular, I want to suggest that the discussion of sustainable development to date has missed a big boat by addressing only one side of this microworld of institutions. The notion of sustainable development is meant to reconcile both the need for intergenerational environmental sustainability and the simultaneous need for economic development (WCED 1987). Yet in emphasizing the need for sustainable *development,* the discussion to date has focused on mitigating the environmental impacts of the private sector and of economic growth. It implies the need for reconciliation of two conflicting forces—economic development versus environmental protection—represented by the invisible hand of private markets on one side and by the visible corrective hand of the state on the other. A standard characterization defines sustainability as asking whether "markets generate incentives for resource conservation sufficient to ensure the welfare of future generations," and, if not, which "policies would adjust market outcomes to secure future welfare" (Farmer and Randall 1997, 608). The premise is that markets are the problem and government policies are the exogenous force that can secure sustainability. Further, the framers of the concept of sustainable development say that a necessary element in its implementation is adherence to the "precautionary principle" (Bergen Declaration of 1990, cited in Cameron and Abouchar 1991, 18; see also Howarth 1995), which in turn is defined as a legal doctrine requiring that "a substance or activity posing a threat to the environment is prevented from adversely affecting the environment . . . [it seeks] a reversal . . . of the current position whereby polluters can continue to discharge . . . [and] marks a re-evaluation of . . . industrial development [that] has severely degraded the environment" (Cameron and Abouchar 1991, 2). The point here is not that the precautionary principle is right or wrong (a topic for another day), but that it has been interpreted in terms of the public-private interface, to require government regulation of private industry unless private industry can demonstrate the absence of adverse environmental effects (see Cross 1996, 852–58). This way of thinking builds on a long tradition, from at least Pigou onward, of diagnosing market failures such as pollution and then imagining that government is a benign exogenous force with the fiat power to impose precisely corrective environmental policies (Wiener 1999, 701–3, citing Pigou 1932, 192–95).

Clearly, ordinary economic markets do fail to protect the environment adequately, and it is absolutely right to say that society should remedy these failures, often through government regulation. What this approach largely neglects, however, are the environmental dysfunctions of the state itself. I argue here that, far from being an exogenous force for sustainability, government is an endogenous and imperfect human institution that systematically generates highly significant adverse environmental impacts. Government can help, but it can also hurt. An analogy to medical care is useful: like a doctor or a hospital, government is a source of remedies for social maladies, but the remedies can be ineffectual or can pose their own adverse side effects (see Wiener 1998; see also Janicke 1990; Moynihan 1993). The range of potentially unsustainable government activities is quite broad, including perverse subsidies for logging and other resource extraction, narrowly conceived public works projects such as dams, regulatory interventions generating new risks, and defects in the very legal system undergirding property and contract rights that shape private markets. In its full scope, unsustainable governance may even be a larger environmental hazard than unsustainable private industry, especially in countries dominated by the state sector but even in market-oriented societies such as the United States.

What is needed is a concept of sustainable governance to proceed in tandem with sustainable development. The marriage of these two concepts would reflect the reality that both private and public activities merit attention and reform. Yet these two concepts are not developing in tandem. Sustainable development, despite (or perhaps because of) its ambiguity, is now moving from a widely accepted norm to an official policy goal and may soon be an enforceable feature of "hard law" (Ruhl 1998). By contrast, sustainable governance, as I elaborate here, lags far behind. Although very occasionally enforced by courts, its application is more often rebuffed or even outlawed by legal rules. It may not have even become a widely accepted norm (the most preliminary of Ruhl's "seven degrees of relevance"). In this case study I explore the reasons for this imbalance, the legal barriers that currently obstruct efforts to promote sustainable governance, and the remedial options for achieving sustainable governance.

Unsustainable Governance

The state plays an enormous role in development and environmental outcomes. In some countries, the state owns much of the means of production. At least until 1989 this was the case in Eastern Europe and the former Soviet Union, countries currently in transition to some form of private market ownership. Even in wealthy Western industrialized countries, the state plays a major role. In member countries of the Organization for Economic Cooperation and Development (OECD), total government expenditures were about 20 percent of GDP in 1960 and grew to over 40 percent of GDP by the 1990s (World Bank 1997, 2). In developing countries, the government share of GDP grew from about 15 percent in 1960 to about 30 percent by 1990 (2). In the United States, the federal government owns about one-third of the land area of the country.

Beyond this role as a direct player in economic activities, the government as regulator wields power over the entire economy and society. At least since the New Deal and the Great Society—including the emergence of modern environmental law since the 1960s—the power of the U.S. (and European) administrative state to regulate private arrangements has become vast (Horwitz 1989; Sunstein 1990). To be sure, many countries place constitutional constraints on state action, such as the separation of powers, judicial review, the Bill of Rights, and the Commerce Clause in the U.S. legal system (see World Bank 1997, 99–109). But these still leave a huge arena of legitimate regulatory power over activities affecting interstate commerce in the hands of the federal government, and additional power over intrastate affairs in the hands of the fifty state governments. In countries without effective legal limits on state power, regulation may be open to nearly unfettered use and abuse (see World Bank 1997, 148–50; Shleifer and Vishny 1998).

Further, the government's role as adjudicator of private disputes and enforcer of private rights makes the state essential to the shape and magnitude of all market activity (see Menell and Stewart 1994, 68, n. 28; Stiglitz 1989; Sunstein 1997, 5; World Bank 1997, 41). In common law countries, the definition and enforcement of private legal rights is performed chiefly by the judicial branch of government, but it also inescapably involves executive and legislative branch actions that finance the effective policing of rights in practice (see Holmes and Sunstein 1999).

The adverse environmental impacts of the state arise in numerous forms, but it seems useful to consider two basic versions of state failure: excessive weakness and excessive strength. By failures of weak governance, I mean the incapacity of the state to protect legal rights and to resist corruptive pressures. By failures of strong governance, I mean the use of unduly caustic measures to achieve well-intentioned policy goals. Of course, there is potential overlap between these types of state failure, and one could adopt a different classification (see, e.g., Wolf 1988), but I use this simple taxonomy here to make a very elemental point: governance can be unsustainable if its medicine is too weak or too strong. Sustainable development, then, is not just a matter of diagnosing market failure and invoking government as a remedy. Given that there can be government failure as well as market failure, the challenge is to minimize the sum of both (Buchanan and Tullock 1962). The challenge of sustainable governance must be to design institutions and policy instruments that optimize governance across the diverse risks of state and market failure.

I have analogized the state to a doctor or a hospital, an analogy I developed in more detail elsewhere (Wiener 1998). This analogy seems apt in the field of environmental protection. When market failures such as externalities—effects on third parties uninvolved in the market transaction that gives rise to the effects—cause environmental harm, they resemble a patient presenting symptoms of an ailment to a physician. The doctor must diagnose the ailment (risk assessment) and then decide what remedy to prescribe (risk management). Yet medical remedies are not perfect, and studies of medical care in the United States indicate that hospitalization regularly injures about one in every twenty-five patients and kills about one in every two hundred—amounting to an estimated 100,000 deaths per year from the adverse side effects of hospitalization (Weiler et al. 1993, 36, 43–55). Similarly, entry into the system of governance can cause injury. Unduly weak governance is like a quack medication that lets the patient die for lack of a real remedy. Excessively strong governance is like a caustic remedy that treats the ailment but also creates new, iatrogenic (care-induced) health problems for the patient (Moynihan 1993; Wiener 1998). Put another way, state intervention to reduce a "target risk" may fail to do so and/or may create new "countervailing risks" (Graham and Wiener 1995). Either weak or strong governance, if extreme enough, can seriously harm or kill the patient.

The environmental injuries arising from weak governance are well-known. These include the worldwide practice of government subsidies for environmentally damaging activities such as logging, grazing, farming, mining, water use, dam construction, and fuel use (see World Bank 1992; Myers 1998). I label these as instances of weak governance because they tend to reflect the lack of state backbone. The state is typically conferring these subsidies as a service to the powerful special interests that dominate politics and distort government activities to favor themselves over the collective public interest (Olson 1965). Subsidies of fossil energy use, for example, have been estimated to account for about 10 percent of observed carbon emissions in 1990 (World Bank 1992), roughly twice the amount sought to be controlled by the Kyoto Protocol.

A second and more general form of weak governance occurs when the state does not or cannot protect security in property rights. The basic inability of the state to underpin market activity not only inhibits economic development in general (World Bank 1997), but is often a prime contributor to environmental degradation. It is the lack of secure property rights (in particular, the right to exclude, whether held by an individual or a group) that yields the "tragedy of the commons" (more accurately, the "tragedy of open access") as a result of a "race to deplete" unowned resources such as fisheries, forests, and the atmosphere (Hardin 1968; Ostrom 1990; Deacon 1994). Indeed, pollution itself—typically seen as a market failure—can also be seen as the failure of governance institutions to create and enforce exclusive property rights that limit access to air, water, and land.

The environmental impacts of excessively strong governance are less well-known but no less worrisome. As a general matter, it is commonplace that well-intentioned government policies can produce undesirable iatrogenic or countervailing side effects (for numerous examples, see Graham and Wiener 1995; Wiener 1998; Cross 1996). Mandating airbags in cars can save adults but kill children (Graham and Segui-Gomez, 1997). Banning drugs can reduce addiction but increase violent crime among gang suppliers (Moynihan 1993). Bombing the forces of despots bent on ethnic cleansing can unintentionally hit convoys of refugees. The same kinds of risk-risk trade-offs occur in environmental regulation. Controlling pollution in one medium, such as air, may simply shift pollution to another medium, such as water, land, or the workplace (Graham and Wiener 1995). Banning one pesticide may result in the use of another pesticide, potentially one that is more toxic or that shifts risk from consumers to workers (Graham and Wiener 1995). Mandating prompt cleanup of oil spills via hot water spraying can cause damage to coastal microorganisms, ultimately injuring the beach more than the spill itself (Lancaster 1991). Restricting carbon dioxide emissions alone can yield increased methane emissions, potentially causing a net increase in global warming (Graham and Wiener 1995).

Understanding the dysfunctions of government requires recognition that government is an endogenous and imperfect institution. Government is not outside of society; it is part of the same society that manifests markets and market failures. The failures of weak and strong governance are *institutional* failures analogous to market failures. Markets may underproduce "public goods" such as environmental protection because "free riders" undermine the incentive of market actors to cooperate in the common interest. Similarly, government institutions may underproduce public goods because free riding in the political arena undermines the incentive of government actors to serve the common interest (see Olson 1965). As market actors respond to the voice of self-interest, so government actors may respond to organized special interests and to internal bureaucratic incentives (Wolf 1988). Subsidies for logging, grazing, mining, dams, and resources left open-access for exploitation appear to reflect the disproportionate influence of concentrated special interests on government policy. Narrow pursuit of "target risks" with inadequate attention to the resultant "countervailing risks" can also be attributed to the excessive political influence of organized groups and the "omitted voice" of the general public, disenfranchised minorities, and future generations (Graham and Wiener 1995, ch. 11). Environmental justice problems can therefore be seen as a form of countervailing risk and of state failure. The central concern of sustainable development—that present development may compromise the interests of future generations—can be traced to the failure of government institutions to internalize the interests of future generations who do not vote in present elections (ibid.; Howe 1997, 604).

These state failures can also be understood as "regulatory externalities" or "government externalities" (for related discussions, see Weisbrod 1978; Wolf 1988). Markets may fail when market actors cause harm to third parties uninvolved in the market transaction. Such market externalities often warrant government response. But government may fail when government actors cause harm to third parties uninvolved in the regulatory decision. Such regulatory externalities are the economic equivalent of what I have called countervailing risks. A plausible hypothesis is that regulatory externalities, like market externalities, arise when the benefit to the causal actor of internalizing the externality is less than the cost to that actor of doing so. In the market context, the net benefits of internalizing externalities would always induce such internalization where the costs of a transaction between victim and causal actor are zero, but because such costs are typically not zero, externalities may persist (Coase 1960). In the government context, the net benefits of internalizing regulatory externalities would induce internalization where the costs of doing so are zero, but because such costs are typically not zero, regulatory externalities may persist. There are costs on both sides of the regulatory externality. The victims of the regulatory externality face costs of political organization and voice; that is why they are saddled with the countervailing risks of policies designed to serve other, more vocal groups. And the suppliers of regulation face costs of analyzing and deliberating about regulatory side effects (see Wiener 1998, 69–78). The costs of such deliberation mean that regulatory agencies may neglect harms outside their mission area, or even within their mission area if it is inconvenient to reconsider a policy that has gained bureaucratic momentum. Ideal regulatory decision making would account for deliberation costs in determining how much attention to give to countervailing risks (ibid.), but government actors, like market actors, may engage in heuristic shortcuts that economize excessively on deliberation costs and yield inefficiently large externalities.

In short, the environmental degradation we observe may be the result of unsustainable governance as much as of unsustainable development. The symptoms of unsustainable governance—perverse subsidies, open-access resources (including pollution in general), and countervailing risk—are widespread and serious sources of environmental hazard. Narrow focus on sustainable development to the neglect of sustainable governance may leave problems unsolved. The concept of sustainable development gives no good guidance on how to resolve conflicts between target risks and countervailing risks, such as cross-media shifts of environmental stress (see Stone 1995, 2). And if the response to unsustainable development is more of the same kinds of governance that we have employed in the past, then stressing sustainable development and precaution, without simultaneous attention to sustainable governance, could actually make things worse.

Legal Obstacles to Sustainable Governance

Constraining the environmental impacts of government is not new to environmental law. It is the central aim of the National Environmental Policy Act

(NEPA), enacted in 1969 as the cornerstone of the modern environmental law edifice and copied around the world. Yet NEPA and subsequent law have not succeeded in achieving sustainable governance. Worse, several legal rules operate to block efforts toward sustainable governance.

The genesis of NEPA was the recognition that "omitted voice" and heuristic limits on analysis, just as they generate market externalities, could also lead government agencies to neglect the harmful effects of their projects on interests outside their mission. In an effort to remedy such government regulatory externalities, NEPA requires every federal agency to include in each proposal for major federal action a statement of the environmental impact of the proposed action, an environmental impact statement (EIS), including discussion of reasonable alternatives to the proposed action (see 42 U.S.C. 4332[2][C]). It also requires agencies to develop methods to take appropriate account of environment values along with economic considerations (42 U.S.C. 4332[2][B]). Although NEPA does not expressly provide that citizens may enforce its provisions against agencies, Judge Skelly Wright held in *Calvert Cliffs Coordinating Committee v. U.S. Atomic Energy Commission,* 449 F.2d 1109 (D.C. Cir. 1971), that NEPA was enforceable under the Administrative Procedure Act (APA; 5 U.S.C. 706). Judge Wright went further, asserting that NEPA contained both a procedural requirement to prepare the EIS and a substantive requirement to balance the environmental costs of the proposed action (read: externalities of state action) against its target benefits. Judge Wright suggested that an agency action could be rejected by the courts as arbitrary under the APA if, after NEPA, it struck an unreasonable balance between environmental costs and target benefits (5 U.S.C. 706).

Subsequently, however, the courts began to limit NEPA. First, the U.S. Supreme Court held that NEPA is purely procedural, requiring only the "stop and think" function of preparing the EIS and not the substantive function of balancing costs and benefits or choosing alternatives that yield greater net benefits (see *Vermont Yankee Nuclear Power Corp. v. NRDC,* 435 U.S. 519, 551 [1978]; *Strycker's Bay Neighborhood Council v. Karlen,* 444 U.S. 223, 230 [1980]; *Robertson v. Methow Valley Citizens Council,* 490 U.S. 332, 353 [1989]). Second, the Court held that NEPA does not apply to appropriations bills—the acts of Congress that finance perverse subsidies (*Andrus v. Sierra Club,* 442 U.S. 347 [1979]). Third, even the purely procedural element of NEPA has been held not to apply to regulatory decisions by environmental agencies such as the Environmental Protection Agency (EPA). The notion is that the EPA's own internal decision-making process would generate the "functional equivalent" of the EIS required under NEPA, so that applying NEPA would be duplicative (see, e.g.,

EDF v. EPA, 489 F.2d 1247 [D.C. Cir. 1973]). Whatever the virtues of this notion as a policy matter, the text of NEPA itself makes no such distinction. Congress then exempted all rule making under the Clean Air Act and Clean Water Act from review under NEPA (see 15 U.S.C.A. 793; 33 U.S.C.A. 1371). The blanket shield provided by the "functional equivalence" doctrine and statutory exemptions, however, relieved the agency of the responsibility to examine the adverse environmental impacts of its decisions even where the agency's internal decision-making process would not generate such analysis. And the problems of "omitted voice," heuristic limits on analysis, and concomitant cross-media pollution shifts suggest that one office within EPA may well neglect the consequences of its rule makings for the environmental media that are under the jurisdiction of other EPA offices (see Hodas 1998).

The Court has further impeded efforts to vindicate the interests of victims of countervailing risks by holding that the "zone of interests" test under the APA (see *Association of Data Processing Service Organizations Inc. v. Camp*, 390 U.S. 150 [1970]) prevents a litigant whose interests are "inconsistent with the purposes of the statute" from challenging agency action under the statute (*Clarke v. Securities Industries Association*, 479 U.S. 388 [1987]). The goal of this doctrine is to limit tangential litigation, but it has the effect of reinforcing the state failure that resulted in the legislation in the first place. It may be precisely because the victims of the countervailing risk were omitted voices in the legislature that they received the brunt of the regulatory externality now visited upon them. If so, it is no wonder that the statute specifies a purpose that is inconsistent with these victims' interests. Thus, the Court's "zone of interest" test ensures that the same people omitted from legislative consideration will be excluded from court. It misunderstands, or worse, condones the problem of regulatory externality.

Meanwhile, the Supreme Court's evolving doctrine of standing makes it difficult for litigants to challenge the environmental injuries caused by government agencies. If a private party causes injury to natural resources, courts have held that the statutes making such parties liable to the state for damages do recognize the state's standing to sue for the loss of "nonuse" or "existence" value (see *Ohio v. U.S. Department of Interior*, 880 F.2d 432 [D.C. Cir. 1989]). But if a government agency causes injury to natural resources, the Court has held that private citizens do not have standing to sue for the loss of nonuse or existence value, at least not on that ground alone. A private litigant must show "injury in fact" (*Sierra Club v. Morton*, 405 U.S. 727 [1972]), which the Court has interpreted to mean that the litigant was actually present on the land affected—an actual user, not merely a valuer of the nonuse status of the land (see

Lujan v. National Wildlife Federation, 497 U.S. 871 (1990); *Lujan v. Defenders of Wildlife,* 504 U.S. 555 [1992]; *Friends of the Earth v. Laidlaw,* 528 U.S. 167 [2000]). The upshot is that the nonuse value losses caused by government actions may tend to be underlitigated and inadequately internalized by government agencies. Nor can the government be sued for its policy choices which are shielded under the Federal Tort Claims Act.

Even presidential efforts to rein in the administrative state have not achieved much progress toward sustainable governance. Executive Order 12291, issued by President Reagan in 1981, directed agencies to perform benefit-cost analysis. But it did not say anything about the adverse health or environmental impacts of health and environmental regulation. Such countervailing risks could be thought of as part of the overall "costs" of a regulation, but the agencies and the Office of Management and Budget (OMB) tend to define costs as the costs of compliance for industry. Again, compared to the well-organized political power of industry to protect its interest, the victims of countervailing risks are an omitted voice. Worse, a focus on industry's compliance costs could actually exacerbate the countervailing risks of the regulation. The reason has to do with substitutes for the regulated product or activity. If switching to substitutes is easy (inexpensive), compliance costs will be low, but countervailing risks (of the substitute product) may be high. If, by contrast, switching to substitutes is difficult (expensive), compliance costs will be high but countervailing risks may be low. (Note that this potentially inverse relationship between compliance cost and countervailing risk also suggests that industry is not a reliable surrogate litigant for the interests of countervailing risk victims whose claims are barred by the "zone of interests" or "injury in fact" tests.)

Executive Order 12866, issued by President Clinton in 1993, made some progress by expressly including adverse health and environmental impacts in the definition of "cost." But it remains to be seen whether OMB and the agencies will pay any attention to this language. And because E.O. 12866 (like E.O. 12291) applies by its own terms only where its analyses are not prohibited by law, the entire admonition to consider *cost* will be irrelevant wherever statutes forbid such consideration (as several do). A better move would be to define countervailing risks as part of the overall calculation of *benefit.* And, of course, the Executive Orders are enforceable only by OMB, not by private litigants in court.

Some courts have held that the harmful side effects of government action must be taken into account for the agency to be practicing "reasoned rule making" that is not "arbitrary" under the APA (see *CEI v. NHTSA,* 956 F.2d 321 (D.C. Cir. 1992)). Others have held that a statute directing the agency to "prevent unreasonable risk" obliges the agency to confront the adverse side effects its actions may have (see *Corrosion Proof Fittings v. EPA,* 947 F.2d 1201 [5th Cir. 1991]). Most recently, a statute directing the agency to "protect public health" against "all identifiable effects" of a pollutant has been held to require the agency to balance both the beneficial and harmful effects of reducing pollution (see *American Trucking Ass'n v. EPA,* 175 F.3d 1027 [D.C. Cir. 1999]). In that case, reducing ground-level ozone could protect our lungs but expose our skin to UV radiation (see Lutter and Wolz, 1998); EPA did not appeal this ruling. New "regulatory improvement" legislation is requiring greater attention to countervailing risks (see the Safe Drinking Water Amendments of 1996, codified in 42 U.S.C. 300g-1[b]; S.746, the Levin-Thompson bill, 106th Cong.). On the other hand, judicial second-guessing of second-order consequences will push agencies into excessive analysis of attenuated side effects and thus excessive delay before regulating target risks, so any such legislation should take care to authorize or require only reasonable, not indefinite, attention to countervailing risks (see Wiener 1995, 76).

As to the problems of state failure to protect property rights and to limit open-access resources, there is little scope for direct remedies through litigation. Reform will require political pressure and institutional change. Similarly, the acute fragmentation of the administrative state and legislative committees into an array of bounded fiefdoms, each interested in its own mission but creating externalities in other mission areas, will not easily be overcome.

Toward Sustainable Governance

A great deal might be said about options for sustainable governance and for internalizing regulatory externality—at least as much as has been written about options for sustainable development and for internalizing market externalities. Here, I only begin this discussion.

First, the choice is not one of "more" government versus "less" government. Government failures derive from both weakness and strength, so that less could be as problematic as more. The real problem is neither markets nor government; it is externality, omitted voice, and flawed institutions. The real question is much more complex than less versus more: it is how to design instruments of government that are both more therapeutic and less caustic. We need not just less or more medicine, but the right medicine for the right ailment at the right time. We need not less or more state intervention, but a state that is sustainable because it is both effective and agile.

Second, we need to see the state as an imperfect institution. Once said, this is trite, but it is continually overlooked in discussions of sustainable development and environmental policy. The implications are significant. Seeing the state as an imperfect institution, that can injure as well as nurture, shows the potential conflicts and opportunities in the precautionary principle. If the PP is understood to apply only to private polluters, it misses the point of sustainable governance. On its terms, the PP could embrace governance as well; it requires the proponent of "a substance *or activity* posing a threat to the environment" to demonstrate its safety (Cameron and Abouchar 1991, 2). Applying the PP to government activity clearly makes sense in the context of dams and logging subsidies. But if the PP is applied to government regulatory activity—given the potential countervailing risks of regulation—it would imply that precautionary regulation of market externalities could not go forward until the proponents of such regulation demonstrated that the regulation would have no ill effects. The PP would swallow itself (see Cross 1996, 861). Either much regulation would be blocked, or the PP's effort to bar every risky action—now understood to include both market activity and the regulation of that activity—would have to give way to a more reasoned balancing of the conflicting risks that may be reduced and created by government action (see Wiener 1998, 59–82). The recognition of state failure and regulatory externality should lead to a new search for "risk-superior moves" that reduce multiple risks in concert instead of replacing target risks with countervailing risks (Graham and Wiener 1995; Wiener 1998, 64–67, 78–82).

Third, much effort is needed to reform perverse subsidies and to bolster the basic legal systems that prevent the race to deplete. Courts should consider expanding NEPA, the Public Trust Doctrine, and the Takings Clause to constrain government giveaways of public environmental assets. Advocates of environmental protection should support subsidy-cutting measures at home and, through the World Trade Organization, internationally. This is one of several areas in which the WTO/GATT regime can assist environmental protection.

Fourth, we should study closely our options for internalizing regulatory externality. One path is to impose on the state financial liability for the injuries it causes, much as the private sector is currently liable. This could involve waiving sovereign immunity, imposing pollution taxes on government agencies, requiring agencies to post security bonds to insure against the errors in their EISs (Hodas 1998), and other measures. A problem, however, is that government agencies are not exactly like market firms; the former do not earn financial profits on their provision of social benefits, so a tax or liability on social harms could induce overdeterrence.

A more promising path is to reintegrate the fragmented decision-making structure of the state. This would involve more vigorous overarching supervision within the executive branch (with the power authorized by Congress to identify and internalize regulatory externalities). Such a function would act as the "primary risk manager" for the state (Graham and Wiener 1995, ch. 11). In addition, agencies, program offices, and congressional committees could be required to notify one another if an action here would cause impacts there. Further, the blanket functional equivalence doctrine (judicial and statutory) under NEPA should be relaxed, as well as the exemption for appropriations bills. Thought should also be given to reviving the substantive cost-benefit test derived from NEPA by Judge Skelly Wright. And some mergers of splintered agencies could be worthwhile, as the U.K. and Mexico have each recently attempted—and as the United States itself undertook when it created EPA thirty years ago. For example, it may be time to study whether bringing the EPA, OSHA, NHTSA, FDA, FAA, and CPSC, among others, into a new Department of Risk Management, would enhance the bottom-line responsibility for the joint impacts of these interrelated agencies and reduce risk-risk trade-offs among them. Congressional authorization should meanwhile be given to agency heads to consider risk-risk tradeoffs in setting regulations, notwithstanding preexisting legislative edicts to the contrary (see the 1996 amendments to the Safe Drinking Water Act, 42 U.S.C. 300g-1[b][5]), which passed the Senate 98 to 0 and the House 392 to 30). These are moves toward a more holistic and sustainable approach to regulation.

Conclusion

It should be clear that focusing attention on the sustainability of government activity is neither antienvironment, antigovernment, nor antidevelopment. If the state is causing environmental harm, advocates of environmental protection should not pretend otherwise. Indeed, constraining the environmental impacts of the state has been a hallmark of environmentalism and environmental law, from at least NEPA forward to the contemporary concern about perverse government subsidies and risk-risk trade-offs. Environmentalism should not equate state intervention with promoting environmental sustainability per se. The goal of environmentalism and sustainability should be protecting the environment against the hazards of all imperfect institutions, private and public, rather than an unbreakable alliance between environmental groups and the state against markets and capitalism. The apparent coalescence of such an alliance, reflected in much of green politics as well as in the academic literature on sustainable development, is both an environmental mistake—because the state can injure as well as nurture—and a public relations mistake—because it gives credence to the cynicism that environmentalism cares more about government control of society than about real ecological consequences. It paints environmentalists as watermelons—green on the outside, red on the inside—a perception that can only undermine the long-term sustainability of environmentalism itself. Moreover, the occurrence of unaddressed countervailing risks will also undermine public confidence in regulation.

Meanwhile, advocates of economic development, in both rich and poor countries, should not see the absence of state intervention as the ideal condition for development. Insufficient governance can be as disruptive of markets and economic activity as excessive governance. If the state does not define and protect property rights, development will wither (World Bank 1997). Nor is advocating attention to sustainable governance, in parallel with sustainable development, a move to downplay the importance of development. Both concepts are essential: without sustainable governance there cannot be successful development.

If it is to succeed, then, sustainable development must go arm-in-arm with sustainable governance. The latter alone could neglect the problem of market failure, but the former alone would target market failure while neglecting state failure. It would risk acting like a patient who feels sick and seeks therapy, but does not care which medicine the doctor prescribes. Because our goal is not just medicine for its own sake, but a healthy planet, we must be informed consumers of state therapy. We must see government as imperfect, endogenous, and improvable. Inasmuch as sustainable development is about redefining development and growth to mean not just more output of goods and services, but better output—more social and environmental value rather than just more quantity—sustainable development must take into account the social and environmental value of the services of all social institutions, including the state. But just tucking this concept into the already sprawling rhetoric of sustainable development seems a recipe for further neglect. It is better, then, to construct a parallel concept of sustainable governance and pursue that goal with equal zest.

Sustainability in the United States: Legal Tools and Initiatives

Celia Campbell-Mohn

The goal of sustainable development is to manage natural systems for the perpetuation of the human species now and in the future. The value judgments underlying sustainability include risk aversity, intergenerational equity, and community. Sustainable development relies on planning and science to govern activities. Beyond ad hoc mentions of sustainability in several statutes,[1] federal agencies are undertaking proactive initiatives intended to move toward sustainable development. In the natural resources field, public land managing agencies are implementing ecosystem management. In the pollution abatement field, the Environmental Protection Agency (EPA) promotes pollution prevention.

The purpose of this case study is to examine present efforts toward sustainable development in the United States and to make suggestions for the future. This study first examines efforts to define sustainable development, and then explores present federal efforts toward sustainable development, namely, ecosystem management and pollution prevention. Finally, it proposes strategies for this emerging area of environmental law. This study suggests that sustainable development strategies incorporate integrating natural resources and pollution abatement law and apply innovative legal tools to capture environmental externalities to private property.

Defining Sustainable Development

The definitions offered to support and implement the concept of sustainable development are numerous and demonstrate the complexity of the concept (see Stone 1994; Hoelting 1994; F. Smith 1994). For example, the EPA compiled a list of sixty-five definitions of sustainable development (Atcheson 1991, 1395). Perhaps the best-known definition was set forth in *Our Common Future* by the United Nations World Commission on Environment and Development: "Sustainable development is development that meets the needs of the present without compromising the ability of future generations to meet their own needs" (WCED 1987, 8).

This definition has been criticized as being too general and of limited use in putting the concept of sustainable development into effect (Lipschutz 1991, 35, 38). It begs the questions: What do we mean by need, present or future, and what limits should consideration of the future place on present decisions and actions (38)? It has also been criticized as considering economics and ethics as competing concepts, to the exclusion of community and the precautionary principles. However, by formulating a general construction of the concept of sustainable development, the Commission invited a worldwide dialogue among members of such diverse fields of expertise as economists, scientists, philosophers, and government policymakers (Stone 1994, 978; Sessions 1993, 15). The goal of sustainability is to manage natural systems for the perpetuation of the human species now and in the future; the value judgments underlying sustainability include risk aversity, intergenerational equity, and community.

Risk Aversity

In terms of risk aversity, or what many scholars are calling the precautionary principle (Hickey and Walker 1995; Fullem 1995), sustainability assumes that the consequences of depleting resources outweigh the likelihood that the resource base will be expanded. In other words, human behavior should maintain natural systems so that these systems are not depleted even though future generations may invent technology that avoids the consequences of resource depletion.

Risk aversity makes sustainable development strategies proactive. Sustainable development seeks to manage environmental resources in the present so that they will not be depleted or degraded in the future. This requires employing a long-term, forward-looking perspective that anticipates future needs and limits future harms (Daly and Cobb 1989, 76). However, because of the inherent uncertainties about the extent of future needs and the potential future adverse impact of present actions,[2] sustainable development also embodies a conservative element to ensure that future generations will not have options foreclosed by present decisions (Weiss 1990, 201–2).

Such proactive management schemes include utilizing environmental resources within the bounds of the Earth's carrying capacity (Postel 1994, 16–19; Daily and Ehrlich 1992, 761; Clark 1994, 655). The purpose is to curb consumption rates of renewable resources so as to not exceed their regeneration rates, and limit pollution within the environment's assimilative capacity. Sustainable development plans usually call for both nonrenewable resources to be used more efficiently and the development of alternatives.

Additionally, sustainable development cannot be addressed without consideration of population pressures and the myriad complex policy and ethical questions it raises (Repetto 1985, 1, 12). Sustainable development seeks to develop policies that protect the viability of ecosystems and prevent the loss of biodiversity, while continuing economic development (Woodwell 1985, 61–63).

Conforming human behavior to maintain natural systems has immediate implications for an economy based on free-market principles and dependent on growth. Economic growth is measured by various benchmarks, such as the gross national product of the level of real income per capita, but may be generally thought of as increasing the standard of living (Pearce et al. 1989, 30).

Sustainable development claims to shift the economic goals from growth to development. In the economic sense, development is based on quantitative expansion rather than qualitative change (Daly and Cobb 1989, 37–38). Rather than trying to increase the standard of living, sustainable development measures the economic success of the system by improvement in the quality of life (Pearce et al. 1989, 17). The tension in this perspective lies in the fact that quality of life is a value judgment, the same value judgment that the free market measures through free choice.

Given this problem, there is very little consensus about how a sustainable economic system should operate. Some commentators agree that our economic patterns should change and hold sustainability as a goal, but the means remain ill defined (Lipschutz 1991, 38). At a minimum, sustainable development has broadened the way we think about economic systems by acknowledging the interdependent relationship between the economy and the environment that supports it (Pearce et al. 1989, 4).

Intergenerational Equity

The values embodied by sustainable development rest inter alia on the principle of intergenerational equity (Repetto 1985, 8). Intergenerational equity is based on the present generation having an obligation or duty to sustain the environmental quality of the world it inherited from previous generations so it may pass the planet on to future generations in a comparable condition. The purpose is to provide future generations equitable access to the planet's resources and assets (see Weiss 1990, 200). In this way, intergenerational equity is conceptually similar to a trust relationship, where net assets must be preserved so that they are equally available to future beneficiaries (200).

The validity of intergenerational equity principles is hotly debated. It has been argued that future generations have no rights that obligate the present generation and must face their own challenges through technology or adaptation (Gundling 1990, 209–10). Other arguments posit that vesting rights in future generations rests on a value judgment that ought to be made by the present generation to ensure the future survival of our species and planet. Still other arguments are based on ethical imperatives. One difficulty with intergenerational equity is that the nature of these reciprocal rights and duties remains unclear. Therefore, it is difficult to determine what obligations in the present must be met to live up to the responsibility to future generations (210–11). However, the pursuit of intergenerational equity provides an important and influential ethical context for evaluating present policies and promoting sustainable development.

Community

Many of the foundational principles of law are atomistic, such as the right to contract and private property rights. The natural environment, however, is built on the functioning of interrelated natural systems. Humans share this natural environment with each other and with other species. Perpetuation of the individual's life as well as that of his or her progeny and all of life within the functioning of natural systems is a common enterprise. Therefore, the natural environment compels humans to focus on principles of community. Virtually all humans share a communal interest in perpetuating natural systems.

One element of the call for sustainable development is recognition of this shared interest with the consequent recognition that communities, from the local to the international, must determine how to respond. Community theory therefore may add to the development of the concept of sustainable development.[3]

Sustainable Development in the United States

Despite the difficulties in defining sustainable development, the federal government initiated several efforts toward achieving this goal. The domestic sustainable development effort can be divided into two parallel arenas: policy discussions and agency actions.

Policy Discussions

In 1993 President Clinton created the President's Council on Sustainable Development (PCSD); Executive Order 12852, 58 Fed. Reg. 35841 [1993]). In April 1995, the PCSD issued its Task Force Policy Recommendations (U.S. EPA 1995, 3). These draft recommendations require the vote of the entire council before forwarding to the president. Proposals include inter alia setting essential learning standards on sustainable development for all students; drafting environmental guidelines for developers, such as building codes and materials guidelines; expanding educational efforts and access to services related to contraceptives; realigning fiscal policies with the aim of shifting the tax burden from labor and investment to consumption; reducing federal subsidies, particularly those for extraction industries such as mining and petroleum; and implementing incentives or requirements for producers and consumers to reduce solid waste. The recommendations have been criticized for being too radical to be implemented (3).

Discussions at the federal policy level are stymied by Congress's efforts to limit federal environmental laws. Proactive moves toward sustainable development have not developed through legislation during recent Congresses.[4]

Federal Administrative Initiatives

At the federal agency level, however, the environmental agencies, while pursuing their substantive mandates, continue to impact the ideas and implementation of sustainable development. Specifically, the land management agencies' efforts to employ ecosystem management and the EPA's stress on pollution prevention should be examined in light of sustainable development efforts.

Ecosystem Management. The U.S. Forest Service and various agencies within the Department of the Interior are implementing what former Vice President Al Gore's National Performance Review (NPR) called ecosystem management, a "proactive approach to ensuring a sustainable economy and a sustainable environment" (Thomson 1995). President Clinton established an Interagency Ecosystem Management Task Force, chaired by Katie McGinty, Director of the Council on Environmental Quality. The Task Force identified seven case studies for a survey team to assess opportunities to assist ecosystem management.[5] Beyond these case studies, there seems to be adequate flexibility within the agencies' legislative delegation of authority to pursue ecosystem management in managing the public lands (Keiter, 1989, 923; 1994, 293). However, application of the concept is proving elusive.

One problem with ecosystem management, like sustainable development, is that the term has been difficult to define. For example, EPA officials are finding that terms such as "community," "ecosystem," and "watershed" may encompass different geographical areas depending on the project.[6] The land management agencies also are having a difficult time defining ecosystem management. The Forest Service alone has four different working definitions of the term, and those definitions differ from the Bureau of Land Management's (see, generally, Grumbie 1994, 27).

Part of the difficulty may lie in the fact that the term ecosystem is less than precise. The term was coined by Arthur Tansley in 1935 (Slocombe 1993, 289, 292). Tansley sought to provide a more precise and more holistic term for the set of biological and physical factors that affect an organism, that form its environment. Ecologists use the word ecosystem to describe a distinct and coherent ecological community of organisms and the physical environment with which they interact. The term thus describes a group of natural systems (see Slocombe 1993). It is a useful shorthand. But applying this term to federal public land management raises a series of applications that the term was not intended to encompass.

For example, there are many severable natural systems within the term ecosystem that do not share common physical boundaries: watersheds, airsheds, wildlife habitat, and vegetative habitat overlap but do not follow consistent geographical boundaries that we can delineate and manage. In addition, each of these natural systems changes over time. So there is no spatial or temporal region we can describe as an ecosystem.

In addition, our public lands are not located along ecosystem lines, nor is division of management responsibility among the various agencies based on ecosystem criteria.[7] Rather, the lands themselves and jurisdiction over those lands are a patchwork. This simple reality requires defining ecosystems by human boundaries rather than by natural systems. "Management" of an ecosystem then becomes a human construct extrapolated by analogy onto natural systems. Like the term sustainable development, ecosystem management quickly degenerates into a malleable concept that is subject to instrumental definition. As the Congressional Research Service (CRS) report on ecosystem management states, "There is not enough agreement on the meaning of the concept to hinder its popularity" (CRS 1994; U.S. General Accounting Office [GAO] 1994).

Pollution Prevention. A second regulatory initiative moving toward sustainable development is pollution prevention. The primary legislative initiative is the 1990 Pollution Prevention Act (42 U.S.C. §§ 13101–09). In response to the Act, in February 1991 EPA issued its Pollution Prevention Strategy (56 Fed. Reg. 7849 [1991]). The two main objectives outlined in the strategy are guidance for internal operations and a voluntary reduction program.[8] These objectives reflected President Bush's reliance on voluntary pollution prevention efforts and "end-of-pipe" reductions.

On Earth Day 1993, President Clinton supported the goal of pollution prevention by requiring all federal agencies to establish pollution prevention strategies.[9] The Clinton administration also expanded the definition of pollution prevention beyond just addressing "end-of-pipe" to include "protecting natural resources through conservation or increased efficiency in the use of energy, water and other materials" (Browner and EPA 1993).

Former EPA Administrator Carol Browner often highlighted pollution prevention in her speeches (Browner and EPA 1993; see also Johnson 1992, 13, 157). Browner elaborated that "EPA is . . . fundamentally shifting U.S. environmental protection strategy towards pollution prevention" and "[the Clinton] Administration is committed to making pollution prevention the guiding principle in all our environmental efforts" (1994).

Under the auspices of pollution prevention, EPA initiated over thirty projects. Of direct relevance to sustainable development, some of these projects included the Sustainable Industry Project (U.S. EPA, Office of Policy Planning and Evaluation 1994), Environmental Labelling and Life-Cycle Assessment (Browner 1994), and the Common Sense Initiative (see U.S. EPA 1994, 17).

The Common Sense Initiative focuses on the regulations governing six industries. In coordination with various stakeholders, EPA-led committees review regulations governing the industry to improve the efficiency and effectiveness of the regulatory controls (Harper 1995, 3). The Common Sense Initiative, however, may be threatened by EPA's decision to target enforcement actions against the participating industries because industries will be unwilling to provide information that will later be used against them (U.S. EPA 1995d, 1).

The pollution prevention effort, and the Common Sense Initiative in particular, are promising efforts moving toward sustainable development. To the extent that sustainable development is definable, it includes concepts of effective and efficient environmental regulation that allocates resources in a manner that does not foreclose future options. At least in theory, EPA's pollution prevention efforts strive to both efficiently and effectively regulate production.

Recommendations for New Sustainable Development Initiatives

Ecosystem management and pollution prevention are both steps toward sustainable development. The purpose of this essay is to recommend two additional steps. These recommendations for sustainable development initiatives proceed on two levels: the federal level and the state and local level.

Federal Initiatives

Although the federal agencies are making impressive strides within the context of a nebulous and ill-defined standard of sustainable development, they are hindered by historical patterns. By approaching ecosystem management on public lands and pollution prevention at the resource manufacturing phase, our present efforts echo the Progressive-era split between natural resources and pollution abatement.[10] These two severable concepts still fail to link resource extraction to pollution abatement.

However, the split of natural resources and pollution abatement into separate camps belies their basic integrated nature and impedes improvement of environmental quality through effective environmental law. Just as natural resources are not separate islands, neither are ecosystem management and pollution prevention. Both are part of one cyclical system in which resources are extracted, manufactured into a product, and then thrown away—back into the system from which they came.[11] In other words, environmental law would be more efficient and effective if analyzed according to the way the natural systems function, rather than the way legislation and regulation happen to have been adopted. We also need to realize that the purpose of environmental laws is to affect human behavior, with the consequent result of affecting the environmental media—that is air, water, and soil. We need to develop a new analytical framework that joins natural resource law and pollution abatement law into one system that more accurately reflects the way the natural system works. The new framework should look at environmental law through the life cycle of the resource from the time we extract a resource from the environment, through its manufacture into a product, to its disposal as a waste. This is called the resource-to-recovery approach. Unlike traditional approaches to environmental law that explain each statute or group the statutes by media—such as air, water, soil—the resource-to-recovery approach reflects the fact that laws govern human activities, not the environment.

The best way to illustrate this analytical framework is by an example.

Applying the resource-to-recovery framework to the paper industry (see K. Rosenbaum 1993), we start with the laws governing timber cutting on public and private land. The National Forest Management Act is the primary statute governing the cutting of trees on national forests. The amount of timber that may be cut on national forest lands is called the "allowable sale quantity" and is set by Congress every year. State and local zoning laws govern cutting timber on private land.

Cutting timber creates environmental impacts, including those on watersheds, fish and wildlife, and recreation. However, looking at the whole framework of our laws, we see that environmental laws limiting harvest may indirectly encourage the use of materials other than wood, and the manufacture and use of these materials may produce more hazardous waste, use more energy, and create more air or water pollution than manufacture and use of wood. This calls for an analysis of the economics and impacts of substitutable products.

In the manufacturing phase, the logs and chemicals are first brought to the pulp mill, then pulped, bleached, and finally rolled into paper. Throughout the many steps in this process, the Comprehensive Environmental Response, Compensation, and Liability Act (CERCLA) may require spill reporting; the Clean Water Act governs storm water runoff and discharges into navigable waters; the Clean Air Act governs the industrial boilers that produce heat and the air-borne chemicals from the pulping and bleaching processes; the Resource Conservation and Recovery Act (RCRA) governs production of hazardous wastes; the Federal Insecticide, Fungicide, and Rodenticide Act (FIFRA) governs biocides used to treat the paper; and the Toxic Substances Control Act governs PCBs and sludge left after wastewater treatment (K. Rosenbaum 1993).

Applying the resource-to-recovery analytical framework to the manufacturing stage, we see how easy it is to shift pollutants from one medium to another. For example, in the pulping process, the paper industry is learning to avoid water quality regulations by freezing water-borne wastes, drying and burning the residue, and shifting the pollution to the air or remaining sludge, where it is not regulated (K. Rosenbaum 1993). Likewise, in many industries, it is easy to take a pollutant that is regulated in one medium, such as air, and switch it to another medium where it is not regulated, such as water. The resource-to-recovery analytical tool allows us to see this media-shifting and more accurately target regulation.

In the final stage of the resource-to-recovery cycle, the paper is discarded; it is generally either recycled or put in a landfill. The recycling of paper can produce its own wastes. As the paper is repulped, clays and other finishes and fillers are separated out and become solid waste. The process water from the repulping and de-inking now contains fibers, detergents, inks, and bleaches (K. Rosenbaum 1993). Applying the resource-to-recovery framework to the disposal phase, the laws encouraging recycling of paper reduce solid waste in municipal landfills but increase the solid waste created in the manufacture of paper. They may also encourage the use of bleaches to remove the last vestiges of ink from the recovered product (K. Rosenbaum 1993).

The resource-to-recovery analytical framework allows us to look at the whole papermaking process from cutting the trees to throwing away the paper. Using this approach, it may be possible to conceive of regulations that would be simultaneously more effective and efficient. For example, when Congress is setting the allowable sale quantity for timber on national forests in the first stage, at the same time it should consider ways to reduce market demand for the final product by both the federal government and the private market. Congress should look beyond the immediate argument of jobs versus the environment to consider the entire environmental impact of producing paper, from cutting the trees to throwing away or recycling the paper.

The paper industry provides an example of how the resource-to-recovery analytical framework applies to one industry, but we can apply this construct to virtually any industry. The resource-to-recovery approach shows how the laws interact with human behavior and how they interrelate. It highlights gaps and conflicts in the law. In other words, it demonstrates how the laws fit together by showing where they apply.

Comparing the resource-to-recovery approach to the present structure of environmental laws reveals similar gains in efficiency and effectiveness. In the resource extraction phase, existing environmental law focuses on just parts of the natural system—wetlands but not wildlands, endangered species but not biological diversity—and not all biological constituents that support the functioning of natural systems. Natural systems function as a unit, in a complex and interrelated process, and ecologists have long recognized that everything is connected to everything else. However, our laws protect certain elements within these systems as if they were distinct from the whole. For example, wetlands are one part of a much larger natural hydrologic cycle, yet we give legal protection to wetlands while ignoring the cycle. Endangered species are likewise part of an evolving cyclical ebb and flow of species populations, yet we stringently protect endangered species without worrying about the basic loss of life itself through diminishing biological diversity. The resource-to-recovery approach focuses on all impacts of human activities on the environment. By using a systems approach and analyzing law with regard to the way natural systems function, the resource-to-recovery approach would come closer to making law mirror the environment.

Our current system of environmental law now impacts the middle phase of the cycle, product manufacturing, most heavily (Breen 1993). In the manufacturing phase, the law focuses on select pollutants emitted into the environment, but not on all pollutants or on siting the facility emitting the pollutants. Again, this is law divorced from the functioning of natural systems. The resource-to-recovery framework analyzes other phases of the cycle to discover where environmental law can produce better environmental and social results. Currently, no laws address all three phases of extraction, production, and disposal; in contrast, the resource-to-recovery approach relies on foresight and planning through all phases of the cycle.

By uniting ecosystem management and pollution prevention, the resource-to-recovery approach may be the next step toward sustainable development. The challenge is that such reform may require legislation, which is unlikely during a Congress seeking to limit environmental law.

State and Local Initiatives

A more likely advance toward sustainable development at this point in history may be at the community level rather than the federal level. Indeed, focusing on the community level seems a natural place to start. Combining ecosystem management and pollution prevention in a resource-to-recovery framework may best begin in the community. Avoiding pollution at the beginning, in the extraction phase, through managing natural systems is a precept of sustainable development.[12]

Environmental law to date, however, has focused on the international, federal, and individual levels, but not the community. Environmental law also has focused on the manufacture and disposal phases, but not on extraction, except on public lands.[13] Perhaps now is a good time to examine how to fill this gap in environmental law. This is where the local community plays the starring role. It is state and local regulation that limits resource extraction on private land.[14] Perhaps the single most important regulatory authority is zoning. Land-use control is exercised almost exclusively by state and local governments. Land-use controls limit what each owner can do on private property (Breen 1993). These controls can authorize what uses can be made of private property and can require mitigation measures and safeguards. These mitigation measures can include measures to prevent soil erosion and to protect standing trees, for example (Breen 1993).

There is a limit on the ability of this regulatory mechanism to impact the rest of the life cycle. Under the Fifth Amendment of the Constitution, the local government cannot regulate the use of land to deprive the owner of all of the land's benefits. There is a reason for this protection. The benefits of mitigation measures to preserve the functioning of natural systems accrue to everyone. For example, we all benefit from safeguards to prevent soil erosion, but the costs fall only on the landowner. The landowner usually had other plans for the land than to operate it for everyone else's benefit. Thus, there is a tension caused by restricting activities on private land to preserve the functioning of natural systems. The concept is that individuals should not have to use their property as an environmental commons for everyone else (Breen 1993). However, there is another way to address property ownership. By using land in a way that destroys the functioning of natural systems, the landowner imposes costs on other property and on the community. For example, destroying a primary dune or wetland may increase the likelihood of flooding on adjacent property. Just as environmental regulations attempt to require industrial polluters to internalize the costs of externalities, so could private property owners be required to pay for their impacts on the community.

This is where community theory comes in. The traditional notion of property is that ownership carries with it a bundle of rights, such as the right to exclude others, and inter alia to use the property in certain ways. The community theorists could argue that property ownership is limited to the right to use only one's own property and not to impact the rest of the community. Under this perspective, property ownership entitles the owner to use the property in any way that does not impact the rest of the community. Impeding the functioning of natural systems that extend beyond the property boundaries are not within the bundle of rights of ownership.

This view of the limits of private property ownership finds some precedent in the law. The Supreme Court has stated: "While the rights of private property are sacredly guarded, we must not forget, that the community also have rights; and that the happiness and well-being of every citizen depends upon their faithful preservation" (*Charles River Bridge v. Warren Bridge,* 36 U.S. [11 Pet.] 420, 548, 9 L. Ed. 773 [1837]). Indeed, the early nuisance cases were based on this premise.

Recent takings bills have brought this debate to a crescendo. As Sax notes on the full compensation provisions, these bills "basically override the notion that the community can demand that property owners accommodate their property use to community values. This approach leads to a renunciation of the very sense that we *are* a community" (1995).

A view that stresses private property responsibilities to the community as well as rights of ownership raises a series of difficult questions. For example, cutting one tree arguably impacts the watershed, airshed, wildlife habitat, and other natural systems on which the community relies. One wonders to what extent a community should be willing to absorb the external costs of private property use versus at what point the private property owner should be responsible for internalizing those externalities. This, however, is a fundamentally different question from that presently being asked. The question also arises as to what tools the community could use to require property owners to internalize the costs of impacting natural systems to the rest of the community.

To assess legal tools that may be used at the community level to require landowners to internalize their costs, we may extrapolate from concepts used at the federal level. There are two legal tools that the local regulating entity may be able to invoke under the police power: the concept of natural resource damage and the concept of environmental assessment. The idea behind applying both of these to private property is that the right to use private property does not include the right to impinge upon the functioning of natural systems that extend beyond that property.[15] Where an action will have extraterritorial effects, the property owner must either pay for the impacts or at least inform the community of them.

Natural Resource Damages. As defined by the concept of natural resource damages, the government is entrusted with the responsibility of managing common lands and resources, such as air, water, and wildlife, to maximize the public's benefit by acting as a trustee over the resources (see Stewart 1995). When these resources are damaged by private parties, in some circumstances, Congress empowers the government to collect damage on behalf of the public for the injury. The money is used either to restore the damage or to purchase comparable properties (see Salzman 1997; Kopp and Smith 1993; Williams 1995). The basic concept is simple: if you damage resources, the government can compel you to clean them up. In broadening this concept one could argue that if you damage natural systems held in common through your use of your property, you must pay for that extraterritorial damage.

Under present law, trustees can recover natural resource damages if the government has some special interest in the damaged resource. Purely private resources are excluded, but the definition of natural resource is not limited to lands *owned* by the government. Indeed, governmental regulation may be sufficient to invoke the trusteeship responsibility. Theoretically, the concept of natural resource damages may be invoked to require private property owners to pay for damages to natural systems from uses of their property that extend beyond their boundaries. Under this theory, one could use property to the extent that such use does not impact the community, natural or human. One would have to pay for any damages private uses inflict on common resources.

Environmental Assessment. A second theory based on the property owner's responsibility to the community not to externalize the costs of his or her property use is environmental assessment. The concept of environmental assessment is codified in the National Environmental Policy Act (NEPA; 42 U.S.C. 4321-4370d [1994]; see 40 CFR 1500-99 [1997]).

President Nixon signed NEPA into law on January 1, 1970. Congress enacted section 102(2)(C) of NEPA to ensure that agencies consider the environmental impacts of their decisions. Section 102, the primary action-forcing provision of NEPA, states:

> The Congress authorizes and directs that, to the fullest extent possible: (1) the policies, regulations, and public laws of the United States shall be interpreted and administered in accordance with the policies set forth in this Act, and (2) all agencies of the Federal Government shall . . . (C) include in every recommendation or report on proposals for legislation and other major federal actions significantly affecting the quality of the human environment, a detailed statement by the responsible official on (i) the environmental impact of the proposed action, (ii) any adverse environmental effects which cannot be avoided should the proposal be implemented, (iii) alternatives to the proposed action, (iv) the relationship between local short-term uses of man's environment and the maintenance and enhancement of long-term productivity, and (v) any irreversible and irretrievable commitments of resources which would be involved in the proposed action should it be implemented. (42 U.S.C. 4332)

These requirements led to the creation of the environmental impact assessment (EIA) process, through which Congress required federal agencies to consider the environmental impacts of the agencies' decisions. The EIA process identifies three classes of agency activities. Activities that may have a significant impact on the quality of the human environment require an environmental impact statement (EIS). When it is unclear whether an action may have a significant impact on the quality of the human environment, the agency prepares an Environmental Assessment (EA), a concise public document that analyzes and provides evidence on whether to prepare an EIS or a Finding of No Significant Impact (FONSI), which justifies the decision not to prepare an EIS (40 C.F.R. 1508.13). The EA contains a brief discussion of the need for the project, reasonable alternatives to the proposal, environmental impacts of the proposal and alternatives, and a list of interested parties. If the EA reveals that there may be a significant environmental impact, a full EIS is prepared. Actions that do not individually or cumulatively have a significant effect on the human environment and for which neither an EA nor an EIS is required are categorically excluded (40 C.F.R. 1508.4).

Where a tiered system is used, the threshold question of when an EIS must be prepared at all is crucial. In the NEPA context, an EIS must be prepared when there is a "major federal action significantly affecting the quality of the human environment" (42 U.S.C. 4332; 40 C.F.R. 1502.3). This provides in-depth consideration of environmental impacts where they are "significant," a term that is defined on a case-by-case basis depending on the context and intensity of the proposal (40 C.F.R. 1508.27). The term "major" reinforces "significant" and does not have independent meaning (*Minnesota PIRG v. Butz*, 498 F.2d 1214 [8th Cir. 1974]).

It would be infeasible and imprudent to recommend that a NEPA process, with its elaborate and expensive EIA process, be applied to extraterritorial impacts from the use of private property. However, theoretically, the idea that extraterritorial environmental impacts from the use of private property should be considered and that information about those impacts should be provided to those affected can be extrapolated from NEPA to the private property context. Rather than require property owners to pay for damage to natural systems from their property use, as the extrapolation from natural resource damages suggests, local regulatory bodies could require landowners to at least document and inform the community of these impacts. The documentation process may inform and mitigate some landowners' actions and inform the community of extraterritorial impacts. This is the same idea as natural resource damages, but requires information rather than payment.

The community level is the part of the resource-to-recovery framework of environmental regulation that is most lacking. It may behoove us to now focus on the community level for environmental reforms. At the community level, the community theorists inspire also asking about the responsibilities of property ownership as well as the rights. These responsibilities may include internalizing the costs of property use through compensation or at least information on the impact on natural systems that extend beyond the property boundaries.

Conclusion

Although sustainable development is an elusive and ill-defined term, risk aversity, intergenerational equity, and community are goals that we can ascribe to it. Federal agencies have begun to incorporate these objectives through ecosystem management and pollution prevention initiatives. This case study challenges us to approach sustainable development on the system rather than the project level. Rather than incorporate new federal policies and projects, the resource-to-recovery approach and innovative legal tools capturing the externalities of private property seek systemic changes to environmental law. Rather than calling for another new federal project, this study challenges us to reach deeper into our legal system and structure and to apply sustainable development concepts in ever widening circles.

Sustainable Development and the Use of Public Lands

Jan G. Laitos

An important natural resource in this country is the land base owned by the federal government. These so-called public lands comprise nearly one-third of the land within the United States (U.S. Department of Interior, Bureau of Land Management [BLM] 1991). Traditionally, these lands have been managed by various federal agencies according to the multiple-use doctrine.[1] This doctrine, firmly affixed to federal land management statutes, requires the simultaneous production of a variety of resources, outputs, uses, and activities through scientific planning.[2] Although the multiple-use doctrine is grounded in the assumption that many uses are preferable to a single use, the reality is that for years two visions of public land use have competed for dominance: the *use* of these lands for development and extraction of commodity resources (primarily minerals, energy resources, timber, and livestock forage),[3] and the *preservation* of these lands for parks, wildlife habitat, and wilderness.

Throughout the latter half of the twentieth century, the debate raged on between commodity resource use and preservationism nonuse (Callicott and Mumford 1997, 32, 34; Cushman 1999). The conflict became so intense that for decades it nearly crippled the environmental law movement (Ruhl 1998, 273, 279). Because multiple use seemed incapable of resolving the use/nonuse conflict, many commentators offered alternative management philosophies.[4] During the past several decades, the debate over competing uses of public lands became even more complex when it became obvious that recreation was becoming, or had become, a force driving utilization of these lands. Indeed, recreation has become so powerful that I have argued elsewhere (and I believe the data support) that public lands are now largely dominated by two nonconsumptive uses: recreation and preservation.[5]

Several important consequences follow from such a transformation on public lands. If these lands are used primarily for just recreation and preservation, this is inconsistent with the statutory mandate of multiple use. Moreover, commodity uses and resource-extractive activities should be able to play some role on the public lands. After all, for nearly a century this country's federally owned lands were chiefly valuable because of the natural resources that were found there, and that were in turn extracted from the lands by private commodity interests (Laitos and Carr 1998, 1–22). It would seem that some management strategy other than multiple use should be applied to America's public lands.[6]

A change in the nature of the public lands debate from use versus nonuse to use versus recreation and preservation suggests that future conflicts about public land use will not be fought along the traditional lines of commodity versus noncommodity use. That battle has already been largely conceded by commodity and resource developers (Laitos and Carr 1998). Instead, the looming conflict in public land use will be between two former allies: recreational and preservationist interests. Recreation *uses* public lands, albeit without extracting natural resources. Preservation seeks to implement a *nonuse* mandate. These two values must inevitably clash.[7] There will even likely be a subset of this recreation versus preservation conflict, as disputes will surely arise between two classes of recreational interests: low-impact, human-powered recreational users, and high-impact, motorized recreational interests (Lofblom 1998; Kelly 1998).

Given the changing uses of public lands and the new types of controversies that have arisen with respect to competing uses (and nonuses) there, the critical question is which land management policy is best able to guide future land use decisions. In this case study I argue that the efficiency criterion is a preferred alternative to the multiple-use land management philosophy (e.g., Leshy 1984, 235). I explain how efficiency principles applied to public lands may deal more realistically with intangible recreational and nonuse values, as well as with tangible resource-extractive commodity uses. Efficiency seems to achieve what multiple use promised yet was unable to deliver: the allocation, development, and maintenance of public lands resources that bring about an overall increase in social value.

I then overlay on the efficiency model the notion of sustainable development. This concept, or philosophy, means many things to many people. At its most fundamental, however, it calls for resource decisions to be made with a planning horizon that does not sacrifice the future for the present, and that defines broadly the beneficiaries of resource policy so as to include a wide range of economic, social, and racial classes (Lash 1997, 83). These time and equity goals add new dimensions to the efficiency goal, which typically is concerned only with present allocations of resources but not with the equitable consequences of attaining economic optimality. The idea of sustainable development is therefore a potentially important component part of public land and resource use management. When combined with efficiency, it would create a comprehensive decisional matrix, where economic efficiency, time, and equity could constitute the three dimensions of the next generation's public lands resource policy.

Economic Efficiency and Public Land Management

The public lands contain a vast amount of resources that have the potential to produce a mix of diverse outputs. These include timber, cattle, extracted hardrock minerals, oil and gas and coal, recreation, and preserved habitat for species, ecosystems, and unique geological structures. A given acre of land may be able to produce multiple commodity products. Under most circumstances, however, recreation and preservation uses are not compatible with the traditional commodity outputs (Clawson 1978, 281, 286; Power 1996, 1–2).[8]

In light of the two uses that now predominate on public lands, as well as the failure of existing policy to rationally accommodate them alongside consumptive uses, two questions arise: Given the 650 million acres of federal land, how much land should be devoted to the production of commodity goods (timber, grazing, minerals), how much of that land should be allocated to recreation, and how much to preservation uses? and How should public land managers answer question 1?

Economic principles suggest allocating land to obtain the goal of efficiency. An efficient allocation means that the current use of resources maximizes the total value of goods and services for a given distribution of income. Mindful of the underlying assumptions,[9] economic efficiency represents a theoretical ideal for allocating resources in a society. This theoretical goal could serve as the benchmark for how policymakers and agencies choose to allocate resources on the public lands.[10] Recent contributions in the economics literature provide an appropriate theoretical framework to determine the optimal allocation of land (e.g., McConnell 1989). To demonstrate the operation of an efficiency methodology, one can begin with the overly simple assumption that the public land base is allocated between just two categories of use: extractive uses that include timber harvesting, grazing, and mining, and nonextractive uses that include recreation and preservation. An efficient allocation maximizes net social benefits from the set of possible land allocations subject to the constraint of the fixed federal land base. The efficiency solution requires that the marginal unit of land yields a marginal benefit of recreation and preservation equal to the marginal benefit of commodity use. Intuitively, this means that the last acre of land allocated to timber production, cattle grazing, or mining should generate the same incremental benefits as the last acre of land allocated for hiking, camping, mountain biking, or wildlife habitat preservation.

Valuing the Benefits of Market and Nonmarket Goods

The efficiency goal seeks to duplicate the result that would be reached if commodity and recreational and preservationist goods and amenities could be traded in a well-functioning market. In such a market, preferences will shift from less valued uses to more valued uses, measured by people's willingness to pay. When markets do not exist for various uses (e.g., for recreational or preservationist uses of public lands), welfare economics teaches us that it is possible to test whether a particular allocation has achieved efficiency by subjecting the allocation to an analysis of costs and benefits. Such an analysis would seek to measure the social benefits of an allocation among commodity, recreation, and preservationist uses, as well as its costs (e.g., Mishan 1976). Although the costs of a given allocation of land uses are significant,[11] what is particularly important in allocating competing public land uses is the measurement of marginal benefits of recreation, preservation, and commodity uses. Unfortunately, it is exceptionally difficult to calculate the social benefits of land used for recreation and preservation purposes, because these uses have no easily discernible market value. I therefore focus on offering both a methodology for valuing recreation and preservation and a general aggregate economic value for each.

The theoretical concept of economic efficiency assumes a full accounting of social benefits of all resources. Social benefits are valued by willingness to pay for a good or service or resource. The social benefits from land allocated to commodity use yields tangible market goods, such as lumber, cattle, metal, and energy products, whose economic value can be calculated. The social benefits of land allocated to recreation include nonmarket activities, such as hiking, camping, fishing, hunting, and birdwatching. These are not easily quantified. It is likewise difficult to put an economic value on land devoted to preservation. The natural ecosystem generates various services outside of the market that are important to humans, such as the collection and storage of drinking water in a watershed, genetic information leading to new medicinal and commercial products, and carbon sequestration of greenhouse gases in a standing forest (Myers 1992, 189–293).

Policymakers must recognize that the full economic value of public lands may extend beyond the traditional use values associated with commodities. The true value of these lands also includes nonconsumptive values, sometimes called passive use values, that may be employed to set the worth of recreation and preservation uses. Although passive use values are more speculative than use values because they are not subject to normal market valuation methods, they are real and valid because they reflect utility derived by humans from a resource. Two generally recognized passive use values are *option value* and *existence value.* Option value measures the amount an individual is willing to pay to reserve the right to use the resource in the future (Weisbrod 1964, 472). Existence value defines the satisfaction an individual derives from knowing a resource continues to exist, even if that person never personally uses the resource and will not likely do so in the future (J. V. Krutilla 1967, 777, 781).

Although option and existence values are extremely difficult to measure, certain nonmarketed resource methodologies are available. One that seems particularly applicable for recreational use in public lands is the travel cost method. This measures recreation benefits indirectly by observing the costs individuals willingly incur to travel to a site (e.g., gasoline, opportunity costs of time) for a recreation activity. Implicit in such behavior is that recreation benefits are at least as great as those travel costs (Clawson and Knetsch 1966). Another methodology for determining option and existence values is the contingent valuation method. This utilizes surveys to directly elicit an individual's willingness to pay for a hypothetical change in resource or environmental quality (R. Davis 1963; Hanemann 1994; Mitchel and Carson 1989). Sophisticated survey methods typically ask respondents whether they would be willing to pay a specified amount of money through such mechanisms as a higher taxes, user fees, or trust fund for an improvement of environmental quality. Both the travel cost and contingent valuation methods can measure use values, such as recreation, but only contingent valuation can estimate nonuse values of natural resources, such as preservation.

The estimation of economic value for nonmarket natural resource use has gained acceptance among policymakers and the courts. The Comprehensive Environmental Response, Compensation and Liability Act of 1980 (CERCLA; 42 U.S.C. §§9601–9675 [1996]) and the Oil Pollution Act of 1990 (OPA; 33 U.S.C. §§2701–2761 [1996]) both impose liability on parties responsible for destroying natural resources. Natural resource damage assessment refers to the process of establishing values for different levels of natural resources lost due to environmental contamination. Both CERCLA and OPA authorize agency regulations that establish protocol methods for natural resource damage assessments (CERCLA: 42 U.S.C. §9651 [C][2]; OPA: 33 U.S.C. §2706 [e][1]) and entitle a plaintiff using such methods to a rebuttable presumption of accuracy (CERCLA: 42 U.S.C. §9607 [f][2][C]; OPA: 33 U.S.C. §2706 [e][2]).

The most controversial features of the regulatory and judicial challenges to natural resource damage assessments concern the reliability of the contingent valuation method and the validity of passive use and nonuse values (Binger et al. 1995; Portney 1993). In the 1989 landmark case *Ohio v. U.S. Department of Interior* (880 F.2d 432 [D.C. Cir. 1989]), the D.C. Court of Appeals instructed the Department of Interior (DOI) to give equal weight to use and nonuse values in assessing natural resources damages (880 F.2d 464 [D.C. Cir. 1989]). The Ohio case upheld the contingent valuation as an acceptable method for calculating option and existence values, and concluded that these two values could constitute acceptable passive use values (880 F.2d 478 [D.C. Cir. 1989]). In 1992, the National Oceanic and Atmospheric Administration (NOAA) organized a blue-ribbon panel of economists and sought recommendations relating to natural resource damage assessment regulations under OPA. After much debate, the NOAA panel concluded that contingent valuation "can produce estimates reliable enough to be the starting point of a judicial process of damage assessment, including lost passive-use values," provided that such studies follow the panel's recommended guidelines.[12] In a separate 1998 ruling on DOI regulations for simplified natural resource damage assessments, the D.C. Court of Appeals upheld the use of older contingent valuation and travel cost studies in the formation of computer model parameters (*Nat'l Ass'n of Mfrs. v. U.S. Dep't of the Interior,* 134 F.3d 1095, 1115 [D.C. Cir. 1998]).

Government agencies have also relied on both the travel cost method and contingent valuation to estimate the value of recreation and nonmarket environmental resources (Loomis 1993, 168). The U.S. Water Resources Council has identified the travel cost method and contingent valuation as the two preferred methods for valuing outdoor recreation (*Nat'l Ass'n of Mfrs. v. U.S. Dep't of the Interior*). The U.S. Bureau of Reclamation and the National Park Service have used contingent valuation to estimate recreation benefits for fishing and rafting in the Grand Canyon under different scenarios of water releases from the Glen Canyon Dam (ibid.). Other state fish and wildlife agencies have also used these methods to value fish- and wildlife-related recreation for the purpose of formulating policy (ibid.).

Measuring the Benefits of Public Lands

If policymakers adopt the principle of economic efficiency for managing multiple-use lands, an assessment of the relative benefits of alternative uses could lead to changes in current management policies. In an effort to discern the possible implications of such a policy, this section develops rough estimates of the aggregate benefits from different uses of multiple-use lands. The following analysis generally relies on quantity data from the year 1995 when possible and utilizes price variables that represent either the clearing price for market commodities or an imputed market clearing price for nonmarket commodities. This analysis relies on many simplifying assumptions and should be viewed as an exercise that explores possible implications of moving toward an efficiency criterion in public land management.

Recreation Benefits

The Forest Service conducts an extensive economic assessment of the benefits of different uses (timber, range, minerals, recreation, and wildlife) on the national forest system under the Resource Planning Act (RPA) program (U.S. Department of Agriculture, Forest Service 1990, Ch. 6, app. B; 1995, ch. 4, app. E; www.fs.fed.us-land-RPA). To estimate recreation benefits, the Forest Service has relied on studies utilizing the travel cost method and contingent valuation.[13] These Forest Service recreation prices are used to derive updated estimates of the benefits of recreation for both the National Forest System (NFS) and BLM lands, based on recreation visitor day numbers for both national forests and BLM lands.

Estimates of recreation benefits in the NFS and BLM lands were derived in the following manner. The quantity of 1995 recreational visitor days for each recreation category was multiplied by the corresponding value of a recreation visitor day.[14] These recreation unit values represent the imputed market clearing price as estimated by the Forest Service and adjusted into real 1995 dollars. The benefits of recreation on BLM lands were derived by multiplying BLM visitor day quantities by the corresponding Forest Service price for recreation. To the extent that Forest Service prices overestimate recreation on BLM lands, the resulting figures would similarly overstate recreation benefits.

In 1995 the total benefits from recreation in the NFS equaled $8.288 billion, and the corresponding recreation benefits on BLM lands were $1.520 billion. Table 1 lists the recreation prices, visitor days, and benefits for the major common recreation activities on Forest Service and BLM lands. The leading

Table 1. Recreation Benefits in the NFS and BLM Lands

Recreation Activity	Price of a Recreation Visitor Day (1995$)	USFS Quantity of Visitor Days (million)	USFS Imputed Market Value (million 1995$)	BLM Quantity of Visitor Days (million)	BLM Imputed Market Value (million 1995$)
Camping and picnicking	12.22	85.8	1,048	34.0	348
Fishing	77.62	17.8	1,381	2.4	186
Hunting	51.88	18.9	983	6.3	326
Hiking and horseback riding	12.92	32.3	417	6.7	350
Mechanized travel	11.64	129.0	1,501	9.9	104
Winter sports	52.38	20.3	1,099	0.7	36
Other	45.44	40.9	1,859	13.4	170
Total		345.1	8,288	73.4	1,520

Source: U.S. Department of Agriculture, Forest Service 1990, appendix B; U.S. Department of Agriculture 1997; U.S. Department of Interior, Bureau of Land Management 1994, 1995.

activities on Forest Service lands are mechanized travel and viewing scenery, fishing, and camping and picnicking. The three most important recreational activities on BLM lands are nonmotorized travel, camping, and hunting.

Preservation Benefits

Measuring preservation benefits raises even more challenging issues than the valuation of recreation. Natural resources that produce preservation benefits are further removed from direct human use and provide various intangible services. Consider some of the diverse characteristics of preservation resources on NFS and BLM lands: millions of acres of wilderness areas (NFS 34, BLM 5.2),[15] miles of designated Wild and Scenic Rivers (NFS 4316, BLM 2032),[16] thousands of miles of fishable streams and rivers (NFS 128, BLM 174),[17] millions of acres of waterfowl habitat (NFS 12, BLM 23),[18] thousands of wildlife, fish, and plant species (NFS 13, BLM 8),[19] and threatened or endangered species (NFS 283, BLM 300).[20] The economics literature contains many studies that attempt to value the benefits of preserving specific natural areas that face proposed development projects (J. V. Krutilla and Fisher 1985). Other studies have estimated the value of specific resources such as wilderness areas (Walsh et al. 1984), wetlands (Thibodeau and Ostro 1981), and endangered species (T. Stevens et al. 1991; Hagen et al. 1992). A recent study by Robert Costanza, Ralph d'Arge, and others takes a new approach (1997, 253): it values entire ecosystems by estimating the various goods and services generated by units of specific types of ecosystems. The authors identify seventeen different ecosystem services (e.g., gas regulation, climate regulation, water supply, waste treatment, pollination, genetic resources) that are performed by sixteen different biomes or types of ecosystems (e.g., coastal estuaries, coral reefs, tropical forests, temperate/boreal forests, grass/rangeland, wetlands, lakes/rivers, desert). Based on a synthesis of over one hundred studies, they develop an estimate of the economic benefit of each ecosystem service for the different biomes in terms of dollars per hectare. The value of the world's ecosystem services is then derived by multiplying the benefit unit per hectare by the total land area for that type of biome.

To estimate the economic value of preserving ecosystem services on U.S. public lands, one can apply the Costanza-d'Arge methodology to specific parcels of federal land that supply these services. Four types of ecosystems characterized most of NFS and BLM lands: temperate forests, grass/rangelands, wetlands-swamp/floodplains, and lakes/rivers. Benefit parameters were converted to acres and adjusted to 1995 dollars. Multiplying these benefit parameters by the corresponding area within the NFS and BLM lands yields the total imputed market value of the benefits of ecosystem services. Table 2 presents the results of this exercise. The total value of ecosystem services amounts to $71.7 billion

Table 2. Benefits of Preservation: Ecosystem Services from the National Forest System and BLM Lands

Type of Ecosystem	Value per acre (1995$/ acre/yr)	USFS Acres (million)	USFS Imputed Market Value of Services (million 1995$/yr)	BLM Acres (million)	BLM Imputed Market Value of Services (million 1995$/yr)
Forests/ temperate	110	136.7	15,036	71.1	7,821
Grass/ rangelands	101	46.2	4,654	167.0	16,824
Wetlands/ swamps	7,923	5.4	42,783	24.0	190,147
Rivers/ lakes	3,431	2.7	9,265	2.2	7,549
Total		191.0	71,739	264.3	222,341

Source: Costanza, d'Arge et al. 1997; U.S. Department of Agriculture, Forest Service 1995; U.S. Department of Agriculture 1997; http://www.fs.fed.us/biology (Wildlife and Fish); U.S. Department of Interior, Bureau of Land Management 1994, 1995, 2000a, 2000b; Zinser 1995.

from the NFS, $222.3 billion from BLM lands, and a total of $294.1 billion for both NFS and BLM lands.

The benefit figures by biome indicate the important role of wetlands on public lands. Despite a relatively small area, the high unit value makes wetlands the most important generator of benefits among the four types of ecosystems. Lakes and rivers provide the second most productive type of ecosystem. Compared to wetlands and lakes, forests and rangelands offer relatively low individual unit value when you consider total benefits.

Comparing the Economic Benefits of Commodity, Recreation, and Preservation Uses

The quantification of recreation and preservation benefits permits comparisons to commodity uses. The benefits of timber, grazing, minerals, and recreation were derived according to two different accounting measures for benefits: government receipts and the imputed market clearing price.[21] Table 3 presents the total benefits of commodity uses and recreation and preservation as defined by government receipts and the estimated market value. These estimates illustrate two principles. First, there is a large disparity between the receipts measure and the imputed market value measure. In the NFS, the traditional commodity uses of timber, grazing, and mining account for 90 percent of the total receipts, whereas recreation amounts to only 9 percent and preservation 0 percent of total receipts. Second, when benefits are calculated by the imputed market clearing price, which includes nonmarket benefits, the preservation benefit share rises sharply from 0 to 88 percent, the recreation benefit share increases slightly to 10 percent, and the commodity use share falls dramatically to only 2 percent of total benefits.

On BLM lands, timber contributes the largest share of government receipts at 44 percent, followed by range and mineral benefits at 17 percent and 16 percent, respectively. Benefits from receipts are virtually nonexistent for recreation and preservation. The imputed market benefits of these different uses convey a very different picture. Mineral benefits become the largest commodity share at 4 percent of total benefits, and recreation remains with a 1 percent share. Timber and range benefits fall to less than 1 percent. But preservation, in the form of ecosystem services, becomes 95 percent of the benefits from BLM lands.

The estimated market value of ecosystem services on public lands overwhelms the dollar figures attributable to commodity benefits. Recreation and ecosystem benefits within the NFS are sixty-two times the size of commodity

Table 3. Benefits from Commodity Uses, Recreation, and Preservation in the National Forest System and BLM Lands

Type of Use	USFS (million 1995$) Receipts to Fed. Govt. 1995 (% of total)	USFS (million 1995$) Imputed Market Value, 1993–95 (% of total)	BLM (million 1995$) Receipts to Fed. Govt. 1995 (% of total)	BLM (million 1995$) Imputed Market Value, 1996 (% of total)
Timber	303.0 (51%)	616.1 (1%)	45.5 (44%)	109.7 (0%)
Range	8.8 (1%)	64.8 [a] (0%)	14.7 (14%)	9,937.2 (4%)
Minerals	221.6 [a] (37%)	605.5 [a] (1%)	14.7 (14%)	9,937.2 (4%)
Recreation	52.0 (9%)	8288.0 (10%)	0.9 (1%)	1,520 (1%)
Preservation: Ecosystem srvs.	0.0 (0%)	71,739.0 (88%)	0.0 (0%)	222,341.0 (95%)
Other	7.3 (1%)		26.8 (26%)	
Total	592.6	91,313.4	103.6	233,997.3

Source: NFS values: U.S. Department of Agriculture 1997; U.S. Department of Agriculture, Forest Service 1995, Table E.2; 1991, appendix B. BLM values: U.S. Department of Interior, Bureau of Land Management 1994, 1995, 2000b. 1993 data from U.S. Department of Agriculture, Forest Service 1995.

benefits, and BLM ecosystem and recreation benefits exceed the corresponding commodity benefits by a factor over twenty. Moreover, because most of the ecosystem benefits arise entirely outside the market, there is no necessary limitation on their potential size.[22]

Policy Implications

The economic efficiency theoretical framework and the above preliminary empirical findings permit four observations concerning the management of public lands. First, empirical estimates indicate that there are significant and sizable benefits from recreation and preservation uses of public lands. A policy that views social benefits solely in terms of government receipts, or otherwise neglects nonmarket benefits, would be economically inefficient (Passell 1998). If government land managers omit nonmarket benefits from their analysis, they will misperceive the demand for recreation and preservation and place it at an unrealistically low level. Indeed, this seems to be what has happened on BLM and Forest Service land, where federal managers have found themselves unprepared to deal with the unprecedented public demand for recreational and preservationist uses of these lands. Their adherence to traditional multiple-use policy has resulted in a quantity of land allocated to the traditional extractive commodities, such as timber, grazing, and mining, that are inefficient compared to the benefits that are derived from nonconsumptive uses. This policy also assumes that an unrealistic percentage of public lands is actually devoted to consumptive uses.

A second observation concerns technological innovation and population growth. Advances in technology generally lead to a reduction in the quantity of natural resources required to produce a given level of manufactured goods in the economy (see V. Smith 1979). To the extent technology dampens the demand for public lands for extractive uses, there is a corresponding increase in the supply of land for recreation and preservation uses. Technological innovation raises the demand for recreation by increasing leisure time, lowering the cost of transportation to federal lands, and creating new recreational pursuits (e.g., mountain biking, inline skating, snowboarding) (Zinser 1995, 3–9). These types of innovations shift upward the demand for recreation and preservation of land. Furthermore, the demand for recreation and preservation of public land will be augmented by a continuation of the growth in population in the Western states, which have the largest holdings of federal lands (8). Over the past two decades, the mountain region states experienced population growth rates that are double to triple that of the nation as a whole.[23] A continuation of Western U.S. population trends and technological innovation in the future will shift the demand for recreation and preservation land even further and increase the optimal allocation of public land allocated to recreation and preservation.

Third, because an efficiency goal would also entail consideration of costs, federal land managers adopting such a goal might consider restricting access to public lands to limit degradation of the natural resources or curtail negative congestion effects for recreational visitors. Land managers could restrict entry by an administrative permitting process based on historical use, random lottery, or some other criterion. Alternatively, a user fee system provides certain advantages for implementing an efficient policy.[24] An appropriately set user fee reflects the scarcity value of public lands and generates a level of use consistent with the efficient allocation of public lands. User fees provide revenue to the federal government to carry out good management policies. Such fees can be adjusted over time to reflect the changing scarcity value of public lands in light of a growing population and technological innovation.

Fourth, although efficiency seems preferable to multiple use as a public land management strategy, it suffers from two deficiencies: it lacks a temporal dimension, relying on uses, and preferences for uses, that now predominate, and it does not factor into its calculus the future value of current uses (or nonuses). In addition, its reliance on market-based preferences for particular uses does not consider whether these preferences are fair and equitable or whether they lead to results that smack of injustice. Social benefits are typically calculated on a willingness to pay basis. Such econocentrism is insensitive to the inevitable societal consequences of a market-driven set of value choices.

One doctrine that does incorporate both time and equity is the notion of sustainable development. Perhaps certain elements of sustainable development should be overlaid on efficiency to ensure that public land management decisions encompass a wider range of inputs. I now turn to this question.

Sustainable Development and Public Land Use

Beyond multiple use, and beyond efficiency, is the increasingly popular idea of sustainable development.[25] The concept was first introduced in the 1987 World Commission on the Environment and Development, usually known as the Brundtland Commission. Sustainable development was defined there as "development that meets the needs of the present without compromising the ability of future generations to meet their own needs" (WCED 1987, 43). A central tenet is therefore *intergenerational* responsibilities. But there is more. In addition to a time dimension, sustainable development also is characterized by a social and equitable component. The Brundtland Commission further identified this component as a "process in which the exploitation of resources [and] the direction of investments . . . are all in harmony and enhance both current and future potential to meet human needs and aspirations" (43).

The new idea in sustainable development is therefore a trio of equal values—environmental, economic, and social—overlaid by a time calculus (the future should not be sacrificed for the present). As such, it introduces two variables not otherwise considered by the efficiency criterion: equity and time. Efficiency is a suitable allocative mechanism with respect to the preservationism versus commodity resource use debate; it is even appropriate as a methodology for refereeing the preservationism versus recreation versus commodity resource use conflict. But because it largely ignores social equity and the distributive consequences of public land use (e.g., Bullard 1993), as well as the future of any reallocative use scheme, it is less comprehensive than sustainable development.

However, one should be sensitive to the downside of sustainable development before it is embraced as part of an allocative mechanism for public land use. First, although the idea of social equity seems a desirable goal in the abstract, it may be quite difficult to define it, make decisions based on it, and then measure whether it has been achieved. Efficiency seeks to achieve an allocation of current uses that maximizes the total value of goods and services for a given distribution of income in a properly operating market. But the initial distribution of income may skew the ultimate racial or socioeconomic allocation of these goods and services. Because efficiency assumes that all valuations reflect the given distribution of income, and because equity may affect, if not control, that distribution, then factors that may make up equity (e.g., the economic condition of persons of color) may drive efficiency.

If an equitable goal is defined as one that benefits all racial and economic groups, then equity may, in theory, call for a nonefficient result. This would be true if some races, cultures, or income groups are either disproportionately disadvantaged by an efficiency goal, or if their hierarchy of values is driven not by market forces but by cultural ones (e.g., Manaster 1995; Shanklin 1997, 333). For example, it was shown above that recreational and preservational preferences can be measured by a willingness to pay methodology. But the social benefits of nonuse, or a noncommodity use, may be undervalued by a particular racial or societal group if historically that group has been prevented from experiencing the benefits of, say, wilderness or recreation. The willingness of such a group to pay for preservational or recreational public land uses might be less than other groups simply because of some previous negative sociological status. If that status is changed, not by efficiency but by equity, then that group's willingness to pay could change as well.

Second, whereas the environmental, economic, and equitable agendas of sustainable development focus on intergenerational, future-looking sustainability, it is unclear how the present value of an efficient and equitable allocation of resources on public lands should be translated, in the present, into a future value. In the case of an efficiency criterion, the benefits and costs of differing use allocations are adjusted by an appropriate discount rate. But if one were to apply the same system to sustainable development, what would that discount rate be, and who would select it? Selection of the wrong discount rate would mean that present sustainability would come at future expense—a result completely antithetical to the underpinnings of sustainable development.

Third, the very idea of sustainable development is so untested that it rests primarily within the realm of academic discussion and government policy, but not hard law. A classic example of the "vision statement" quality of the concept can be found in the 1997 report of the President's Council on Sustainable Development.[26] The idea of sustainable development has also become a term of art within federal agencies, which employ it in various programs as if its content and scope were accepted as a norm. However, the harsh reality is that there is no consensus about the meaning of the term, and no new laws have been enacted that further define it or mandate sustainable development (Ruhl 1998, 286–87). As a result, to base a land use policy on a concept called sustainable development is to rely on something that is neither universally understood nor extant in any current form other than commentary or vague government policy (288, 292–93).

With no coherent overarching and widely accepted definition in place, and with no implementation of new initiatives based on it, one should be somewhat wary of grounding resource use allocative decisions on sustainable development. This is not to say that it should be entirely abandoned by public land managers. To the contrary, it might be quite useful to factor into public land use decisions its two central components: (1) a balancing of economic and environmental and social equity interests in light of (2) intergenerational sustainability. This allocative methodology would seem to improve on the current multiple-use paradigm and be a useful adjunct to an efficiency criterion.

The Impact of Political Institutions on Preservation of U.S. and Canadian National Parks

William Lowry

National parks are political entities. They are created through political processes and managed by political agencies. This simple fact has an immense significance for the future of national parks. Visions and plans for the protection of these areas have been and will continue to be affected by the political systems in which they exist. Specifically, preserved lands are affected by political institutions. Political institutions are frameworks of rules and procedures that shape and constrain the behavior of policymakers. In the context of national parks, institutional relationships exist among various actors involved with preservation efforts, shaping their actions and ultimately affecting these lands.

The empirical context for examination of this argument is planning for the future of national parks. In discussing the current state of parks, many policymakers and observers suggest that these revered lands currently face severe challenges that make the future quite uncertain (e.g., Frome 1992; U.S. National Park Service [NPS] 1992). To attempt to reduce that uncertainty, policymakers have offered plans for the attainment of certain preservation goals and created agencies to pursue those ends. This case study considers the U.S. and Canadian experiences to illustrate the impact of institutional relations on that pursuit.

This empirical context enables an intriguing assessment of the concept of sustainability. The simplest definition of sustainable development is use of a resource that meets the needs of this generation without compromising the ability of future generations to meet their needs (Daly 1996). The fact that sustainability generally emphasizes human uses of a resource whereas preservation emphasizes natural needs of the environment suggests that preservation goals may be at odds with sustainable development (Gillroy 1998). The context of long-term planning for parks, however, retains the focus of both sustainability and preservation on future outcomes. As the concept of preservation has evolved, the "needs" to be met in this context are the restoration and maintenance of natural conditions. Sustainable preservation thus necessitates natural conditions for current visitors as well as for visitors in future decades. Pursuit of such a goal depends on planning.

The case study is organized as follows. The first section specifies preservation goals for parks in the twenty-first century and suggests why progress toward those goals will be difficult. The second section describes how political institutions have shaped implementation of plans enacted to pursue those goals. The third section offers some recommendations to address the impact of political institutions in preservation efforts.

The Challenging Task of Preservation

The preservation of national parks constitutes one of the highest ideals to be pursued by modern governments. Pursuit of that ideal is inevitably difficult, the intergenerational focus of preservation inherently at tension with present demands.

The Evolution of Preservation Goals

The ideal of preservation is integral to national parks. National parks are lands to be preserved in relatively natural condition for perpetuity. This definition is used in the legislative mandates of both U.S. and Canadian national park systems, statutory language mandating that users "leave them unimpaired for the enjoyment of future generations."[1] This definition is consistent with that offered by the International Union for the Conservation of Nature (IUCN), which defined national parks at the second World Conference on National Parks in 1972 as "not materially altered by human exploitation and occupation" (1972, 389). What exactly does that ideal mean in terms of goals?

The precise goals of preservation efforts have changed over time. In part, shifts of priorities were inevitable given a mandate calling for use and preservation simultaneously. Since the establishment of Yellowstone in 1872 and Banff in 1885, both U.S. and Canadian national parks were managed for decades in ways that satisfied only a fairly diluted version of what it means to be "preserved." In essence, policymakers tried to preserve the views of magnificent scenery for visitors and tourists (Bella 1987; Bryan 1973; Frome 1992; Runte 1987; Sellars 1997). So long as the views of tall trees, thick forests, snow-covered peaks, and clear lakes were maintained, other natural conditions could be sacrificed to roads, hotels, golf courses, and other forms of development. Occasionally, questions were asked, but the major goal of preservation efforts through the late 1960s in Canada, the United States, and most other nations was to provide scenic vistas to attract tourists.

In recent decades, the goals of preservation efforts have changed. The seeds for those changes were sown throughout the world in the 1960s. For one example, the IUCN established the Commission on National Parks in 1961. The Commission facilitated communication of ideas about preservation of natural areas and fostered support for systematic criteria as to what constitutes a national park. By the time of the second World Conference on Parks in 1972, delegates tempered their pride at a century of growth since Yellowstone's establishment with admissions of nonsystematic expansion and protection. Historian E. M. Nicholson stated, "To this day, piecemeal and opportunist aims have remained uppermost" (1972, 33). The delegates called for more parks, more representation of diverse biogeographic areas, and greater protection of existing parks through attention to the biological needs of entire ecosystems. These calls were consistent with and reinforced by the demands for greater attention to environmental issues throughout the world.

The increased attention to a more scientific preservation has evolved into two major goals for many park systems. The first is representation of diverse natural environments within park systems. Policymakers in many nations now use sophisticated techniques of mapping and GAP (Geographical Approach to Planning) analysis to identify biological regions in need of protection (Agee 1996; Keiter 1996; Noss 1996; R. Wright and Scott 1996). The second major goal of current preservation efforts is the restoration and maintenance of conditions in park units that approximate the natural state of the park before significant alteration by humans (Freemuth 1991; Keiter 1997). In 1991, the chairman of the Commission on National Parks summarized these objectives by citing "increasing recognition of . . . protection of representative samples of natural regions and the preservation of biological diversity to the maintenance of environmental stability in the surrounding country" (Lucas 1991, vii).

This mandate has been endorsed by many national governments throughout the world as well as by the scientists and international policymakers cited above. Subsequent efforts to pursue these goals, however, have been affected by the political environments in which they occur.

The Impact of Political Institutions

Ideals of setting aside lands as national parks to remove them from politics cannot be sustained. Rather, parks exist within and depend on political systems. As political entities, they are subject to impact from the frameworks of rules and procedures that constrain the behavior of actors within the systems. Those institutional relations involve inherent tensions between those pursuing preservation and other political actors. Several inevitable sources of tension are posited below, both in generic terms and in the context of park preservation.

Planners and Authorizers. Planning is one thing; authorization of the plan is something else. If those making plans are not the ones who authorize them, the institutional relationship between these two sets of actors can contain significant tension. This tension is manifest in democratic political systems through a temporal dimension. Within democratic systems, authorizers are elected officials whose presence is determined by open and frequent elections. Elections create pervasive incentives for politicians. To be or remain in office, they must be attentive to electoral needs. This necessitates a focus on time periods of electoral intervals of at most six years. Planners must necessarily be attentive to trends and developments that occur over decades. To oversimplify somewhat, planners focus on the long term, but the electoral incentives of authorizers are focused on the short term (Heclo 1977; Moe 1985; Nathan 1983).

As political entities, park systems are ultimately controlled by political authorities. Parks are established by legislative acts and funded by government budgets. Implementation of expansion plans depends on the actions of elected officials. The short electoral time frame shaping the incentives of these elected officials contrasts dramatically with the consistent, sustained, long-term focus necessary to implement plans for the preservation goal of representation of diverse ecosystems.

Agencies and Financiers. Public agencies are severely limited if they do not control the funds with which to pursue their goals. If the financiers of implementation processes are separate from those who actually put the plans into effect, these two groups of institutional actors may have dramatically different priorities. In the public sector, the financiers are generally the same elected officials described above. Their priorities are to provide immediate benefits for their constituents, not future generations whose demands and votes cannot be counted or counted on. Implementers are generally employees in public agencies, personnel whose incentives are focused on agency priorities that may be less focused than financiers would prefer (Derthick 1979).

National park units are created out of public funds. Though the processes for establishment of parks differ between the United States and Canada, institutional tensions exist in both nations between those planning for new units and those financing them.

Between Agencies. The public sector contains considerable potential for interaction between different agencies. Those agencies may be located at either the same level of government or between levels. Different agencies generally have different goals, thus leading to complications and institutional tensions (Bardach 1982; Pressman and Wildavsky 1984).

Parks do not exist in a vacuum. Rather, they are often surrounded by or bordered by lands managed by other federal or state agencies. Many aspects of national park management involve different levels of government. The impact of interagency and intergovernmental relations is becoming increasingly evident to national park systems throughout the world, if for no other reason than the recent emphasis on ecosystem management. In recent years, many observers have encouraged greater attention to ecosystem needs, such as the protection of areas large enough to sustain native species (Agee 1996; Keiter 1996). National parks are neither large enough nor are their boundaries drawn to sustain such a role. To move toward ecosystem management will require coordination of protection efforts on adjacent lands. Often, these lands are managed by other federal or state/provincial agencies.

Empirical Assessment

The impacts of institutional relationships in the United States and Canada on preservation goals have been immense. The following pages describe these impacts regarding recent efforts to attain representative expansion and ecosystem restoration at the flagship parks of each system, Yellowstone and Banff.

Representative Expansion in the United States

In Theory. The U.S. National Park Service (NPS) authored a plan in 1972 for systematic expansion over the coming years. The *System Plan* mandated definition of a complete representative system of historic and natural sites, identification of existing "gaps" in representation, and filling of those gaps through creation of new units (U.S. NPS 1972, vii). The plan categorized American history into nine major themes (such as westward expansion) and forty-three subthemes, many with their own specific niches. Agency personnel then fit existing units within the total number of niches, concluding that 85 of the 281 possible spots were represented (pt. 1, 81). The planners undertook a similar procedure with natural conditions such as tundra and boreal forest and then concluded that seventeen of forty-three natural regions were not represented at all, with many others underrepresented (pt. 2, 14).

Expansion was then to occur as follows. The Division of Park Planning, through the Office of New Area Studies, would determine the qualifications for a proposed new unit. Criteria could include representation of a natural resource or historic theme as well as recreational opportunities and the relatively unspoiled nature of the proposed land. The NPS would then make a recommendation to either the Secretary of Interior (for historic sites) or to Congress (for parks).

In Practice. The institutional relationship between planners and authorizers has minimized the actual effect of the 1972 plan. Rather, elected officials, most notably members of Congress, have added questionable sites to the system and ignored proposals from the NPS for their own recommended sites. This has been true for historic and natural units (Foresta 1984, 77–78; King 1988; National Parks and Conservation Association [NPCA] 1988). The Office of New Area Studies was even dismantled during much of this period. Since then, NPS officials admit that "the Park Service has lost some control over the process of establishing new parks" (U.S. NPS 1992, 135). Many former NPS officials have denounced their inability to pursue systematic expansion (Hartzog 1988; Rettie 1995; Ridenour 1994). In 1995 congressional hearings, a variety of witnesses agreed that the expansion process has been increasingly political and dominated by congressional manipulations.[2]

Why has this occurred? In short, authorizers have supplanted the planned process for adding new units with their own procedures and priorities. Political incentives for adding new units to the system often differ from reasons of ecosystem representation. Units are added due to "sufficient constituent appeal and/or economic development benefits in selected regions" (U.S. NPS 1992, 12). Agency requests for inclusion of areas like grasslands go unheeded for decades, while more politically popular units such as the jazz historic site in New Orleans are added. As NPS Director of Planning Denis Galvin told me, "The 1972 plan is great as intellectual exercise, but it's not the way the system is built. The system is built by individual acts of Congress" (interview, 21 December 1995).

The impact of the institutional relationship regarding financing the expansion process has led to more units for the system, but often units that the NPS does not desire and can hardly afford. The most notorious example is Steamtown National Historic Site. Steamtown became an NPS site in 1986 when

Table 4. NPS Construction Funding (in millions of dollars)

Fiscal Year	Requested	Appropriated	Add-ons (% of App)
1983	69.1	92.5	23.4 (25)
1984	61.6	95.2	33.6 (35)
1985	61.7	92.6	30.9 (33)
1986	50.0	94.8	44.8 (47)
1987	19.3	72.0	52.7 (73)
1988	19.3	77.1	57.8 (75)
1989	6.6	141.3	134.7 (95)
1990	23.2	156.7	133.5 (85)
1991	52.5	156.7	104.2 (67)
1992	84.2	216.6	132.4 (61)
1993	94.0	172.5	78.5 (46)

Source: U.S. National Park Service, 1995, 16.

Scranton Representative Joe McDade (R-PA), ranking minority member of the House Appropriations Committee, amended an Omnibus Appropriations Bill to channel funds to the creation of a site commemorating steam locomotives. McDade's manipulation was criticized by NPS officials and historic experts, but the influential representative managed to funnel over $70 million for its renovation (Hinds 1992; U.S. General Accounting Office [GAO] 1991b). More recently, the NPS has been given the responsibility of managing the Presidio site near San Francisco. Maintenance of grounds and buildings at the centuries-old military installation are expected to cost at least $45 million per year and the NPS estimates annual appropriations for the site not exceeding $25 million (U.S. GAO 1993). Overall, according to testimony from former NPS Director James Ridenour, the NPS received $1.4 billion more than it had requested for construction and acquisition between 1983 and 1993, most of it for congressional projects.[3] Table 4 displays congressional add-ons to line-item funding for the years 1983–1993 as a percentage of total appropriations.

Funding through the Land and Water Conservation Fund (LWCF) is particularly frustrating for NPS personnel. The LWCF was created in the mid-1960s to funnel revenue from sales of federal properties and user (especially offshore oil) fees to create a steady flow of cash for purchase of parklands. The fund has since been abused in the 1970s, virtually cut off in the 1980s, and nearly ignored in the 1990s. Though the annual pot today is supposed to be about $900 million, congressional appropriations toward parklands constituted only about $140 million per year in 1996 and 1997 (McManus 1998).

The expansion process will continue to be determined by the institutional relationships between NPS personnel and members of Congress for the foreseeable future. Some agency personnel continue to try to revive systematic planning for expansion, but as Galvin said of one recent effort, "Our list [of proposed sites] doesn't move as fast as it might because our studies are being displaced by lists from Congress."

Representative Expansion in Canada

In Theory. The Canadian parks agency, then titled Parks Canada (PC), also produced a systematic plan for expansion in the early 1970s. The PC planners identified nine marine and thirty-nine terrestrial natural regions differentiated according to "biologic, geologic, physiographic, geographic, and oceanographic" features (Parks Canada [PC] 1979, 38). This plan was later expanded to include priority historic themes, and the number of marine regions warranting representation increased to twenty-nine (Canadian Environmental Advisory Council [CEAC] 1991, 55). When the plan was first initiated, only 40 percent of regions were adequately represented (Carruthers 1979).

Efforts to add new units are concentrated on regions that remain unrepresented. Agency planners would conduct studies to identify possible new areas that represent natural features of various regions and display minimal human impact. One "worthy" area would then be selected as a park site. Parks Canada then would do a full feasibility study with direct involvement from provincial authorities and representatives of local communities. If this assessment revealed feasibility and support, then the federal government would negotiate transfer from current owners, usually provincial or territorial governments (PC 1990, 7–8).

In Practice. Implementation of the plan started quickly but has since slowed perceptibly. By 1978, the number of parks had increased from eighteen to twenty-eight. Between 1978 and 1992, eight new parks or reserves were added, all but one in previously unrepresented regions. That growth was deceptive, however: several of the new units had already been in the pipeline. After this spurt, growth dropped off dramatically. In 1990, PC admitted that expansion was "becoming increasingly complex, expensive, and time-consuming" (1990, 9). By the late 1990s, eighteen terrestrial regions and twenty-seven marine regions still awaited completed units. Observers became increasingly skeptical of completion (Canadian Nature Federation 1991; Walker 1992). In their 1995 review of expansion, the Canadian World Wildlife Fund awarded the federal government a grade of C– for terrestrial parks and D– for marine sites (WWF 1995). Figure 2 shows the status of the system plan as of the start of 1996.

Completing Canada's National Park System
Western Mountains
1. Pacific Coast Mountains
2. Strait of Georgia Lowlands
3. Interior Dry Plateau
4. Columbia Mountains
5. Rocky Mountains
6. Northern Coast Mountains
7. Northern Interior Plateaux and Mountains
8. Mackenzie Mountains
9. Northern Yukon
Interior Plains
10. Mackenzie Delta
11. Northern Boreal Plains
12. Southern Boreal Plains and Plateaux
13. Prairie Grasslands
14. Manitoba Lowlands
Canadian Shield
15. Tundra Hills
16. Central Tundra
17. Northwestern Boreal Uplands
18. Central Boreal Uplands
19a. West Great Lakes – St. Lawrence Precambrian Region
19b. Central Great Lakes – St. Lawrence Precambrian Region
19c. East Great Lakes – St. Lawrence Precambrian Region
20. Laurentian Boreal Highlands
21. East Coast Boreal Region
22. Boreal Lake Plateau
23. Whale River
24. Northern Labrador Mountains
25. Ungava Tundra Plateau
26. Northern Davis Region
Hudson Bay Lowlands
27. Hudson-James Lowlands
28. Southampton Plain
St. Lawrence Lowlands
29a. West St. Lawrence Lowland
29b. Central St. Lawrence Lowland
29c. East St. Lawrence Lowland
Appalachian Region
30. Notre Dame – Megantic Mountains
31. Maritime Acadian Highlands
32. Maritime Plain
33. Atlantic Coast Uplands
34. Western Newfoundland Highlands
35. Eastern Newfoundland Atlantic Region
Arctic Lowlands
36. Western Arctic Lowlands
37. Eastern Arctic Lowlands
High Arctic Islands
38. Western High Arctic
39. Eastern High Arctic
Quttinirpaaq
Northern Bathurst Island
Aulavik
Ivvavik
Vuntut
Sirmilik
Auyuittuq
Tuktut Nogait
Kluane (R)
Wolf Lake
Nahanni (R)
Ukkusiksalik (Wager Bay)
Torngat Mountains
East Arm of Great Slave Lake
Wood Buffalo
George River
Mealy Mountains
Gwaii Haanas (R)
Wapusk
Lac Guillaume-Delisle
Terra Nova
Gros Morne
Mingan Archipelago (R)
Jasper
Glacier
Mount Revelstoke
Elk Island
Prince Albert
Pacific Rim (R)
Gulf Islands
Banff
Yoho
Kootenay
Waterton Lakes
Riding Mountain
Manitoba Lowlands
Forillon
Cape Breton Highlands
Prince Edward Island
Kouchibouguac
Fundy
Kejimkujik
La Mauricie
Grasslands
Pukaskwa
Georgian Bay Islands
St. Lawrence Islands
Bruce Peninsula
Point Pelee
Legend
National park or reserve (R)
Region represented
Region with interim protection
Region not represented
Lands withdrawn for a future national park / reserve
National park area of interest
Land assembly underway
Kilometres
200 0 200 400
Operational Services
National Parks Directorate
July 2001 ian joyce
Parks Canada
Parcs Canada

Figure 2.

In large part, the disappointing recent progress on system expansion is attributable to the institutional relations between PC planners and provincial officials. Whereas U.S. NPS personnel must deal with federal politicians, the Canadian system of federalism ascribes a greater significance to subnational authorities. Most public land in the United States belonged originally to the federal government, whereas according to the Constitution Act of 1867, most Canadian land belonged to provincial governments. At the end of the 1980s, less than 6 percent of U.S. land was owned by state governments, whereas provincial governments owned more than 50 percent of Canada's land (Leman 1987, 29; Weaver 1992, 44).

Negotiations between the parks agency and provincial authorities over expansion plans have become increasingly complex, costly, and often contentious (CEAC 1991, 55; McLaren 1986; PC 1990, 9). Negotiations for desired sites are often slow and lengthy, creating time lags in which the price of lands may escalate and natural conditions on the desired lands may be reduced. In addition to time, a second major consequence is cost. Land values become even more inflated when provincial negotiators realize that certain areas are targeted for expansion. For example, provincial leaders of British Columbia virtually held for ransom lands on Moresby Island in the 1980s when they realized the strategic worth of the land to federal planners (Munro 1987, A7). The province "settled" for over $100 million in 1987, more than nearly all the other national parks had cost together. Even if acquisition is achieved, new areas come with substantial compromises, such as continued logging and development, that make the land less suitable for inclusion in the park system. Finally, provincial negotiators can stymie the process altogether, keeping the land in provincial hands or responding to local concerns about federal restrictions on current activities.

Like their NPS counterparts, Canadian planners have been somewhat frustrated at the impact of political institutions on planned expansions. When I asked PC Chief of Planning Murray McComb about the WWF review, he admitted sadly, "Their assessment is right," and then added, "We haven't done much good except for a lot of studying and a lot of negotiating" (interview, 5 February 1996).

Ecological Restoration at Yellowstone

Restoration of natural processes in the Yellowstone ecosystem depends on long-term planning, sufficient funds, and cooperation with neighboring authorities. As the earlier discussion on institutional relations suggests, all three of these aspects of restoration efforts involve tensions between the NPS and other political actors. Ecological restoration at the world's first park has thus been difficult.

History of Planning. The history of planning at Yellowstone is one of frustrated efforts to restore natural conditions in the park's ecosystem. In the 1920s and 1930s, biologists Milton Skinner and George Wright deplored the lack of scientific input into previous park decisions (Chase 1987, 235; G. Wright et al. 1932, 85). Their recommendations led to proposals to adjust park boundaries and manipulate herd sizes of park species, but momentum for those efforts died with Wright in a car crash in 1936. Ecological planning was rejuvenated by the Leopold Report of 1963 and increased environmental awareness in the late 1960s. This led to the Yellowstone Master Plan of 1974. Specific parts of this plan, such as removal of built structures at Fishing Bridge and the development of a transit system in the park, were subsequently overruled or ignored by political authorities (Chase 1987, 206; Freemuth 1989, 280; Lowry 1994, 158; U.S. NPS 1974).

Fire Policy. More recently, NPS efforts to restore the use of natural fires in Yellowstone's ecosystem were questioned after the intense fires of 1988. The agency had adopted a let-burn policy for many naturally occurring fires in the 1970s but stopped short of endorsing a significant increase in prescribed burning. The fires in 1988 resulted from the dry fuel buildup in the forests, the driest summer on record, and twice the normal amount of lightning in the park. After the August fires burned roughly 36 percent of the park, local commercial interests and their congressional representatives skewered the NPS for its let-burn policy, ignoring the effect of too little prescribed burning. The resulting plan for the fires called for much more pressure on NPS personnel to avoid the risks of letting natural fires burn without suppression. For example, Senator Wallop (R-WY) boasted on the release of the new fire policy, "All the words about natural fire are in there, but the fact is that they're now going to have to suppress the fires" (quoted in Reid 1989, A3).

Establishing Long-Term Priorities. When I last visited Yellowstone in 1996, I began all interviews with the superintendent, two planning officials, and an official with the Greater Yellowstone Coalition with the same question about long-term planning. The initial response of all four was to laugh. Long-term planning at Yellowstone is difficult due to changes necessitated by political circumstances and the short-term focus of most expenditures. Park officials track various projects with a priority list that categorizes according to time frames and deadlines. Unfortunately, most of the long-term projects are neglected due to necessary attention to short-term situations. As Park Planner Beth Kaeding told me, "We go from crisis to crisis and can't, with government downsizing, devote resources to long-term planning" (interview, 16 July 1996). One classic example involves the planned transit system mentioned earlier. By 1991, only 60 of the 410 paved road miles in the park were in good condition. The NPS thus devotes nearly all of the $1 million/year in the Cyclic Road Program to immediate repairs rather than pursuit of long-range goals. Further, with a backlog of over $28 million in repair work, future prospects for funding for a transit system are unlikely any time soon (U.S. NPS 1991, 49).

Ecosystem Restoration. Today, many planning efforts involve the larger ecosystem of which Yellowstone is the base. The greater ecosystem includes eighteen to nineteen million acres. This ecosystem includes six national forests, three wildlife refuges, BLM lands, state lands, private lands, geothermal drilling, logging, ranching, and commercial development.

Efforts to address issues within the ecosystem have been slowed by political obstruction. Congressional hearings in 1985 acknowledged that current federal efforts in the area were deficient but did little to solve coordination problems (U.S. Congress 1985). A 1989 symposium called for management of the area as "an integrated ecological entity" but carried little political clout (Keiter 1991, 5). The NPS plans for the ecosystem were contained in a 1990 document entitled *Vision for the Greater Yellowstone Area* that drew criticism from Senator Alan Simpson (R-WY) and White House Chief of Staff John Sununu. These political authorities demanded revisions. Ultimately, the plan shrank from sixty to ten pages, and the primary author, Lorraine Mintzmayer, was transferred. Recent efforts to address ecosystem issues provide evidence of a continuing lack of coordination even between federal agencies. When the NPS was opposing a proposed gold mine just northeast of the park in 1996, their efforts received little assistance from Forest Service officials on whose land the project would have been completed. Superintendent Mike Finley called the process "one of the poorest I've ever seen in terms of analysis or cooperation with surrounding agencies" (interview, 15 July 1996).

In summary, success in restoration efforts is hard to achieve. Even the most promising project in recent years, the reintroduction of wolves, is now under severe challenge in federal court. As a compendium on ecosystem research published in 1991 concludes, "This evolutionary change in federal natural resource management policy is not coming easily in Yellowstone" (Keiter and Boyce 1991, 381).

Ecological Restoration at Banff

As at Yellowstone, ecosystem restoration efforts at Banff depend on a long-term focus, funding priorities, and cooperation with authorities in the larger ecosystem. Also similar to the experiences of NPS personnel, Canadian planners are affected by the institutional tensions inherent in agency relations with other political actors.

History of Planning. The history of planning at Banff is one of changing priorities. For decades after its creation in 1887, the focus of planning was to encourage commercial activity and tourist access. Plans encouraged commercial development of the hot springs, growth of the townsite, and cultivation of activities such as golf and skiing to make the park a year-round attraction. When park boundaries were redrawn, they were done so to allow the province of Alberta to use more land for mining, grazing, and other activities (Bella 1987, 163; Hildebrandt 1995, 9). Roads such as the Icefields Parkway to Jasper were built. Beginning in the 1960s, planning priorities began to change. Though PC was not always in the lead on these changes, various plans were questioned. These included staging the Winter Olympics in Banff, encouraging more townsite growth, and twinning the Trans-Canada Highway. Eventually, park planners took the lead in several efforts, notably opposing an airstrip in the park and resisting expanded operations at some ski facilities. In those cases, PC personnel often found themselves at odds with provincial and federal politicians who were responding to the well-connected residents of Banff and the surrounding province. As Perry Jacobson, a park manager at Banff for eighteen years, told me, "If you did something someone there didn't like, you could bet you'd be hearing from Ottawa a half hour later" (interview, 30 June 1992). Park planners increasingly found themselves "muzzled" in their restoration efforts (Bergman 1986; Hildebrandt 1995, 79; Howse 1989; Lowey 1985).

The Townsite. The most prominent obstacle to restoration efforts was and remains the town of Banff itself. Because the town existed before the park, PC officials have never even had the option of removing it from the ecosystem. Rather, the town has grown to nearly ten thousand permanent residents and at least as many units for overnight visitors. The pressure to grow in Banff and its neighbor park of Jasper remains enormous. As one historian notes, "Banff and Jasper have felt unrelenting pressure from developer and business interests to exploit commercial opportunities and expand tourist facilities" (Hildebrandt 1995, 79). After extensive negotiations, Banff was formally recognized as a town in 1990, responsible for its own services and with the power to levy taxes. Some PC personnel are optimistic about this change, but others admit that the planning process will inevitably become trickier.

Ecosystem Restoration. Planners with PC have moved toward ecosystem restoration in the past decade. In 1986, the agency released a planning document for the four mountain parks together (Banff, Jasper, Kootenay, Yoho), signaling a shift toward managing them as one unit. The subsequent management plans released two years later again stressed coordination among the four parks. The plans specify use zones and restricted activities, but specific restoration proposals were cautious and boundary limits on the Banff townsite indeterminate (Canadian Parks Service [CPS] 1988, 145). Implementation of the plans was subsequently sidetracked by the incorporation of Banff and by the emergence of an even larger project.

After years of criticism of overdevelopment at Banff, the Banff–Bow Valley Study was commissioned in 1994 to provide a long-term vision for the park and the surrounding ecosystem. Outside planners and consultants, not PC personnel, conducted the study, a conscious decision to incorporate diverse viewpoints and encourage public confidence. Director of Planning for the Mountain Parks Judy Otton said, "We stayed out to keep it objective. We thought there would be more faith in an outside study" (interview, 22 July 1996). The Study Task Force deliberated for months, soliciting input from many sources, receiving hundreds of submissions and deputations. Released in 1996, the study concluded that Banff's "ecological integrity has been and continues to be increasingly compromised" (R. Page 1996, 14). To remedy the situation, the study offers five hundred specific recommendations; Table 5 provides summaries of several key areas. Underlying all of them are the need to think long term about ecological integrity and for PC to take the lead in those efforts.

In general, the tone at Banff regarding ecosystem restoration is one of acute awareness. Those involved with managing the park and concerned about its future are aware of past mistakes and of current opportunities. Putting that awareness into effect will be a challenging task.

Recommendations

Attainment of preservation goals remains challenging. Goals involving representation and restoration of natural ecosystems have been the focus of verbal support and substantial planning efforts. Progress toward those goals has been significantly affected by the tensions inherent in institutional relations be-

Table 5. Banff–Bow Valley Study Recommendations

Aquatic ecosystems	Set specific limits on pollutants from specific sources. Monitor and reduce nonpoint source pollutants. Identify species needed to restore biodiversity. Integrate research for entire ecosystem. Parks Canada take lead for integrated management.
Terrestrial ecosystems	Set specific limits on pollutants from specific sources. Monitor and reduce nonpoint source pollutants. Identify species needed to restore biodiversity. Integrate research for entire ecosystem. Parks Canada take lead for integrated management.
Human usage	Combine rail and highway corridors into single multiuse transportation corridor. Close sections of highways where animal mortality is high. Reduce speed limits on other highways. Convert some existing roads to hiking/cycling trails. Develop more public transit for commuters from cities. Restrict periods of use of ski areas. Restrict number of skiers on ski areas. Stop plans to expand golf course and, if need be, reduce the number of holes. Provide incentives for tourism operators to use ecologically sound practices. Adopt current zoning system to site-specific limits. Establish ecological carrying capacity limits for trails, campgrounds, waterways. Impose limits on use in sensitive areas.
Towns	No commercial expansion in Lake Louise hamlet. No commercial expansion in Banff town. No residential expansion in Lake Louise hamlet. Only residential fill-ins within existing Banff town. Plan model ecological communities for Banff and Lake Louise.

Source: Parks Canada 1996.

tween agency planners and other political actors. How can further progress be facilitated?

Proposing specific recommendations to moderate the impact of electoral institutions on park planning is not easy. Changing electoral incentives is not a viable option, as it would call for drastic constitutional revisions. A second, less dramatic possibility involves requesting official declaration of priorities. Worthwhile as this may be, real impact can be minimal. Politicians could simply respond with rhetorical promises. A third, and perhaps more substantial, option involves finding means of institutionalizing long-term priorities. To be more specific, mechanisms could be created that mandate a permanent focus on future generations. For example, Costa Rica has achieved an impressive level of education and literacy by requiring that at least 10 percent of the national budget for each year be spent on education. Perhaps a similar arrangement could be established for national parks agencies by instituting a focus on the long-term future in terms of financial resources. Currently, very small percentages of park service budgets (less than 5 percent in the United States and Canada) is spent on scientific research. Mandating a minimum of 10 percent would force politicians to include an emphasis on the long term in funding priorities for park service agencies.

If not done wisely, delegation is a double-edged sword. Left to politicians, plans would lack expertise and a long-range perspective. Delegated entirely to civil servants, plans can lack the political authorization needed to be implemented. A different form of delegation is one that perceives plans, once formulated, as more authoritative. In recent decades, the Canadian parks agency has moved toward this model by having plans approved at the highest levels of government. To put the contrast in specific terms, management plans at U.S. national parks are approved by regional directors within the agency, whereas Canadian plans receive approval from the Minister of the Environment. One institutional means of making implementation more likely is to give general management plans, once approved, the power of law that can be referred to in legal processes should subsequent conflicts or questions arise.

Park agencies simply need more resources with which to pursue restoration goals. In the U.S. National Park System, until 1997 the money raised from entrance fees at national parks went back into the General Treasury fund to be spent at the discretion of the U.S. Congress. Parks were then fiscally dependent on congressional appropriations. In 1997, Congress approved an experimental program allowing many national parks to keep the revenue they raise. Many state park systems have utilized a similar system but have gone even farther in demanding that their parks become self-supporting. The current experiment regarding entrance fee revenue within the NPS is a step in the right direction. The previous system allowed too much discretionary behavior by Congress; significant amounts of money were spent on parks and programs that were preferred by individual politicians but were not a high priority for the NPS. Still, the current experiment of keeping entrance fee revenue in individual parks may eventually lead to unsustainable demands for self-sufficiency and may foster unhealthy competition among individual units. Further, needy parks that may not be able to raise much revenue, in remote locations, for example, will suffer. I recommend instead that all money raised by national parks be kept in the national park system but in a central trust fund to be allocated by central agency planners.

Another means of increasing resources available to the parks involves concessions operations. Returns to the government from concessionaires operating in national parks are notoriously low, estimates consistently measuring returns at less than 5 percent (U.S. GAO 1991a). Low returns are a consequence of past efforts to lock in facilities within parks and the present political clout of concessionaires. As the popularity of parks continues to increase, fears of little interest from entrepreneurs are no longer justified. Finally, national parks are moving to more competitive bidding and shorter-term leases. In this respect, they could learn from state park systems that generally get a higher rate of return on their concession operations.

Parks are important to many people. One dramatic piece of evidence for this fact comes from a 1995 survey of Americans on national parks. Over 96 percent of those answering the survey said that it was at least somewhat (11 percent) if not very (85.2 percent) important that parks remain protected in the future (Deruiter and Haas 1995, 14). Obviously, views on what constitutes "protection" can vary widely, but the fact remains that parks inspire strong feelings. Individuals and interest groups thus want to have a say in how parks are to be managed, for now and for the future. The political institution used to account for these feelings is participation in planning processes. Planners in both Canada and the United States have moved aggressively to involve public participation in park planning, especially in recent years. Some prominent plans in the past have inspired an impressive number of comments, notably ski expansion proposals in Banff, and policymakers have increasingly attempted to involve as many relevant participants as possible and to actually incorporate their feedback in making decisions. The recent Banff–Bow Valley Study and the current planning efforts underway at Yosemite are prominent examples. Indeed, the Banff study relies on outside planners precisely to encourage involvement by people outside of Parks Canada (R. Page 1996).

Diverse viewpoints must be incorporated into plans because, if not part of the planning process early, they will demand changes later. As one Yosemite planner told me about the impact of subsequent opposition to previous plans in that park, "History is littered with plans that went nowhere." Park planners must continue to aggressively solicit and utilize outside input into the planning process, but they can also institute mechanisms to keep those external individuals and groups involved in park management and implementation of plans. One means of pursuing this is to make interested individuals stakeholders, perhaps by creating advisory boards or boards of directors that continue to serve in an institutional capacity to park managers. An example of this is currently underway at a recently created unit in the U.S. park system. The Tallgrass

Prairie National Preserve Advisory Committee consists of thirteen members who represent the NPS, the National Park Trust, and various local interests. Incorporating these interests in a permanent relationship gives them a stake in the future of the park.

Federalism eventually will affect the implementation of plans; therefore, those pursuing a vision for the future of protected areas must incorporate this institution in their efforts. Doing so could take a variety of forms. One modest but straightforward approach is to establish regional conferences with overlap between state/provincial and federal officials that will at least facilitate open lines of communication. A more dramatic approach has been attempted in Australia with varying degrees of success. Some national parks in Australia are governed by joint management authorities combining federal and state officials in one governing body. For example, the top level of the Great Barrier Reef managing authority consists of one federal representative, one state representative, and one scientist. Decision making is thus somewhat complicated, but the Australians view it as a necessary complication given the importance of the federalism institution.

Summary

Sustainable development can take many forms. One of the most challenging involves using national parks even while pursuing the restoration and retention of natural conditions for this and future generations. The attainment of preservation goals for national parks will continue to be affected by institutional arrangements in political systems. To effectively address those traditional procedures will require corresponding responses in the creation or modification of institutional arrangements that emphasize future conditions in parks. Proposed potential changes include mandated allocations of funds for scientific research, formal legal recognition of plans, incorporation of diverse outside viewpoints in planning and subsequent management efforts, regional councils of federal and subnational officials, and retention of revenues in park systems.

Global Environmental Accountability: The Missing Link in the Pursuit of Sustainable Development?

Robert V. Percival

Sustainable development has become one of the guiding principles of global environmental policy during its remarkable recent ascent. Yet the more widely the concept has been embraced, the more elusive its meaning seems to be. Despite broad agreement that we should leave future generations at least as well off as our own, sharp disputes persist over the proper path toward achieving this goal. Efforts to define sustainability or to apply the concept in policy debates evoke memories of the late Justice Potter Stewart's famous comment about obscenity. Rather than "trying to define what may be indefinable," he applied a simple test: "I know it when I see it."[1]

This case study argues that sustainable development, regardless of how it is defined, can be advanced by designing policies that increase the accountability of individuals, governments, and corporations for the environmental consequences of their actions. In this context, accountability means not only internalizing external costs, but also assigning burdens of proof and the risks of loss in a manner that encourages the development of better information about, and more realistic forecasts of, environmental consequences. It means ensuring that laws and regulatory standards are implemented and enforced and that risks controlled in one jurisdiction cannot easily be transferred to another.

This case study begins by reviewing the concept of sustainable development and why it is so popular yet difficult to apply in practice. It then discusses the challenges posed to sustainable development by the new global economy, where vast private capital flows can rapidly alter the economic fortunes of entire nations. The study then explores the accountability theme, reviewing institutional mechanisms for holding actors accountable for the environmental consequences of their actions. It concludes by exploring some examples of how law is extending its global reach to promote greater environmental accountability in the pursuit of sustainable development.

The Concept of Sustainable Development

The roots of the concept of sustainable development can be traced to the 1972 Stockholm Conference on the Human Environment (see Hodas 1998, 1, 8), but it became a prominent theme of global environmental policy following publication of the 1987 report *Our Common Future by* the World Commission on Environment and Development (WCED, the Brundtland Commission). This report articulated the goal of sustainable development as "meet[ing] the needs of the present without compromising the ability of future generations to meet their own needs" (WCED 1987, 43). Five years later, when representatives of 178 nations gathered at the U.N. Conference on Environment and Development, known as the Rio Earth Summit, they adopted sustainable development as the centerpiece of the Rio Declaration's statement of global environmental principles.[2]

Sustainable development is now widely embraced as the central goal of global environmental policy, even though it remains highly vague and difficult to apply to resolve specific environmental controversies. As economist Herman Daly has observed, "Sustainable development is a term that everyone likes, but nobody is sure of what it means" (1996, 1). Professor David Hodas observes that "when one attempts to implement the concept, one quickly discovers that, except in its most obvious applications, sustainable development is, if not meaningless, an oxymoron" (1998, 2–3). Indeed, the popularity of the concept may stem from its very malleability, as it combines elements that appeal to both environmental and business interests.

Efforts to implement sustainable development have been monitored by the U.N. Commission on Sustainable Development, created as part of Agenda 21 adopted at the Rio Earth Summit. Agenda 21 recommended that each country establish a national council for sustainable development, as the United States did in June 1993 by creating the President's Council on Sustainable Development (PCSD; Lash 1997). This Commission, which included government officials, CEOs of major corporations, and leaders of environmental, labor, and civil rights groups, issued its report in March 1996. The report outlined the Commission's principles for promoting sustainable development (PCSD 1996). These included the statements that "economic growth, environmental protection and social equity should be interdependent, mutually reinforcing goals" and that decision makers "should consider the well-being of future generations, and preserve for them the widest possible range of choices." The report concluded,

"Sustainable development requires fundamental changes in the conduct of government, private institutions and individuals." It was short on specifics concerning what these changes should be, but it emphasized the importance of market strategies, the precautionary principle, citizen participation in the policy process, and public access to information. Despite confusion over the meaning of sustainable development, it is not difficult to determine that certain trends reflect the consequences of unsustainable policies.

For instance, inadequate sewage treatment and industrial pollution have precluded a large portion of the world's inhabitants from having reliable access to safe drinking water, posing an enormous threat to human health. Water-borne diseases account for 80 percent of human health problems in developing countries, causing the death of 4 million children per year (U.N. 1996, 259). In 1980 the United Nations established as a global goal that clean water and adequate sanitation be provided for all the world's population by the year 1990 (U.N. General Assembly Resolution 35-18, 10 November 1980). But more than a billion people remained without access to safe water supplies in developing countries years after the 1990s had commenced. Even with improvements in sanitation and water supply services, it was projected that more than 750 million people would not have access to safe drinking water in the year 2000 and the number without access to sanitation would be more than three billion people (U.N. 1996, 260).

To transform sustainable development into a more useful policy goal, the United Nations Commission on Sustainable Development is trying to develop indicators for monitoring progress toward sustainable development. The goal of the Commission is to have an agreed set of indicators available for use by all countries in the year 2001 (U.N., n.d). Although these indicators are not sufficiently developed to provide a comprehensive picture of progress, when the United Nations met in June 1997 to review progress toward implementing Agenda 21, its assessment was largely negative: "(1) an increase of 450 million more people on the planet; (2) an increase of 4% in global carbon emissions; (3) the loss of another 3.5% of global tropical forests; and (4) a decline in development aid to 0.27% of donor GNP," its lowest level in fifty years and far below the U.N.'s target of 0.7 percent (A. Miller 1998, 287, 290).

The New Global Capitalism

With the collapse of communism, the explosive growth of world trade, and the shrinking of capital controls, the majority of the world's population has been thrust rapidly from state-controlled economies into a new world economic order. Shortly after the U.N. declared its glum assessment of progress since Rio, the economies of developing countries in East Asia were devastated by a financial panic following the devaluation of the Thai currency. International investment banks that had made enormous investments in these emerging markets suffered heavy losses as the panic spread from country to country. The rapid capital flight away from Asia rocked the world economy, prompting the observation that global capital flows had become "the new neutron bomb" that can devastate a nation's economy virtually overnight. Although the world's financial markets weathered this "Asian flu," it demonstrated a new economic reality: that private capital flows now dominate international investment in developing countries. This was not true at the time of the Rio Earth Summit, where developing countries insisted on increased development assistance as a condition for accepting new international environmental commitments. Yet, as Alan Miller notes, "in the five years since Rio, all forms of public sector aid have been declining, while private sector flows have increased dramatically." As a result, in 1997, "private investment flows to developing countries totaled approximately $260 billion, more than five times greater than the $50 billion in multilateral aid" (1998, 291). This has important implications, both positive and negative, for sustainable development policy as private multinational corporations increasingly influence environmental and economic conditions in developing countries.

This is not an entirely new phenomenon, to be sure. In 1842, the United States suffered the wrath of some of the world's richest investors when Louisiana, Maryland, Mississippi, and Pennsylvania defaulted on bonds they had sold to London bankers (Kristof 1998, A18). But what is unprecedented here is the scale of the private capital flows, which are now so vast that they are capable of devastating the economies of entire nations.

Describing this as the new "global capitalism," Professor Jeffrey Sachs argues that it poses new challenges to legal systems to foster "the strong yet law-bound state" combined with a new international regime of law "fit for our global capitalist society." As Sachs notes, the former remains a problem in many developing countries where governments are either too weak or too corrupt to meet basic infrastructure needs, and the latter is complicated by the absence of any overarching political authority (n.d., 8–10). These are daunting challenges, yet legal processes are responding to them and the contours of a new global legal order are beginning to emerge, as described below.

Mechanisms for Enhancing Environmental Accountability

Economists have long explained environmental problems as the product of market failures due to the absence of property rights in common goods. Their prescription to protect the environment is for some form of collective action to internalize externalities by making actors bear the full social costs of their actions. This can be accomplished by holding parties who cause environmental damage liable for the cost of remediation or by creating property rights in goods formerly held in common. Both solutions involve the use of law to create mechanisms for holding humans accountable for the environmental consequences of their actions. Recent developments in U.S. environmental law also are stimulating regulatory innovations designed to increase the fairness and efficiency of the modern regulatory state, promoting the kinds of values served by long-standing principles of private law. These developments include increased emphasis on informational approaches to regulation that help harness market forces to prevent pollution, the creation of marketable emissions allowances that reduce compliance costs and combat strategic behavior, and experiments with new regulatory approaches such as environmental contracting and challenge regulation.

Liability

Prior to the emergence of the modern regulatory state, common law liability was the principal vehicle for holding actors accountable for environmental damage. But liability is no longer the primary legal tool for promoting environmental protection. In the past three decades regulatory legislation has largely eclipsed the common law because the difficulty of proving causal injury and tracing it back to individual agents made common law liability an inadequate tool for protecting the environment. As a result, environmental law has come to be largely dominated by statutes that authorize administrative agencies to set regulatory standards limiting emissions or that require agencies to consider prospective environmental effects as a condition for approving projects likely to have significant environmental consequences.

Recognizing the obstacles to common law recovery, Congress extended liability for environmental contamination beyond the common law model with the enactment of the Comprehensive Environmental Response, Compensation and Liability Act (CERCLA) in 1980. This law, commonly known as the Superfund legislation, broadened liability by extending strict joint and several liability for the costs of remediating environmental contamination to much broader classes of parties, including current owners of contaminated property and generators of hazardous substances who arranged for its disposal. The law creates powerful incentives for businesses to prevent releases of hazardous substances to avoid future liability.

Regulation

In the regulatory context, environmental accountability requires mechanisms to ensure that government agencies implement and enforce statutory directives. To enhance the accountability of government agencies, most U.S. environmental laws authorize citizen suits against agencies and officials who fail to perform nondiscretionary duties. These provisions have been widely used to force agencies to implement regulatory programs, particularly those for which Congress has specified deadlines for agency action. Citizen participation in the regulatory process is another important mechanism for increasing the accountability of government agencies. U.S. laws give private citizens the right to participate in rule-making proceedings and guarantee citizen access to information held by government agencies.

Some provisions of regulatory legislation enhance accountability by generating better information about potential environmental risks that can be used to inform government policymakers and private citizens. In a few cases the laws require demonstrations of safety before certain substances—pesticides and therapeutic drugs—can be marketed. Pesticides can be marketed only if they are licensed following a demonstration by the manufacturer that they will not cause unreasonable harm. Therapeutic drugs can be approved only following a demonstration that they are safe and effective for their intended uses. These requirements have generated far more extensive test data than are available for most other chemical substances for which the environmental laws generally place the burden on regulators to act.[3]

Informational Approaches to Regulation

U.S. environmental policy is placing increasing emphasis on informational regulation. Informational approaches seek to enlist the marketplace in controlling environmental problems by providing information that will affect consumer choices in environmentally positive ways. They are particularly attractive to those concerned about the high costs of command and control regulation. The Emergency Planning and Community Right-to-Know Act (42 U.S.C. 11001–11050), enacted in 1986, requires industries to report annually the volume of their releases of hundreds of toxic substances. The Act creates a national inventory of toxic releases that is made accessible to the public, using information as a tool for mobilizing public pressure to reduce toxic emissions. This approach has been so successful that EPA has dramatically expanded the list of chemicals and industries subject to reporting.

One of the most innovative developments in U.S. environmental law has been California's Proposition 65. This law, which was adopted by voter initiative, combines a duty-to-warn approach with a shifting of the burden of demonstrating the safety of emissions of carcinogens and reproductive toxins. The law prohibits companies from exposing anyone without warning to any substance known to be a carcinogen or reproductive toxin unless the person responsible for the exposure can show that it poses no significant risk to human health. This legislation is founded on the simple, ethically irresistible notion that no one should intentionally expose another to a significant risk of harm without warning. Proposition 65 avoids most of the problems of defining regulatory targets by specifying that it will be applicable to those releasing any substances that have been identified as carcinogens or reproductive toxins. Avoiding one of the most serious pitfalls of conventional regulatory programs, Proposition 65 makes it unnecessary for regulators to decide how stringently to control the great variety of discharges by simply leaving it up to the source. If the discharger thinks it is too expensive to stop exposing others to chemicals on the list, it can continue its discharges, but only if it provides a clear and reasonable warning to the exposed population.

Tradeable Emissions Allowances

Title IV of the Clean Air Act Amendments of 1990 provided the first large-scale experiment with emissions trading approaches long advocated by economists as a more efficient means for reducing pollution. While mandating significant reductions in sulfur dioxide emissions, the law creates emissions allowances that may be bought and sold to ensure that the reductions are obtained in the cheapest manner possible. Experience with Title IV's emissions trading program produced some surprising results. The allowances have been selling at prices that look astonishingly low compared to estimates made when the 1990 Amendments were enacted. Emissions reductions also are occurring at a faster rate than EPA had anticipated as utilities switch to low-sulfur coal to reduce emissions at a time when the price of such coal has dropped substantially. In the first five years after enactment of Title IV, sulfur emissions were reduced by 1.7 million tons from 1990 levels at a dramatically lower cost than Congress had anticipated (U.S. EPA 1995a). The low prices for which emissions allowances have sold appears to confirm that industry estimates of the cost of complying with Title IV were greatly exaggerated. When the 1990 Amendments were debated, it was projected that allowances could sell for $1,000 to $1,500 per ton because of the high cost of installing pollution control equipment to achieve the emissions reductions required by Title IV. However, emissions allowances have been selling at prices ranging from $100 to $200 per ton because the cost of reducing sulfur dioxide emissions has been much lower than expected. The price of low-sulfur coal has fallen due to improved mine productivity and reductions in the cost of transporting coal by rail. The costs of installing scrubbers also has fallen substantially ("Cyprus Amax" 1995, 1). In addition, many early cost estimates probably were deliberately inflated as a familiar form of strategic behavior by a regulatory target seeking to avoid regulation. Thus, one benefit of marketable allowances is that they make the true costs of compliance more transparent.

How Much Accountability? Improving the Fairness and Efficiency of Regulation

The first generation of U.S. regulatory programs made dramatic progress in reducing certain kinds of air and water pollution, particularly emissions from large industrial point sources, and created powerful incentives for more careful handling of hazardous substances. But they also came under considerable fire from regulated industries who argued that they were inefficient and inequitable mechanisms for improving environmental quality. These criticisms have intensified as environmental law's regulatory tentacles have been extended to embrace smaller and smaller entities and to affect development decisions by individual property owners.

Some of the apparent backlash against environmental regulation represents ordinary strategic behavior by the regulated community. But it also has some roots in tensions that the common law has long struggled to reconcile. These include the tension between the common law's insistence on individualized proof of causal injury and the inherently probabilistic and uncertain nature of environmental consequences. Regulatory legislation sought to overcome this problem by dispensing with the common law's causation requirements and endorsing preventative regulation by expert administrative agencies. But a major factor contributing to the apparent backlash against federal environmental regulation is the failure of national regulatory programs to deal with some of the very genuine concerns for fairness and efficiency that animated common law requirements. The difficulty of tailoring nationally uniform regulations to respond with flexibility to local circumstances and the perceived unfairness when changing regulatory standards benefit some groups and burden others help explain persistent complaints from the regulated community that the burdens of environmental regulation are distributed in an unfair or inefficient manner. Another major focus of fairness complaints has been CERCLA's extension of strict joint and several liability to broad classes of parties who may be far removed from the activities that result in releases of environmental contamination.

Although not yet widely employed in the United States, there is considerable interest in more flexible regulatory tools. Proposals for greater regulatory flexibility reflect the belief that environmental objectives could be achieved at far less cost if industries are given the latitude to reduce emissions plantwide and without the constraint of medium-specific requirements (e.g., Rice 1994, 15). In one approach, called "challenge regulation," the government establishes a clear environmental performance target while allowing the regulated community to design and implement a program for achieving it. Unlike purely voluntary programs, the government specifies regulatory responses that will take effect if the target is not met. Another approach, called "environmental contracting," involves an agreement between a government agency and a particular source to waive certain regulatory requirements in return for an enforceable commitment to achieve superior performance. Under this approach, companies are offered the flexibility to design their own program for meeting regulatory goals in return for a commitment to achieve specified levels of environmental performance.[4]

Efforts to provide greater regulatory flexibility were endorsed by the PCSD in its March 1996 report. The Council concluded that basic regulatory standards contained in existing environmental laws have been successful and should not be relaxed. However, it found that the efficiency and effectiveness of the current environmental management system could be improved by developing new approaches to regulation that emphasize performance targets rather than prescribing the means for achieving them. The Council strongly endorsed the notion of giving companies "greater operating flexibility, enabling them to *reduce their costs* significantly in exchange for achieving superior environmental performance" (PCSD 1996, 27). But it also stressed the importance of maintaining accountability, a goal that may be in tension with flexibility given the greater monitoring difficulties that more flexible approaches may entail.

Expanding the Global Reach of Law

During the past three decades there has been a remarkable surge in the enactment of environmental laws throughout the world. This is the product in part of the very forces that have fueled the explosive growth of global economic activity. Trade liberalization and advances in communications and information technology have contributed to pressure to harmonize environmental standards, as have bilateral and multilateral environmental agreements (see Kimber 1995, 17). In addition to enacting environmental legislation, countries that recently have revised their constitutions are now incorporating environmental rights or duties into them.

Yet it is far easier to put laws on the books than to develop the capacity and political will to implement and enforce them. Most countries have established specialized environmental agencies, but in many developing countries these agencies lack the resources even to begin to implement national regulatory programs. As a result, in many countries environmental legislation exists largely in the law books with little implementation or enforcement.

Inadequate resources represent an important barrier to increasing environmental accountability in many developing countries. In former Soviet bloc countries, where the government has owned most production facilities, lack of funding has been a crucial obstacle to investment in pollution control technologies. As industries are privatized, many countries are requiring that pollution control issues be addressed as part of privatization agreements, which provide financial incentives for environmental improvements. Environmental improvements also can be accomplished through policies that promote more efficient management of natural resources, such as improved planning and pricing of public water supplies to ensure that they are used where they provide the greatest social benefits.

To hold agencies accountable it is important to mobilize citizens to participate actively in the process of implementing and enforcing regulatory standards. In June 1998 the Convention on Access to Information, Public Participation and Access to Justice in Environmental Matters was signed in Aarhus, Denmark. This treaty represents the first time that citizen participation has been the focus of an international environmental agreement. It requires signatory nations to change their national laws to give citizens access to environmental information, including the right to know about toxins in their communities and the ability to bring actions to enforce the environmental law. This treaty ensures that more countries will adopt the U.S. model of citizen involvement as a means for ensuring that regulatory legislation is implemented and for supplementing government enforcement.

Most countries have adopted some form of framework environmental legislation that requires environmental impact assessment. Indeed, an environmental impact assessment requirement has become the primary and most universal element of environmental law around the world, having been adopted by more than 130 countries. The United States was the first country to adopt such a requirement, but some countries have improved on the model of our National

Environmental Policy Act by requiring that forecasts of environmental impacts be embodied in enforceable commitments. In these countries the environmental impact assessment process serves as a framework for applying all environmental standards in a single licensing process with follow-up audits to ensure that predictions of environmental consequences prove accurate. By holding the proponents of development projects accountable for their projections of future environmental consequences, this approach creates incentives for more realistic forecasts, while relieving the public of some of the cost of remediation if forecasts prove too optimistic (Hodas 1998, 45–48).

Although few countries approach the United States in terms of tort liability, civil liability for damages caused by pollution is beginning to appear in the laws of several countries. Some of the more egregious forms of health and environmental damage are generating tort claims by foreigners against multinational corporations. In 1996 three thousand workers on banana plantations in the Phillippines settled a lawsuit against U.S. chemical companies who manufactured or used dibromochloropropane, a pesticide banned in the United States for causing reproductive harm. In October 1998, another group of more than twenty thousand banana workers filed a similar lawsuit ("Philippine Banana Workers" 1998). A $1.5 billion class action against Texaco on behalf of thirty thousand residents of an Ecuadoran rain forest ravaged by oil drilling was reinstated by an appellate court in October 1998 (*Jota v. Texaco, Inc.*, 157 F.3d 153 (2d Circ. 1998)). Some Central and South American nations have sought to facilitate lawsuits on behalf of their citizens in the U.S. courts by closing their own courts to such claims to reduce the possibility of a forum non conveniens dismissal (Newman 1999). When successful domestic tort claims are made against companies for marketing hazardous products that also are sold abroad, it is not surprising now when lawsuits by foreigners seek to follow. For example, in the wake of the $206 billion settlement by the tobacco industry of the class actions brought against it by states, the governments of Guatemala, Nicaragua, and Panama filed similar lawsuits against the industry (Torry 1999, A12).

As international megamergers increase the size and influence of multinational corporations, international cooperation will become increasingly necessary to hold such corporations accountable for their actions. Coordination of national regulatory strategies can be a useful means for avoiding a regulatory race to the bottom. For example, when one Caribbean island nation sought to impose a waste disposal tax on cruise ships that call at its port, a major cruise line threatened to drop the island from the itinerary of its ships. This was successful in scuttling the tax until the Association of Caribbean States adopted a coordinated policy in which several nations adopted the same tax simultaneously. Faced with the reality that it could not play off one island nation against another, the cruise line backed down and agreed to pay the fee.

Conclusion

Speaking to the World Economic Forum in February 1999, U.N. Secretary-General Kofi Annan invited multinational corporations to enter into a formal "compact" with the United Nations. The terms of the proposed bargain were that the companies would agree to promote sustainable development, human rights, and fair labor practices in whatever countries they do business in return for the U.N.'s political support for open global markets. The secretary-general explained the problem as follows: "Globalization is a fact of life. But I believe we have underestimated its fragility. The problem is this: The spread of markets far outpaces the ability of societies and their political systems to adjust to them, let alone guide the course they take. History teaches us that such an imbalance between the economic, social and political realms can never be sustained for very long." Annan warned, "We have to choose between a global market driven only by calculations of short-term profit and one which has a human face." Pledging to use the United Nations to "help make the case for and maintain an environment that favors trade *and open markets,*" Annan argued that "unless those values are really seen to be taking hold, I fear we may find it increasingly difficult to make a persuasive case for the open global market" (quoted in Cowell 1999, A10).

The secretary-general's comments reflect a growing appreciation that additional institutional mechanisms for holding multinational businesses accountable will be required in an era of unprecedented global capital flows and trade liberalization. A variety of initiatives are underway to increase environmental accountability at every level of government and in the private sector. For instance, the International Standards Organization is involved in efforts to establish voluntary standards for corporate environmental management systems and environmental auditing that could have a significant influence on the environmental performance of businesses. Individual nations are adopting and upgrading national environmental laws, implementing environmental assessment requirements, and incorporating citizen participation and access to information provisions. Indigenous tribes and workers injured by toxic chemicals in foreign countries are taking their own actions to enforce environmental laws, filing tort actions against multinational corporations and linking human rights claims with environmental protection concerns (see Bearak 1998, A26).

Thus, although the newly globalized economy poses important challenges for sustainable development, the legal system is not retreating from its underlying commitment to global environmental protection. No single approach to regulation is appropriate for every underlying legal framework (see Wiener 1999), yet we have sufficient experience with a variety of regulatory approaches to make wise choices in the selection of policy instruments for promoting environmental accountability and sustainable development.

PART III

Moral Principles and Sustainable Environmental Policy: An Analysis of Ends and Means

Introduction

Our discussants (Buck, Paehlke, Kassiola, Bowersox, Norton, Sagoff, Gillroy, and Taylor) now turn their attention to the ramifications for environmental policy of the case studies in sustainability from Part 2. This provides a unique opportunity for them to test, amend, confirm, or refute their positions in Part 1. Perhaps not surprisingly, most of the authors find a mixture of positive and negative developments within the case studies that they utilize to concretely illustrate and expand on their earlier work. Specifically, the discussants' positions in Part 3 reflect a common concern that concrete policies, whether they are sustainable or not, seem to show little attention to prior normative commitments or the fulfillment of goals arising from systematic belief systems. At one level, it appears the authors don't find this too surprising: it is as if they too assume that value debates will have little impact on the present processes utilized to form, implement, and enforce policies. However, although they are unsurprised by this negligence, the authors uniformly are quite concerned about its consequences.

The authors suggest that even the most well-intentioned and thoughtful of the sustainable policy reforms outlined in the case studies significantly undermine their long-term efficacy and legitimacy by leaving critical assumptions unstated, unexplored, and untested. Kassiola, Bowersox, Taylor, and Gillroy are particularly concerned that sustainability be clarified as a policy principle. Although they clearly disagree with many of the case studies' *implicit* definitions and concepts, they suggest that the primary failure of the case studies is to provide internal normative standards for sustainable policy by which citizens, policymakers, and scholars can realistically evaluate their recommendations and analysis of the case under consideration. Hence, such conceptual ambiguity, though it may make the policies more palatable in the short term, ultimately may undercut policy legitimacy.

Another strategy is taken by Paehlke, Buck, and Norton. Each is less concerned with the particular meaning of the term sustainability or sustainable development in the essays of Part 2, and more concerned about the term's practical implications and its influence on contemporary policies. Paehlke and Buck both utilize additional examples to suggest ways in which sustainability or sustainable development motivated or have been incorporated into both domestic and international environmental planning programs. Both argue that sustainability ideally provides a forum in which science and policy converge. Similarly, Norton's argument in this section notes the weaknesses and ambiguities associated with the term sustainability, but he refuses to propose the term's abandonment. Taking a conventionalist view, Norton suggests that the term remains useful in part due to its plasticity and the ability to further define and make its meaning concrete via its ongoing use in environmental discourse and the policy process. Like Paehlke and Buck, Norton argues that the term's utility lies primarily in its potential use as a nexus for science, policy, and social values.

Mark Sagoff serves a function different from that of the other authors. Rather than examining the concept of sustainability and its theoretical and practical consequences, Sagoff questions the utility of unitary explanatory theories in general. Suggesting that we may wish to preserve some natural entities due to their religious or aesthetic value (as perceived by humans), Sagoff finds the totalizing visions of the world outlined by some sustainability advocates misleading, deceptive, incorrect, and even immoral. Further unimpressed with the ethical and policy implications of such essentialist positions in science, policy, and ethics, Sagoff argues for a pluralist, pragmatic, and technologically optimistic response to environmental dilemmas. Hence, although Sagoff's depiction of environmental "foxes" and "hedgehogs" may be controversial, his message is a fitting "reality check" for environmental policy and ethics generally.

While clearly marking out their own paths, our discussants exemplify an emerging and divergent trend in environmental policy and ethics: some choose as their level of analysis the rationalizing tendencies within policy debates, examining the consequences of the more sophisticated use of cost-benefit analysis, risk assessment, and policy innovation for the creation of workable and efficacious policies; others tend to look at the implications of value determination, policy formation, and program implementation for democracy (C. Foreman 1998). In one sense, these trends demonstrate the extent to which the theoretical analysis of environmental policy has matured in the past twenty years, during which time such analyses often completely ignored concerns about democracy or effectiveness. However, because the authors here focus on either rationalization or democratization, it might appear that the two trends are exclusive and incompatible. A careful reading of the essays should lead one to question that presumption, which will be further challenged in the conclusion to this volume.

Issue 1: Science and Sustainability

Sustainability, Sustainable Development, and Values

Robert Paehlke

I prefer to distinguish between sustainable development and sustainability, though both are concepts that advocate balancing the needs of today against the needs of tomorrow and integrating economic activity and environment protection. Sustainability might be seen to emphasize the well-being of future generations and the environment, whereas sustainable development is more thoroughly ambivalent as regards any balancing between economy and environment. Sustainable development advocates, then, place equal weight on each pole, whereas some environmentally oriented commentators (sustainability advocates) see the concept as an oxymoron, an impossibility. They argue that present levels of energy and materials use and environmental damage already place impossible and nonsustainable burdens on the future—that human economies are already "living on borrowed time." Any additional increments of "development," in this view, only compound our troubles and place additional burdens on the environment and on the future.

At issue here is a fundamental difference in worldview that is part science (and faith in science) and part value preference (compounded by political prognostication). In value terms, perhaps in too moralizing a tone, some environmental advocates ask what is more important: the diversity of species on the planet or the price and availability of forest products (woodpeckers or a new sunroom, salmon streams or multiple editions of the same yellow pages)? Air and water quality or the profit levels of the producers of chemical products? RVs or our great grandchildren's having an affordable way to avoid freezing in the dark? But are moral choices at the heart of the matter? Can there not be both profits and clean water, both economic growth and reduced environmental impacts and resource demands (through recycling and energy efficiency, or more services and fewer goods, for example)? Can science and technology not give us both, or failing that, can we not forgo some products and reject some technologies without "damaging" the economy as a whole?

Advocates of sustainable development believe that additional economic growth is essential and desirable and can be sustained through technological and policy advances—by doing things smarter, cleaner, and more efficiently (in terms of energy and materials use). Green critics of sustainable development see this, often sincere, faith operating as a cover for unrestrained economic growth, the primary benefits of which go to those who least need them. Advocates emphasize examples of wastefulness that can be corrected and situations whereby poverty forces unwise (and blatantly nonsustainable) resource use that might have been avoided through the alternative means of survival provided by "green" development. The critics emphasize growing economic inequality and the obscene waste inherent in such things as sports utility vehicles and private jets. They calculate that it would take three to six Earths to bring all the present population of the world to a North American living standard (Wackernagel and Rees, 1996).

The essentials of this difference are captured in two quotations juxtaposed by Rogers (1998, 70) to contrast these opposing views. These assessments followed the 1980 publication of *World Conservation Strategy,* a document that was a precursor to the articulation of sustainable development in *Our Common Future.* The quotes effectively articulate the distinction between sustainable development and sustainability. The first quote, essentially the perspective of sustainable development advocates, is from Canadian geographer Bruce Mitchell:

> When the environmental movement was reaching its initial peak in the late 1960s a situation developed in which those concerned about protecting the natural environment became the opponents of those concerned with economic development and growth. This polarization of views led to many confrontations between the two groups . . . as time went by, those supporting environmental quality issues created a credibility problem for themselves by consistently opposing development. . . . During the 1980s, a significant shift in thinking appeared. The idea was presented that sustained regional economic growth and ecological integrity were complementary. This idea appeared as the core of the *World Conservation Strategy.*

The contrasting view, from Donald Worster, interprets the same transformation from a different perspective, one that might be seen as nearer to the view of sustainability advocates:

> Back in the 1960s and 1970s, when contemporary environmentalism first emerged, the goal was more obvious and the route more clear before they came to be obscured by political compromising. The goal was to save the political world around us, millions of species of plants and animals, including humans, from our own technology, population, and appetites. The only way to do that, it was easy enough to see, was to think the radical thought that there must be limits to growth in three areas—limits to population, limits to technology, and limits to appetite and greed. Underlying this insight was the growing awareness that the progressive, secular materialist philosophy on which modern life rests . . . is deeply flawed and ultimately destructive to ourselves and the whole fabric of life on the planet . . . since it was so painfully difficult to make this turn, to go in a diametrically opposite direction from the way we had been going, however, many started looking for a less intimidating way. By the mid-1980s such an alternative, called "sustainable development," had emerged.

Worster sees the fundamental cleavage between sustainability and development as one of values and philosophy, as much as it is one of science. As Campbell-Mohn put it: "The value judgments underlying sustainability include risk adversity [the precautionary principle], intergenerational equity, and community." For Worster's 1960s environmentalists, nonsustainability itself was at least in part the product of science and technology. In contrast, prior to the rise of environmentalism as a successful social movement, unabashed advocates of development would leap at any and every technological and growth opportunity without hesitation. After environmentalism they would sometimes look again if it was clearly demonstrated that there was a net loss to society, especially in dollar terms, but might expect to be compensated for undoing their errors (see Campbell-Mohn on "takings" claims, this volume).

Sustainability advocates approach every new technology skeptically in terms of its costs to other species and to environmental quality. Some have also concluded that because the planet is finite, perpetual economic growth is straightforwardly impossible. Indeed, some assume that much of today's economic activity is likely temporary, given that energy supplies (fossil fuels) accumulated over countless millennia appear destined to be exhausted within another century (see, e.g., Catton 1980). Sustainable development, from this perspective, is but an attempt to continue on this nonsustainable path by adding additional economic activity (and resource extraction), some of which is only marginally more benign, utilizing marginally more ecologically efficient technologies.

Many, however, are caught in the middle of this debate, somewhere between the poles of strong sustainability and sustainable development. Some, however green their perspective, find it impossible to deny that the majority of humans desperately need more by way of economic well-being and guess that science and policy can buy us some additional "environmental space." Perhaps in the context of continuing growth, they hope, those at the bottom can obtain a proportional share of the increment. Those who come to such conclusions fit Worster's view of how "painfully difficult" is the logic of sustainability advocacy: in the end it may cost human lives (though arguably fewer lives than would any more abrupt resource/sustainability failure). Some of these morally "intimidated" greens would also acknowledge that the redistribution of the holdings of Bill Gates and his ilk would fall short of resolving poverty on a global scale, and may be politically improbable and morally doubtful in any case. Others might argue that Microsoft and Wal-Mart money might be a good place to start, but opt not to take up the issue in a short paper.

Clearly, then, sustainability and sustainable development are value-laden forms of inquiry. However, despite the fact that in the end moral judgment, political savvy, and best guesses are inevitably necessary, scientific analysis has much to offer. Several recent analytic approaches assess just how far (and how) the global economy might be shifted toward sustainability. These include concepts and measures such as metabolism, MIPS (material intensity per service unit), dematerialization, life cycle assessment, and footprint analysis (Fischer-Kowalski 1997; Schmidt-Bleek 1994; Robinson and Tinker 1997; Paehlke 1999; Wackernagel and Rees 1996). One recent detailed effort has been published under the title *Factor Four* (von Weizsäcker, Lovins, and Lovins 1998). The subtitle of this work carries its essential theme: *Doubling Wealth, Halving Resource Use.* The authors make the case that there is at least some room to maneuver. Technologies exist, they contend, that, if fully utilized, would allow global economic output to double, while at the same time energy and materials use could be halved, to levels that might well be sustainable on a renewable basis.

The authors of *Factor Four* assess both the technologies and the policy prescriptions necessary to accelerate the introduction and wide adoption of those technologies (e.g., shared, superefficient cars; reshaped cities; expanded public transportation; superefficient appliances). These authors are, interestingly, explicit advocates of capitalism, albeit capitalism guided by a fundamental transformation of tax and subsidy systems and by newly created markets in such things as saved (forgone) energy use. As the authors put it: "So effective is the profit motive that perhaps markets headed for nonsustainability can best be redirected through creative use of market forces themselves, to harness their ingenuity, rapid feedback and diverse, dispersed, resourceful, highly motivated actors. That is, after all, the strategy of threatened biological systems that evolve new feedbacks and defenses against a threat, often turning the threat's own strengths against it—a common ploy of the immune system" (von Weizsäcker et al. 1998, 142–43). All in all, this view captures the political middle ground by blending a faith in technological innovation and a hope of continued economic growth with a recognition that increases in overall energy and materials extraction must be rapidly halted and then reversed over a period of fifty years.

My view of *Factor Four* is that its technological optimism is plausible and its policy advocacy generally sound, but its political optimism is excessive, especially for North America. The case studies of sustainable development policy in this volume give us a sense of how far away we have been, and in many cases still are, from sustainability of the sort imagined by Worster or envisioned by *Factor Four.* As Lowry notes with regard to national parks (and wilderness habitat, a dimension of sustainability largely long since lost in Europe), "The major goal of preservation efforts through the late 1960s in Canada, the United States, and most other nations was to provide scenic vistas to attract tourists." We have moved more toward an attempt to sustain the diverse habitats of flora and fauna, toward ecosystem management. But we have moved slowly, restrained in Canada by the pro-development orientation of most provincial governments and in the United States by the dominance of politics over ecological science in Congress. In both nations, some national parks are threatened by the uniqueness of their success on a global scale; some, such as Banff, are overwhelmed by excess tourist traffic. Worster might remind us here about too many people, too much affluence, and being overwhelmed by our own technologies (in Banff the trains, automobiles, and campers threaten to exterminate the very animals their drivers are traveling to see).

At present in North America we are still struggling to establish the basics of sustainable development. Much of the third world is a very long way short of that standard. Consider the issues raised by Schwab and Brower regarding natural hazards. Most particularly, consider the possibility that some of the damage and death from mud slides associated with Hurricane Mitch are not unrelated to excessive clearing of forests in Central America and that the intensity and/or frequency of such storms may be related to climate warming. In many cases of this sort, cause and effect are not straightforward, but it is clear that nonsustainability comes in many unexpected and indirect forms. Like the *Factor Four* authors, Schwab and Brower make clear the power of tax shifts and spending powers to promote more sustainable practices.

Laitos gets to the heart of the issue when he notes that "econocentrism is insensitive to the inevitable societal consequences of a market-driven set of value choices." And later, "The new idea in sustainable development is therefore a trio of equal values—environmental, economic, and social—overlaid by a time calculus (the future should not be sacrificed for the present)." Getting to equality among those three values is a very long way from where we are now, but what *Factor Four* suggests is that the market is a tool, not a proper—or exclusive—basis for human values. Laitos's calculation and conclusion that the recreational and preservation-related benefits of public lands overwhelm commodity uses is important. It suggests that sustainability is, as Worster saw, bound up with less by way of material demand, especially when "material" is taken literally as in more contemporary calculations of energy and material throughputs and the "dematerialization" of economies, that is, the argument that GDP has considerable room to grow without an expansion of energy and material throughputs—a possibility that tends to move sustainability from the realm of values to the realm of science.

Rabe's examination of New Jersey's approach to pollution prevention presents another dimension of how materials balances and analysis are important. Closely examining and regulating toxic inputs and outputs within products and into all media is clearly a superior way to proceed. This is very much the comprehensive approach of life cycle assessments of all environmental effects, from cradle to grave. Sustainability analyses, such as MIPS or metabolism, do the same thing, only they do so with regard to all material throughputs rather than just toxic throughputs. They do so on the grounds that there is environmental damage and sustainability loss associated with all extractions and, although the amount of damage per unit weight varies considerably with the quality of extractive practice and environmental circumstance, the damage is nonetheless proportional to the amounts of material extracted (and energy used). Sustainability is about getting more economic well-being per unit of energy and material, just as pollution prevention is about reducing the amount of by-product per unit of output.

What is worrisome in all cases is how slowly we in North America are taking to this way of thinking. As Rabe notes, New Jersey's approach is widely praised but rarely imitated. Campbell-Mohn makes clear broadly what is needed in legal terms: "We need to develop a new analytical framework that joins natural resource law and pollution abatement law. . . . The new framework should look at environmental law through the life cycle of the resource from the time we extract a resource from the environment, through its manufacture into a product, to its disposal as a waste." But we North Americans are a very long way from seeing the extent of change that is necessary to achieve a truly sustainable economy. Europeans and the Japanese get far more economic output per unit of energy input—their energy productivity is enormously higher. The United States gets about $2,000 of GDP per ton of oil equivalent; Germany gets more than $4,000 and Japan more than $6,000 (Carley and Spapens 1998, 110). In achieving energy efficiency, pollution levels are correspondingly reduced. Over and above these differences there is more political support (and intellectual work) in Europe for new sustainability initiatives, including reductions in work time. As well, detailed national studies of sustainability have been prepared and widely read (see Carley and Spapens 1998, 192–93).

In North America thus far, the analytic and political steps taken are more modest and tentative despite the fact that we have so much further to go. The challenge to environmental policy analysts in North America is to set the outlines of a more sustainable society and economy and to assess the economic costs and benefits of moving in that direction. Also essential is an articulation of the policy tools that would move our society and economy in the direction of sustainability. I close with some of the policy possibilities. Each of the three measures mentioned here would establish structures and rules within which the power of markets would be marshaled to creatively and efficiently execute the detailed changes associated with sustainability. All are globally oriented, as that is the level at which economies are (the economy is) now structured.

The three measures are (1) the wide introduction of an energy and materials throughput tax, perhaps as a replacement in wealthy nations for taxes on annual incomes below $10,000 and/or payroll taxes; (2) the introduction of an international penalty-and-incentive fund for compliance (and noncompliance) with crucial environmental treaties; and (3) a global minimum wage linked to WTO-supervised North-South trade (perhaps pegged to national GDP/capita). The first of these might seek to gradually bring North America nearer to European levels of energy taxation and to broaden such taxation to all extractive commodity production other than food (Durning and Bauman 1998). The second might provide an urgently needed compliance carrot for, for example, the 1997 Kyoto Agreement on climate warming (the predecessor to which was a thoroughgoing failure for want of enforcement). Both measures would, arguably, help to establish a broad political and economic basis for many of the sustainability initiatives discussed in this volume. Some agreement regarding the third item would spur the increment in equity that sustainable development advocates assume is both necessary to sustainability and an important objective of development.

Saving All the Parts: Science and Sustainability

Susan Buck

"Breakfast," said Aldo Leopold to his daughter Nina, "comes before ethics" (quoted in Meine, 1988, 504). What are we to make of this? At one level, Leopold may be saying that practical concerns—food, security, a good night's sleep leading inevitably to a tidy breakfast in a cozy kitchen—are the stuff of real life. Make sure, he seems to imply, to take care of the practical before worrying about the abstract. We could also see this pronouncement as recognition that we must be adequately prepared before engaging in the hard work of ethical debate. Leopold does not relegate ethics to the postprandial fireside, an intellectual indulgence after the day's real work. Breakfast first, *then* ethics! Wake up, drink your coffee, *then* buckle down to business!

My business is to address the relationship between science and sustainability. In my earlier essay, "Science as a Substitute for Moral Principle," I argued that during the implementation phase of the environmental policy process, bureaucrats properly use their own technical and scientific expertise, tempered by legislative mandate and constitutional obligation, to make administrative decisions. The focus in that argument was *administrative outcomes.* In this essay, the focus is *policy formulation.* My argument here is that in policy decisions about sustainability, the role for science is a curiously indirect one.[1] When policy decisions about sustainability are made, scientific input is so filtered through economic imperatives and political feasibility that the environmental result often falls short of sustainability. Political decision makers seek to sustain resources with immediate economic value (or those with high emotional value, such as sea otters), and the price they are willing to pay is determined by short-term economic and political factors. The harm from this approach is incalculable. It is not possible to ensure true, long-term sustainability without also maximizing biodiversity. It may be possible in the future to know enough to pick and choose among species, habitats, and levels of pollution when we strive to sustain particular species. However, the scientific ability to do so does not currently exist, and by ignoring the imperatives of biodiversity, we may lose the opportunity forever.

Sustainability is a political term. It is framed by human needs and desires and defined by economic criteria (Turner, Pearce, and Bateman 1993, ch. 15). It is inevitable that scientific inputs would be filtered through social processes, thus diluting the role of scientific information in decision making. Science itself is, in part, a political discipline. Research projects are often chosen in response to funding opportunities or accidents of scholastic propinquity. Scientists bring political or policy biases to their research and their conclusions; they may be in dual roles as researchers and policy advocates (W. Rosenbaum 1998, 137). Regardless of the weight of these criticisms of the supposed "neutrality" of science in policy decisions (Alm and Simon 1999), the accumulated weight of scientific observation is beginning to force political and public awareness of failing ecological systems, disappearing species, and the terrifying adaptive capacity of diseases and insects.

The essays in Part 2 of this volume follow the definition of sustainability established in the 1987 report of the World Commission on Environment and Development (the Brundtland Commission): "development that meets the needs of the present without compromising the ability of future generations to meet their own needs" (1987, 8). I prefer the definition written by Helen Kolff in *Beyond War:* "the ability to support, provide for, nurture the total life system in our bioregion and all other bioregions on the planet; [and] the ability to supply in perpetuity all life forms with the necessities of life" (quoted in Chiras 1994, 6). This is a biocentric approach that places humans squarely in the midst of their ecological communities. It is also the definition that drives the new disciplines of conservation biology and ecological economics.

Assuming we believe that proportionately more attention to scientific inputs is desirable, we need to change two things: first, "science" itself (i.e., the hypotheses being tested, the kinds of data gathered, and the paradigm in which conclusions are made), and second, the effects of the political and economic filters. To some extent, this is already being accomplished. For example, the science of ecology has changed over the past several decades; the "new paradigm" that is emerging perceives ecological systems as open systems in a regular state of flux in which the human species is an integral part (Pickett and Ostfeld 1995, 261–78). This translates into a new academic discipline known as *conservation biology* (Knight and George 1995, 279–95). Conservation biology moves away from the species or ecosystem approach and recognizes what pollution specialists would call multimedia approaches: species and habitat are linked within and across political boundaries and cannot be successfully managed separately (Nassauer 1997; Knight and Landres 1998, esp. Pt. 1).

Changing political and economic filters is more problematic. In 1949, Leopold wrote the following: "To sum up: a system of conservation based solely on economic self-interest is hopelessly lopsided. It tends to ignore, and thus eventually to eliminate, many elements in the land community that lack commercial value, but that are (as far as we know) essential to its healthy functioning. It assumes, falsely, I think, that the economic parts of the biotic clock will function without the uneconomic parts" (1966, 251).

The "dismal science" of economics has made strides since 1949. The emerging field of *ecological economics,* although unavoidably anthropocentric, defines sustainability as follows: "Sustainability is a relationship between dynamic human economic systems and larger dynamic but, normally, slower-changing ecological systems in which (a) human life can continue indefinitely; (b) human individuals can flourish; (c) human cultures can develop; but in which (d) effects of human activities remain within bounds, so as not to destroy the diversity, complexity, and function of the ecological life support system" (Costanza 1995, 332–33). Although this new approach to economic analysis is not yet the accepted norm, a substantial literature in ecological economics is developing. The economic logic of including such factors as natural capital in the global ledger sheets alone is enough to initiate change in government accounting systems.

The political realm moves more slowly; technocrats have begun to advocate broader ecological concerns. This approach is not news to them; it is just that the political climate has become more accepting. Elected officials are still strongly motivated by interest group pressures and the constant, unrelenting, routine demands of the electoral process. In many countries, political pressures in favor of conservation, preservation, and pollution control are beginning to change the political filters. Elected officials have significant and powerful interest groups that encourage environmentally sensitive legislation. Interest groups, issue networks, and epistemic communities have become more sophisticated in their lobbying efforts and more strategic in their lawsuits. In addition, decades of public education by these same groups have made the mass public more attentive to environmental issues. However, the rate of change in the political policy streams is slower than in economics.

Thus the conceptual streams from scientific, technical, and (albeit, to a lesser extent) economic communities collide with more conservative and entrenched political forces. The result is a Charybdis of epic proportions and a series of ecological shipwrecks such as the loss of the Atlantic cod fishery. That policy change is occurring at all is encouraging, but the slow pace of change is a cause for dismay. As Rachel Carson wrote in 1962, "Time is the essential ingredient; but in the modern world there is no time" (1964, 6).

There are some examples of policy arenas in which science and the scientific community have played a leading role in placing the policy issue on the government agenda and in formulating at least the initial policy positions. This situation is the exception rather than the rule, and it may be useful to examine a few of these arenas to find commonalities. Two areas with which I am familiar are the Antarctic Treaty System and ongoing scientific discussions about global climate change (Buck, 1998, 117–21).[2]

The Antarctic Treaty System (ATS) is an example of a policy decision process relying, at least initially, on scientific input. The ATS is also unusual in the active role played by the scientific community itself. The policy story begins with the two International Polar Years (IPY) in 1882–1883 and 1932–1933. The effect of the first IPY on the scientific community was immense. It was the first time that scientists of different nations and disciplines had merged their efforts to study global phenomena, and it was also the beginning of public funding for cooperative, extranational scientific research (Roots, 1986, 171–72).[3] Although neither of these cooperative efforts had an Antarctic component, the IPY in 1932–1933 demonstrated that permanent polar bases could be established and maintained.

In 1957–1958, the International Geophysical Year (IGY), which did have an Antarctic component, provided an opportunity for the scientific community to stake its claim on polar research. The Antarctic scientists were absolutely clear that their work should be free of the taint of politics; at the end of the first IGY planning session in July 1955, they passed a resolution affirming the temporary and apolitical nature of their activities.

Inevitably, the IGY was colored by cold war concerns. The Soviet Union conducted extensive photographic surveying and mapping, although cartography had not been an official IGY discipline because of its political implications. The Soviets were eager for an extension of the scientific endeavors and announced plans to maintain an Antarctic research presence after the end of the IGY in 1958. Other nations chose to follow suit.

The Soviet launch of *Sputnik* in 1957 was an epiphany for the rest of the international community: "*Sputnik* made the Southern Hemisphere allies nervous that the Soviets might install missiles in their Antarctic bases, putting their countries within range. Suddenly, their internecine disputes in Antarctica were dwarfed by the common fear of a Soviet military presence there. A formal accord among nations, including the Soviet Union, renouncing military activities [in Antarctica] seemed the only way to forestall a costly arms race" (Shapley 1985, 90).

In 1958, the scientific community seized a unique opportunity. Clearly, the old bureaucratic and political disincentives to cooperation were either reduced or eliminated, especially in comparison with the advantages of cooperation. This was not a matter of economic profit; the costs of cooperation were balanced by the mutual benefits of closing the Antarctic to a Soviet military presence while continuing to provide shared access to research for all participants. There was also the potential benefit of shelving the issue of territorial sovereignty, thereby avoiding political conflict and perhaps violence.

However, a less utilitarian view of the scientific community finds them seizing their opportunity to promote peaceful, scientific research in one of the last uncharted places on the planet. They were fortunate that the Antarctic was also virtually devoid of economic interest. No one foretold the Antarctic tourist boom or the discovery of new, accessible sources of energy. The scientists had the experience of the IGY behind them, the public support generated by the publicity on the IGY, and a triggering event in the launch of *Sputnik*.

Also in 1958, the scientific community took advantage of national interests to protect the gains already made in the region and established the Special (later Scientific) Committee for Antarctic Research (SCAR). This committee was an important precursor to an Antarctic treaty because, in addition to its scientific functions, it provided a formal organization to coordinate national activities on the continent. Cooperation was justified as a mechanism to sustain the investments of the various national governments. A scientific organization rather than a governmental one, SCAR has its own constitution and governing rules, and its "territory" now includes the area out to the Antarctic convergence (where the cold Antarctic waters meet the warmer northern waters). The constitution for SCAR prohibits political activity except to provide technical advice to the nations involved. This prohibition on political activity has been scrupulously observed for the most part. Although SCAR itself provides only scientific advice, in practice many of the scientists involved with SCAR are also policy advisors for their governments. In theory, the roles are kept separate.[4]

The role of the Antarctic scientific community is unique in the annals of the international commons. They were the policy entrepreneurs who had already invested financial, personal, and institutional resources in building a cooperative regime. Even more important, the scientific culture required the scientists to be politically impartial. Like the purity of Caesar's wife, it was not enough to be politically neutral; it was imperative to be seen as politically neutral as well. In fact, the appearance of neutrality might be an acceptable substitute for its reality.

A second case in which the scientific community has had a prominent role is climate change. Climate change (global warming) is an anticipated increase in the average planetary temperature caused by three major sources: increased atmospheric carbon dioxide and nitrous oxides (by-products of fossil fuel combustion); methane, primarily originating in agricultural processes, especially animal wastes and paddy farming; and chlorofluorocarbons, which are also implicated in the loss of stratospheric ozone. Increased carbon dioxide levels are also affected by destruction of the rain forest (where trees absorb carbon dioxide) and the reduced capacity of the world's oceans to absorb carbon dioxide.

The scientific community has still not reached consensus on the *causes* of global climate change, although scientists now generally agree that a warming trend is occurring. Some scientists assert that the planet moves through warming and cooling cycles independent of human activity and that the current warming trend is simply part of the planet's natural cycle. Recent research points toward a complex relationship between climate and massive oscillations in ocean currents in the North Atlantic and Pacific. If the global warming trend continues despite cooler ocean currents, the likelihood that human activities are the main culprit is increased (W. Stevens 1997). Although the cumulative impact is uncertain, changes in patterns of agriculture and in sea levels seem likely, and the problem comes not from climate change alone but from the *rate* of change.

The greenhouse effect was first noted in 1827 by Baron Jean Baptiste Joseph Fourier, although he did not draw an explicit connection between human activity and climate change (Fourier in Paterson 1996, 17). In 1908, Svante Arrhenius was the first scholar to connect human industrial processes, carbon dioxide emissions, and world temperatures (1908, cited in Paterson 1996, 19–20). Climate change remained a minor item on the world environmental agenda until well after World War II. Weather data collection had become more sophisticated to provide information for military and commercial aviation, and cold war fears of a possible "nuclear winter" led scientists to examine climate data carefully. By 1979, a scientific consensus was emerging that carbon dioxide and other greenhouse gases were inclining to dangerous levels. In 1985, an international conference at Villach, Austria, concluded that climate warming appeared inevitable, and participants began to talk of a climate convention.

After the Villach Conference, the pace of policy formation picked up considerably. The same year, British scientists discovered a large "hole" in stratospheric ozone over the Antarctic; this provided a dramatic underscoring to the idea that human activities could damage Earth's atmosphere. In 1987, the United Nations Environment Programme and the World Meteorological Organization created the Intergovernmental Panel on Climate Change (IPCC) to study global warming. However, after this point, political and economic filters began to change the impact of scientific research.

Politicians jumped on the bandwagon: U.S. President Reagan signed the Global Climate Protections Act (January 1988), British Prime Minister Thatcher discussed global warming in a public speech (September 1988), and presidential candidate George Bush made global warming a campaign issue. However, once elected, Bush proved less enthusiastic, and in an example of spectacular political misjudgment, his administration ordered National Aeronautics and Space Administration scientist James Hansen to alter his testimony to Congress to suggest scientific uncertainty about the effect of greenhouse gases on global climate. Hansen went public, and the ensuing publicity brought global warming to the political forefront. Despite increasing international pressures, the United States continued to oppose a climate change convention.

In December 1988, the United Nations General Assembly passed the Resolution on Protection of the Global Climate for Present and Future Generations of Mankind (A/Res/44/207). In 1990, the final report from the IPCC's first Working Group presented a relatively united front of the world's atmospheric scientists that increasing levels of carbon dioxide were having a deleterious effect on the world climate. Many nations took this conclusion very seriously: between May and December, a majority of the Organization for Economic Cooperation (OECD) member states had begun to curb greenhouse emissions (Rowlands 1995, 78).

In the same year, the second World Climate Conference met in Geneva to address global climate change issues. Representatives from over 130 nations called for an international convention on global warming; this call eventually resulted in the 1992 Earth Summit (United Nations Conference on Environment and Development, UNCED) in Rio de Janeiro, where the Framework Convention on Climate Change opened for signature. Movement toward a climate change convention was extraordinarily susceptible to domestic politics.[5] The same coalitions that had opposed acid deposition policies were also opposed to climate change programs because the same domestic economic interests were threatened. For example, the OPEC nations did not support the Convention because they feared the development of alternative energy sources (especially nuclear power), which would cut into their exports. The United States relies heavily on cheap fossil fuels, and so the potential economic costs of the Convention were high. The developed nations did not present a united front either.

The newly industrialized countries were willing to accept emissions limits only in exchange for the most current technology, whereas some smaller states, especially the Alliance of Small Island States, wanted strict international controls at any price because they are so vulnerable to even the smallest elevation in sea level. Parties to this convention agreed to limit and mitigate carbon dioxide emissions and to provide technology transfer and financial assistance to developing nations. As a concession to U.S. policy, the convention contains no quantified emissions targets.

What do these two cases have in common? The most obvious characteristic is the international flavor of the policy arena. The resources at stake are multi- and interjurisdictional, and public rhetoric in the United Nations usually reflects a broad concern for the public good. These are relatively unexplored resource domains, so there are no preexisting bodies of national scientific knowledge about the resources or the environmental problems. Research is conducted in the glare of international peer review and often public review as well. Probably most important is the fact that at the beginning of the policy issues, virtually no one perceived any economic value in the resource domains. The continent of Antarctica has no extractable resources and until after WWII was uninhabitable for most of the year. Only the scientists were interested. The issue of climate change altered radically as soon as the potential development and economic costs became apparent.

I could make a similar case for the other global commons (high seas, deep seabed mining, and outer space). The lesson to be drawn is that science will always take a back seat to politics and economics when market forces at work dominate political discourse. This explains why most of the essays in Part 2 do not address scientific information explicitly. In most environmental policy areas, science is only one of many inputs into an economic and political calculus. In an anthropocentric world, this orientation is to be expected and is easily defended. In a biocentric world, it is less defensible.

For Aldo Leopold, the practical world of scientific observation and management are inextricably linked with ethical concerns for the environment. Probably his most famous and influential statement is in "The Land Ethic": "A thing is right when it tends to preserve the integrity, stability, and beauty of the biotic community. It is wrong when it tends otherwise" (1966, 262). Equally well-known and, to my mind, more memorable, is the following from *Round River:* "If the biota, in the course of aeons, has built something we like but do not understand, then who but a fool would discard seemingly useless parts? To keep every cog and wheel is the first precaution of intelligent tinkering" (1966, 190). Fifty years ago Leopold articulated the fundamental precepts of conservation biology: the centrality of a biocentric worldview and the precautionary principle (see O'Riordon and Cameron 1994; Hey 1992). Both are essential for sustainability.

Discussion

Gillroy: Scientists sometimes begin discussions of environmental issues, like the preservation of the Antarctic. But, if oil or gas is found, for example, they are eclipsed by economists and politicians who seek the resource regardless of the environmental consequences. I wonder if Susan Buck could say something about the difference between the economic forces and the political forces as they interact with science. Could we not say that it's the responsibility of the political system to control or to balance (that's a wonderful word), economic and scientific interests, and protect the kinds of policies that scientific evidence can support?

Buck: What I meant by political forces were really the electoral forces, including special interest groups, that pressure decision making. I think it's a luxury for a "lame duck" representative to be able to truly consider nothing but the public good when she makes political decisions. So, to the extent that economic forces drive the special interest groups and the economy and election results, that directly influences the role of science in making policy. Then there are the market forces themselves, the multinational corporations that have budgets and internal structures that rival the largest of the developed countries.

Taylor: I'd like to follow up on that. Once the gas is found, of course, the only real hope the scientists have is going through political channels . . . that is, trying to find a common balance to the economic interests that are so powerful without political regulation.

Buck: Well, the environmental interests have gotten much more politically astute. For example, in the Antarctic Treaty Regime, it used to be that only nations that were members could have research stations. But in the past couple of decades, the environmental groups have started moving into the administrative structure, and now Greenpeace maintains a year-round research station in Antarctica. So they are working their way into the institutional structures . . . they're getting more astute.

Paehlke: The comment I would make is that all along, in terms of environmentalism, much of the lead has come from scientists, from Rachel Carson on, who decided to step back from the technical scientific work and express it in such a way that it has public appeal. People who had detailed technical knowledge felt an obligation not only to place it in scientific journals, but also to convey their findings through op-eds and by writing a popular book, and I think that's an important factor. It's enormously encouraging that environmental groups like Greenpeace are funding science themselves. It's not so much that when corporations fund science they're going to distort what the scientist does, but at the same time they can distort the scientific process by picking and choosing among scientists who are doing things that are of interest to them. If there is no one else putting money into the game, well, then science as a whole is moving in that direction. They're not lying, they're just emphasizing one project rather than another.

Buck: But there are certainly a number of very highly publicized occasions when the companies and the government did lie. James Hansen, the NASA scientist who was ordered by the Bush administration to mislead Congress in his testimony on global warming, went public with it and there was this big brouhaha. I'm from North Carolina, and the tobacco industry has been lying about the scientific research they report on tobacco from the get-go. My guess is that that kind of distortion of scientific information happens very frequently. We just don't always know about it.

Bowersox: Just hearing you talk about pure science and the use of science by different interests, both corporate and environmental, causes me to ask . . . Is the only good science applied science, as it is out there to solve real problems? Should we be giving up on theoretical science and simply directing science toward specific applied problems?

Buck: Well, if we didn't have the pure science, applied scientists wouldn't know what to apply.

Paehlke: Science can ask the objective "ecological" question that does not favor one side or the other . . . that just gives basic background information. The basic theoretical models are the only way to know if things in the field have been environmentally altered. Basic knowledge underlies observation in biology. One may not be approaching a question from a conservation biology perspective, but that theory underlies the investigation regardless.

Buck: In my research, we had a contract with BLM [Bureau of Land Management] to look at the Baltimore Canyon when they were talking about doing oil leases out there. My lab was doing the soil analysis, and the science that undergirded the analysis that we were doing had been done in almost pure laboratory conditions. The interesting thing about that, by the way, was that our preliminary results indicated that it was much more ecologically dangerous to drill in the Canyon than we had anticipated. We gave BLM our preliminary results in

October, and in December, they canceled the contract. But by March, they had preliminary leases out in the Canyon anyway. . . .

To understand the role of the scientist, I think we need to define our terms. For policy, this means we must distinguish the four or five stages of the policy process running through agenda setting, formulation, legitimization, and implementation. Implementation is done to a large extent by the street-level scientists, the guys on the ground walking around counting the deer, doing the game warden stuff. My argument in the first essay was that for those people who are actually implementing policy, their primary responsibility is to do what they have been told to do under statutory and administrative guidelines within the constraints of the Constitution. And science there plays a substantial role. For a government policy implementer to follow his or her own moral principals when they are in opposition to statutory mandate is improper. This is a narrow focus. . . . What I'm talking about here is that beginning level of agenda setting and formulation in which I see the roles of science much differently. I am using science in two different ways: as an instrumental source of data in policy implementation and as a broader agenda generator in policy formulation.

Norton: At the level of legislation, for example, would you argue for biocentric legislation?

Buck: I'd like to see that. I'm not sure if it's practical.

Norton: I can't imagine what it would look like, but I'm beginning to understand what you're saying, once the legislation is in place. But then, of course, if you had biocentric legislation, it might change an awful lot of other things, some radically.

Buck: And that's exactly the problem that you run into with the Forest Service and BLM, when multiple use and sustained yield came through, especially for the Forest Service. They are the classic example of the professional culture that knew what they were supposed to be doing. They were supposed to be raising trees so that there would be a reliable and inexpensive source of timber for American industry; that is what Congress told them to do and that's what they were doing. If you are a forester, letting a tree grow beyond its maximum economic worth is like saying: Let the corn rot in the field, it makes the raccoons happy. So when multiple use and sustained yield came in they had tremendous trouble making the shift. I suspect that if we get scientifically sensitive biocentric legislation, the administrators on the ground would have a terrible time doing it. But they ought to do it because not doing it is, at least in our political system, I think, immoral.

Kassiola: But the critical moral question is whether biocentric or anthropocentric policy is superior. Then the political concern is: How do you implement it? I think your comment about how the civil service employees on the ground need to behave is really a product of how those individuals see their responsibilities and what oath they took in carrying out their profession. If they did have a moral conflict between congressional policy and what they are supposed to enact, then, as individual moral actors, they have to confront that, for it will influence policy. Even if Congress passes a law and their supervisor orders them to carry out the act they might have the moral obligation to leave the position rather than implement the policy. But I think the more interesting question is: What does biocentrism mean when you leave the moral/theoretical level to the policy implementation process, whether or not it's superior to anthropocentrism?

Paehlke: I just have one comment here. On any level of the policy process, with whatever legislation exists now, you could come to the conclusion, through the combination of the Endangered Species Act and the prudence principle, that we should be doing no more old-growth cutting in the Pacific Northwest of Canada and the United States. We are scientifically running a risk of losing all kinds of species within those areas if we don't restrict ourselves to cutting in second-growth forests. In fact we do, at least in the United States, have biocentric legislation. Canada does not have a national endangered species act—for ten years they have been trying—but the United States does.

Buck: But in Section 7 of the Endangered Species Act there's the God squad in there, which is a government-appointed commission that gets to recommend exceptions when the economic damage of enforcing the Endangered Species Act is too great.

Kassiola: There is also a pretty strong movement now in Congress to try to further weaken the Endangered Species Act, precisely because it has "biocentric" protections that harm economic interests.

Buck: That's why we haven't ratified the International Biodiversity Convention. Bush refused to sign it, Clinton signed it immediately after he took office, but it never made it out of the Foreign Relations Committee because Jesse Helms won't let it out. Helms won't let it out partly because of states rights, partly because there are implementation problems, but also because Jesse's making an antibiocentric statement. The Biodiversity Convention institutionalizes at the treaty level a lot of the same things that are in the Endangered Species Act. Helms has been trying to gut the ESA forever and ever. If we get it in a treaty, that would override state and national legislation. So, as long as we have policymakers like Helms in charge of the Foreign Relations Committee and a conservatively dominated Senate, we're not going to be part of the Biodiversity Convention, no matter what the scientific community might say.

Audience: But is the Endangered Species Act biocentric or anthropocentric legislation?

Norton: It's clearly anthropocentric if you look at the list of goals set out in the preamble: they're all anthropocentric. Things like scientific, aesthetic, educational, and so forth. There's no mention of biocentric values. The fact that it puts species at a very high level for protection doesn't entail that it has a biocentric motivation because there are many human motivations for protecting at the species level. One of them being the purely instrumental argument: What else are you going to count? Are you going to count subspecies? So I think there are any number of explanations for what was going on in the development and writing and passing of the Endangered Species Act.

Paehlke: There were clearly some biocentric thoughts at the time it passed.

Norton: Sure, we can talk about the motivations of individuals who voted for it, and I'm sure there might have been five, six, or eight Congresspersons who thought biocentrically, but if you look at the Act they didn't explicitly put it in there. My point is that biocentrism has not had any impact on policy, in Aldo Leopold's day or now. Leopold quite sharply separated his philosophical speculations, which were important and interesting, from his policy recommendations. He quite carefully bracketed them because he knew they wouldn't fly politically.

Buck: You have to remember the political message that is getting sent. Congress needs utilitarian economic justifications, regardless of what individual Congresspersons think. If they're going to get the thing passed, they've got to be able to justify it in the larger public arena. Looking at something with the impact of a statute like the Endangered Species Act, we must say that it has a very heavy biocentric component. It's like the argument that there's no such thing as altruistic behavior. If you do something altruistic, it's because it gives you warm fuzzies, and if it doesn't give you warm fuzzies it's because you think you are going to Heaven if you do it, therefore, there is no such thing as altruistic behavior. Well, that is an unanswerable argument because it revolves on internal motivation. One says Yes, I am altruistic, and someone says that can't possibly be, and then you want to throttle him and say, Who do you think you are to tell me what my motives are? So you look at something like the Endangered Species Act, which is a human artifact, and it's really hard to think that the United States Congress is going to engage in a public-spirited act that's going to get 98 percent of them defeated by their challengers next time around, but that might be the case.

Taylor: I don't think fighting over the biocentrism or anthropocentrism of the Act is the key issue. I think the Endangered Species Act clearly has certain values that cannot be reduced to economic utility. I think the legislative history of America is full of nonutilitarian values. We banned child labor, for heaven's sake, while economic efficiency requires it. Mark Sagoff makes this point in his book, that more often than not, economic utility does win, but there are certainly traditions to draw on that are nonutilitarian that also encompass moral values that respect the dignity of other living things or nonutilitarian human goods. I just don't know how far it gets us to say, Well, are those biocentric, utilitarian, or anthropocentric? I think the question is: How broad and how deep are the moral traditions that we draw on in our public life? And they are broad and deep, I think.

Bowersox: To answer that, I think we could debate about the origin of the Endangered Species Act. There was a lot of debate on the floor of Congress signifying that it was both biocentric in some ways but also very anthropocentric about saving all the species. If you look at the way it is being applied, particularly, I would argue, since 1991 with the Northern Spotted Owl, critical habitat designations have become more important—a thing that was previously left out in our exclusive focus on species extinction. At minimum, we're starting to get surrogates for biocentric thinking . . . I think it's been coming in from the Northwest, where we've now moved beyond forests to fish. The recent listing of various steelhead and salmon runs are moving the debate from public lands to what the effect is going to be on private urban communities. This is really getting us to question what we're doing with the structure of our societies and what we're doing when we're building neighborhoods. This is going to raise a lot of questions about the relationship between private land ownership and the Endangered Species protection. It's going to make the forest experience look like a cakewalk. Questions are going to come up: How far do we sacrifice economic efficiency to protect a species that we like to eat but that also represents our culture?

Audience: I have a story and a question about the difference between anthropocentric and biocentric. In the Philippines, the desire to keep the rain forest viable was an issue in the Marcos administration, a keynote one that took on added importance after 1992 and the Earth Summit, when the University of the Philippines was assigned the task of getting the indigenous people, who had lived as long as they could remember in that context, to move out of the rain forest because biocentric policy wanted humans out of the picture. Then a group of younger scientists, social and biological, persuaded the university that these people were actually a part of the rain forest, not *apart from* the rain forest. So which policy is anthropocentric and which one is biocentric, when native communities are part of the picture?

Buck: I bet it depends on how technologically sophisticated those people get. As soon as they start using internal combustion engines and gasoline-powered harpoons, or whatever they fish with, they're suddenly going to get labeled as artificial perturbations in nature. The label is going to change.

Norton: But I still don't think we have a clear definition of biocentrism, and this example points it up as clearly as possible. Do you mean to say that they're nonhuman until they become technological?

Buck: No, I'm just saying that if there had been a technologically advanced group of indigenous people nobody would have been making the argument that they were part of the ecosystem. There has been international law on this subject. As part of the law of the sea negotiations it is argued that certain inland and inshore fishermen ought to be protected from international fishing regulation because they are indigenous fishermen and part of the ecosystem.

Gillroy: Okay! Maybe we need to confuse the terminology a little more.

Taylor: I don't think it can be more confusing.

Gillroy: Maybe the policy question is whether the human dimension of the human/nature interface harms its natural functioning or not. Are humans conducive to mutual support of whole natural systems or not? In that sense, I think the policy questions start to take over the science. Do we have alternatives to leveling all the old forests? Can we establish critical habitat for certain classes of species to put us more in harmony with their long-term survival? There used to be a difference between eco- and biocentrism, and maybe the question you have to ask is what the effect is of having humans in natural systems. In the Philippine example, they are conducive to the continuation of that natural system; therefore, they should be left there as they are part and parcel of it, they are in harmony with it, they are evolving with it. Whereas, if they did, for instance, call Shell Oil and say, Could you come in here and flatten some trees for us and drill some oil wells in this wetland?, then they may no longer be considered in cooperation with "their" natural system and are then subject to regulation like the rest of us. I think that's a more important policy question, and also a more important moral question.

Norton: Those questions can be completely debated without bringing in biocentrism.

Gillroy: But I think they cannot be discussed without moving past a strictly human definition of the importance of nature. In other words, if you see nature as strictly of use to us and as nothing else, then species protection has but one principle: present or future use. I'm just saying that these kinds of policy questions require us to look at both kinds of consideration: economic utility and other things. Most people consider the movement toward a consideration of nature in noninstrumental economic terms as moving toward some kind of eco- or biocentric policy that will ruin us economically. That may not be true, Bryan, but I'm just telling you what the popular perception is, which I think was what Susan was talking about a few minutes ago. Whether or not nature is of utility to us is the essential point. Do we just see nature as part and parcel of some kind of giant materials balance, or is it important in other, more fundamental ways?

Kassiola: I agree! The critical point is not the terminology but the implicit assumption, within anthropocentrism, that humans are superior, and the inherent equality of humanity and nature in eco/biocentrism. How you start out is important: it determines the final policy. I think there's more to the distinction and it should be looked at very practically. If one point of view helps advance a particular problem, then use it, and if it doesn't, or if it confounds it, then you should get rid of it. I think it depends on what your problem is, whether the anthropocentric or the biocentric distinction helps make policy.

Norton: Well, it may help, but the point is that it doesn't draw the distinction that John was just suggesting, because all you have to do is look at Genesis, where it says that humans were given dominion. Okay . . . There's good dominion and there's bad dominion.

Kassiola: But "dominion" in itself is a very problematic concept.

Buck: And dominion came just before expulsion!

Norton: The obligation comes in chapter 2 where man was put in the garden to dress and to keep it. So the stewardship tradition does not conflict with dominion in the technical sense at least. Now we have bad interpretations of dominion that conflict with the stewardship obligation. The point is that within an anthropocentric tradition that gives way to stewardship, you do have the same kinds of obligations that people are talking about, so biocentrism versus anthropocentrism does not get at the key point. Just as it does not get at the key point about whether the natives in the rain forest are human or not. They are human and they are dressing and keeping it, therefore they are not doing anything wrong.

Taylor: There's another way that science has played a role in the growth of all these debates. It's true that scientists were at the forefront of bringing all the environmental problems to light. Many of these scientists were also responsible for discrediting science in important ways: using it badly and therefore discrediting environmental arguments. When people should have been making moral arguments they instead were trying to make scientific arguments so that they wouldn't have to make the moral argument.

Buck: But scientific arguments, in our culture, have more legitimacy, whether they are accurate or not . . .

Kassiola: . . . and moral arguments should, but do not!

Taylor: But science does not always want to have a democratic argument with the people. Scientists want to put an end to public argument.

Paehlke: They want to play the scientific trump card and stop the debate.

Buck: And they have that power. . . . Science is factual, science is true. If I can convince you that I have the science, I win!

Gillroy: Really?

Issue 2: Environmental Policy, Sustainability, and Social Justice

Why Environmental Public Policy Analysis Must Include Explicit Normative Considerations: Reflections on Seven Illustrations

Joel J. Kassiola

The conclusion to my earlier contribution to this volume "Why Environmental Thought and Action Must Include Considerations of Social Justice," is a good source, I think, to introduce reflections on the seven public policy essays included herein. I wrote: "A morally sound environmental political theory, an effective environmental social change movement, and the implementation of a desirable environmental public policy can be achieved only with careful and thorough consideration of social justice." In reading the case studies, I was struck by the uneven recognition of the importance and specific discussion of normative considerations—both value and obligation issues, in general, as well as social justice, in particular—in the authors' various policy foci and analyses.

Normative discourse—clarifying, judging, and defending prescribed social values and obligations regarding the environment—is totally absent in the Campbell-Mohn, Lowry, and Rabe essays. Schwab and Brower, in contrast, assert: "Sustainable development may be seen as a moral imperative that we must pursue more thoughtful ways, must change our values and assumptions, must consider beyond the here-and-now." To be comprehensive, Laitos does write about the social value of public lands in addition to the competing commodity use versus recreational and preservationist values embedded in different public land management policies. Sustainable development, according to Laitos, contains the value of social equity. However, at the end of the discussion, the author seems to raise skeptical doubts and questions about this purported value *but does not address them:* "Although the idea of social equity seems a desirable goal in the abstract, it may be quite difficult to define it, make decisions based on it, and then measure whether it has been achieved." Clearly, if the important normative concept of sustainable development based on the value of social equity is going to be a credible rival, or even "a useful adjunct" to the dominant economic value of efficiency, such fundamental issues as these must be specifically raised and satisfactorily analyzed and, certainly, at a minimum, discussed; alas, neither occurs in this essay.

In his case study on global environmental accountability, Robert Percival continues the discussion of the popular concept of sustainable development, with the former concept claimed as the missing link in the pursuit of sustainable development. He argues that "sustainable development, *regardless of how it is defined* [emphasis mine], can be advanced by designing policies that increase accountability of individuals, government, and corporations for the environmental consequences of their actions." However, mirroring the trend of these authors' omitting normative considerations, Percival's focus on accountability is only aimed at, according to the author, "better information about and more realistic forecasts of" environmental consequences, *not* normative questions at all.

Percival's discussion attempts to explicate the concept of sustainable development, but, like most other users of this now central concept in environmental policy, he does not recognize the Brundtland Commission's famous definition of sustainable development as meeting "the needs of the present without compromising the ability of future generations to meet their own needs." This definition of the concept emphasizing the needs of future generations requires value analyses and discussion of present priorities when limits abound and sacrifices must be made. Therefore, he fails to recognize that this central concept in contemporary environmental discourse essentially functions as a normative concept.

Typical of this normative myopia is Percival's reference to the U.S. President's Council on Sustainable Development, which held that "economic growth, environmental protection and social equity should be interdependent, mutually reinforcing goals." Anyone who has studied these goals—or values—knows that they are contradictory and competitive, making their mutual reinforcement impossible. A normative analysis of these social goals would easily demonstrate this. Regrettably, Percival's remaining discussion about various legal mechanisms (liability, regulation, and tradeable emission allowances) of greater environmental accountability omits any reference to any of these three values, let alone discussion of how their "mutual reinforcement" would be achieved. So, once again, we have a normative conclusion that is not pursued in the case studies.

Jonathan Baert Wiener makes the claim in his case study that government can be part of the problem regarding environmental protection and regulation, as well as part of the solution. Therein, he emphasizes the concept of "sustainable governance" regarding the state's activities and their impacts on the environment. Yet, once again, values are not explicitly mentioned until the last paragraph of the essay, where the author says: "Sustainable development must take into account the social and environmental value of the services of all social institutions, including the state." Throughout his argument, Wiener uses a medical analogy between the state and the medical doctor: both are imperfect, such that each "can injure as well as nurture." This occurs for the doctor when the medicine is either too weak or too strong. Wiener perceptively adds: "We need not just less or more medicine, but the right medicine for the right ailment at the right time." What Wiener overlooks in his health analogy is that good health is normally an unchallenged value. It is just the opposite when we discuss the "health" of a society. What precisely the health of a society comprises requires normative inquiry and defense. Expanding this issue to planetary level is the author's final recommendation: "Because our goal is not just medicine for its own sake, but a healthy planet, we must be informed consumers of state therapy." This merely seems to expand the necessity to pursue normative questions about what a "healthy planet" may be like, or the nature of "state therapy" to treat what ails Earth. Alas, support for the case study's main prescription—in this instance, the concept of sustainable governance—is, once again, flawed by the lack of normative sensitivity and inquiry.

I have argued that environmentalism (within which I include environmental public policy and analyses) must include social justice. In this current work, I approach the seven case studies with the following essential questions in mind: How do the adduced facts by the authors about making and implementing public policies in the selected environmental issue areas produce a normatively desirable or improved society? How are the values and obligations of humanity advanced and met by the policies that the authors propose and analyze?

An abstract discussion of the necessity for normative considerations and discourse in environmental thought, action, public policymaking, and implementation can and should be applied to actual environmental problems and policies in order to resolve or mitigate them. This must be done to demonstrate how the importance of normative considerations regarding environmentalism are useful in making tough environmental policy decisions and implementing them in the rough-and-tumble world of a special interest–dominated, liberal political system within modern industrial, capitalist cultures, such as the United States, with our consumerist and economistic values.[1]

I shall restrict the following remarks, because of space limitations, to comments on several of the case studies' treatments of normative issues and their impact on the arguments' overall rational quality.

Jan Laitos's case study addresses the important environmental issue of how U.S. public lands should be used, or, which public land management policy should be adopted: maximum commodity extraction of natural resources, or the nonconsumptive use of recreation and preservation of the natural environment? The author attempts to apply the criterion of "social value" to assess each of the examined public land management perspectives. These are actually competing value positions, as Laitos appears to recognize early in the case study; curiously, he does not return to this critical observation during the remaining portion of the study.

Laitos *describes* different values purportedly held by the American people, but does not engage in any value analysis, judgment, or defense of these American social values. In short, he overlooks normative discourse. The author, therefore, talks *about* values but does not conduct normative argumentation, that is, provide an assessment of the comparative merit of selected values by judging one value superior and defending this judgment with reasoned evidence. Instead, most of the essay is an empirical statement of the claimed social benefits and economic consequences of the three competing American land management approaches identified earlier.

Significantly, the author omits the unequal distribution consequences of commodity extraction by very large private corporations with great internal income disparities because of the hierarchical corporate structure between upper management and lower-level workers. In addition, the unequal distribution of the profits earned by the corporation through these extractive activities, with the lion's share going to the relatively few large shareholders of the corporation, is not discussed. Furthermore, the devastating environmental impact of such extractive activities as mining or logging is glaringly omitted.

The social value of recreation at the heart of the recreational policy for public land management is also unequally distributed. Not all Americans are able to purchase the means to enjoy the recreational activities on public lands to go camping, fishing, hunting, or biking, or to absorb the costs for horseback riding or equipment rental. In addition, the leisure time required for such recreation and the costs for travel to the recreational site are available only to upper-middle-class and upper-class Americans (see Laitos, Table 1).

Preservation value orientation and policy yield economic benefits as well (see Laitos, Table 2), yet the author does not consider the inherent value of such ecosystem components as scenic views, waterfalls, and wildlife and plants. An ecocentric view may be controversial, but a normatively sensitive analysis would at least raise this issue to achieve a full presentation of the noneconomic perspective on the policy question of how we value our public lands. A more productive discussion, I believe, would prescribe one of the policy orientations and its accompanying values as normatively superior—in this case, morally superior—to the others.

To his credit, Laitos does mention the value of intergenerational equity in the brief section on sustainable development, but, as quoted earlier, he seems to undermine the value of specifically addressing a normative concept by noting the definitional and measurement problems of the value of social equity. The discussion ends with the author's observing the "legal vagueness of the concept of sustainable development," yet Laitos does little to reduce or even analyze this imprecision within normative discourse. Public policy analyses as well as the formation of actual public policy must cope with the definition and measurement of human values and discourse. Such values as equity, justice, and human rights are contested in their meanings and measurement, but this should not mean that we abandon such key normative terms in policy deliberations and analyses.

When we reflect on the proper or desirable use of public lands, it is really an allocative normative question. Thereby, we are raising a question of how the immense value potential inherent in the (public) land is to be fairly distributed: as a source of natural commodities for human use; as a venue for human recreational activities and enjoyment; or as a source of aesthetic and spiritual enlightenment by being left alone and free of human use except contemplation.

Asking what type of social order will be produced from each public land management policy *and* which one is normatively superior (and why) is my prescription of how to integrate normative considerations into Laitos's policy discussion of the most desirable public land management policy. Laitos does recognize, at the end of his essay, the need to supplement the criterion of economic efficiency with noneconomic values by "a balancing of economic and environmental and social equity interests in light of intergenerational sustainability." But I pose the question: How and why, from a normative perspective, specifically morality, should this "balancing" occur? This normative heart of

Laitos's inquiry—the choice of criterion (or criteria) on which to base our public land management policy: economic efficiency, environmental sustainability, intergenerational equity, or other values—is fundamentally a value choice that requires explicit consideration and reasoned defense. Failure to address this decisive issue for public policy and its analysis is a damaging omission, in my judgment, weakening both the usefulness and persuasiveness of Laitos's contribution.

Among the case studies under review, the policy analysis by Schwab and Brower incorporates normative analysis most incisively and usefully in its discussion of sustainable development and natural hazards mitigation. In this respect, it stands alone among the seven case studies contained in this volume.

Schwab and Brower start off by raising the important and inescapable normative question for environmental inquiry: Is there "a duty imposed on all living things to live within the carrying capacity of their environment," as many analysts claim? The authors admit the social impact of the moral doctrine of anthropocentrism within the dominant social paradigm of industrialism. They point out that by focusing, as we normally do, on geophysical processes and natural cyclical events and calling them "hazards," producing periodic "disasters" for humans and their settlements (e.g., floods), adversely affecting human lives and property, we are distracted from examining our own behavior and its consequences for the environment. I would add that such a necessary examination of human behavior must include the values of society on which that behavior is based.

The authors are to be commended for acknowledging the important normative component of sustainable development: "Sustainable development may be seen as a moral imperative that we must pursue more thoughtful ways, must change our values and assumptions, must consider beyond the here-and-now." All too often, this now fashionable concept of sustainable development is used by policy analysts omitting entirely this moral and transformational element articulated by Schwab and Brower. If it is mentioned at all, it is not adequately developed or applied (as in the Campbell-Mohn and Rabe case studies).

My objection to the Schwab and Brower study occurs where they declare that sustainable development advocates "recognize that our economic structure and the natural environment are not necessarily in conflict." My own research on and thought about the environment and values have led me to the conclusion that the modern industrialist capitalist economic structure is, indeed, predicated on a denial of limits—as all modernity is. The environmental crisis and the accompanying morally embedded, transformational concepts such as sustainable development present our social order, as globally hegemonic as it currently is, with its most severe challenge. This is so, I maintain, because the environmental crisis highlights the conflict between the limit-denying modern civilization with its supreme value of endless economic growth and the unavoidable environmental limits, made clear and distinct by environmental crises (e.g., global warming and famine) that constitute a fatal threat to the social order that denies limits to human desires. The current environmental crisis is the result, despite attempts at ideological disguise and distraction by modern social institutions such as the mass entertainment media. Once the value bases of industrial civilization are made clear, the necessary conflict between the capitalist endless-growth economic structure and the finite environment is indeed demonstrated, *pace* Schwab and Brower.

Once this important point about the inevitable conflict between the modern economic structure and the natural environment is granted, it supports the authors' own paradoxical claim (given their denial of this necessary conflict) that "sustainable development implies a change in values and speaks in terms of *needs,* not desires" (emphasis in original). Their conclusion here fits the conflict view better than their own: If no conflict between the economic structure and the environment existed, why then seek a change in values and the additional important shift from infinite desires to limited human biophysical needs?

Moreover, a change of values is called for, as I see it, by the real utopians in this debate, not the critics of the dominant growthmaniacal economic ideology but the ruling advocates of such ideology, who must deny the existence of limits, a proposition that is required by the continuous, limitless economic growth doctrine on which the current global economic structure is founded. The concept of human needs is centered on the ideal of inherent limits that make them capable of being satisfied, in marked contrast to the industrial society's endlessly created desires through corporate-dominated mass media and the mass marketing of new products. This economy can create only more unmet desires. Such an economy and social order, predicated on continuous dissatisfaction and endless consumption, depletes resources and increases pollution such that it can only be in conflict *necessarily* with the finite or limited environment.[2]

I agree with the authors' profound assertion: "We must cease to view ourselves as merely consumers of the world's goods; instead, we must recognize our role as stewards of the planet." However, this important claim is only rationally persuasive if the essential value differences between endless consumption and stewardship of natural goods are discussed and adequately defended. One basic difference between such consumption and stewardship in today's hegemonic consumer capitalism is that limitless consumption is the essence of our identity and value structure, the ultimate end toward which all other values are means, whereas stewardship is the recognition of the inherent value of natural goods, a biocentric or ecocentric viewpoint that recognizes the need to limit the human use of these goods.[3] In addition, the emphasis on the primary significance of limits reinforces the authors' point about the central value of intergenerational equity included in the concept of sustainable development. The acceptance of later generations' rightful needs places limits on human actions today, such as consumption, in order to preserve natural goods for the humans and other living creatures that will follow us on the planet.

Schwab and Brower are to be commended for recognizing and discussing our moral obligation to future generations through, among other values and actions, natural hazard mitigation. However, it is not sufficient merely to announce that there are competing values in environmental public policy, as the authors do here; they must show why public health and safety eclipse or override protection of property as a determinant of policy. Normative analyses and arguments are required to defend the putative moral obligation prescribed by Schwab and Brower: nonimmediate health and safety over protection of property.

The authors' admirable prescription of a "wider moral community" of neighborhoods, towns, states, regions, and, possibly, the entire planet must be further delineated, clarified, and defended to be rationally useful. "Our 'neighbors' are citizens of the world, some of whom are not yet born," is both true and, unfortunately, quite controversial. Therefore, it requires discussion and defense of the practical policy implications that acceptance of this proposition would produce.

Schwab and Brower deserve praise for a normatively and morally sensitive essay containing several profound principles and claims. Their argument would be greatly enhanced, however, by further discussion and reasoned argumentation in support of their views.

The words "values" and "value judgments" occur on the first page of Celia Campbell-Mohn's case study, and, as far as I can discern, are never mentioned again. Although the author appears to recognize that the concept of sustainable development includes risk aversity and intergenerational equity, no delineation of these key values or the role they should play in environmental policymaking or analysis is provided by the author.

Clearly, the concept of private property and state regulation of the human use of such property by its owners will necessarily involve values and value judgments. What these values are will determine the purposes of public policy and the appropriate means to achieve them. I realize that the author is concentrating on environmental law and legal decisions in this case study; nevertheless, a value analysis of the nature of private property within society could inspire an *explicit* normative theoretical perspective to be used by the courts in their adjudication of conflicts involving property.

The law, as we know, is a tool to achieve and maintain social order. For that social order to be morally sound, laws that regulate human behavior must themselves be morally informed. I respectfully recommend that legal analyses like Campbell-Mohn's include a clear statement of the pertinent social values—either implicit in past relevant legal decisions or in the author's own prescription—and then examine how the courts' current decisions advance or detract from these values. Furthermore, such a normative approach can show how these values are "implemented," or realized by public policy, public administration, and court decisions beyond "planning and science." More than planning and science are needed to analyze and defend the vital normative elements of the concept of sustainable development and, more important, the creation of a social order that is compatible with this social value.

Although William Lowry's title does not contain the term sustainable development, the author appears to want to examine the sustainable development concept and value applied to the selected policy issue area of national park preservation in the United States and Canada. (Perhaps the word "preservation" in the title is the clue for this intention of the author?) Nonetheless, we are faced, once again, with a policy analyst who offers no explicit value analysis, although the policy goal and value of preservation is emphasized and contrasted with the human resource use perspective. In my view, Lowry begins the essay correctly by telling the reader, "National parks are political entities. They are created through political processes and managed by political agencies." What is missing is an appreciation of normative issues and thinking in public life and policymaking.

The author establishes a value conflict between the anthropocentrism of the human use approach versus preservation, which is claimed to involve "the natural needs of the environment." This value conflict between human use anthropocentrism and biocentric or ecocentric preservationism should be specifically addressed and discussed thoroughly, given its central importance to the author's theme in this case study: management policy for national parks in the United States and Canada. This issue turns on the management goal or ultimate value in creating and operating national parks. At bottom, the subject of this study, the conflict between commodity use of natural resources found in national parks and preserving and not consuming these park resources, is really a value conflict that drives policy debates and should inform the behavior of public agencies charged with overseeing the parks. This value conflict should be addressed, judged, and defended by the author. Once this fundamental issue of national park policy is decided, the political institutional implementation of the selected value choices can be effectively addressed. Not to do so seems like putting the cart before the horse!

Regulatory fragmentation in the United States is the theme for the study by Barry G. Rabe. This topic essentially concerns how policy goals—social values such as good environmental policy—can be implemented in a multijurisdictional American political system characterized by federalism and intra-agency and interagency conflict. Also, the multidimensional nature of pollution caused by the existence of many toxins conveyed by various media, including water, air, and soil, produces daunting challenges to the achievement of effective environmental policy. How can environmental public policy goals be implemented by bureaucratic governmental agencies if the latter are unable to organize themselves and deliver the needed services to accomplish such goals and the social values embedded within them?

Facilitywide state permitting seemed to make significant ecological improvement because it enabled inspection agencies to identify emissions previously undetected by fragmented and disjointed inspections and enforcement by environmental agencies. Environmental protection was enhanced, according to the author, as a result of the technique of comprehensive facilitywide permitting and inspections. Yet, the values underlying this environmentally superior management approach were not discussed, nor were the values that underlie the highlighted obstacles to this ecologically improved system of environmental protection (called "political impediments" by the author). What were the values involved in the legislative challenges? In the "loss of jobs" or "personnel cuts"? Or, finally, in the "very nasty politics" of agency reorganization?

In my earlier theoretical essay, I attempted to show why social justice is essential to environmental thinking and action. The discussion was necessarily abstract and theoretical in nature, given its theme. In this essay, I have tried to apply my theoretical position about the importance of normative considerations to the illustrations of environmental public policy in our seven case studies. The goal of this discussion is to argue persuasively that public policy analysts must thoroughly address normative—especially moral—values and issues in their work, especially if they rely on the heavily normative environmental concept of sustainable development, as do all seven.

Unfortunately, only one case study even comes close to this aim (Schwab and Brower). The others are weakened by failing to provide value-oriented analysis, as I discussed above. I strongly urge students of environmental public policy to include, along with scientific political descriptions of nature and public agencies, careful treatment of the role of values in the creation of environmental problems (as part of the modern industrial consumerist ideology), as well as their role in the transformation of individual behavior and social institutions if we are to achieve a sustainable and just global social order. I appreciate that environmental and political scientists may feel uncomfortable and unprepared to include such value-based discussion in their works, given their scientific training, preoccupation with factual inquiry, and the skeptical view toward normative discourse that is pervasive in contemporary culture. This very understanding of the barriers to normative environmental inquiry serves to accentuate the need for multidisciplinary and interdisciplinary study of the environment necessitated by the multidimensional nature of this subject matter that includes, I contend, not only scientific elements but inescapable normative ones as well. As I have tried to show, we cannot talk incisively about the environmental crisis unless we focus on the normative foundation of modernity and industrialism that provides the values on which our ecologically damaging behavior and institutions are based. Furthermore, we cannot prescribe the need for societal transformation of our current social practices and institutions without engaging in normative assessment, prescription, and defense, especially when we prescribe alternatives to the status quo.

When I was in high school, I heard a teacher say: "No matter what subject we teach, we are all English teachers" (a view seldom heard today, with perhaps some causal connection to the declining reading and writing skills of current students). To reinterpret this remark about teaching, I would like to conclude by asserting: No matter what our special field of expertise concerning the environment, all policy analysts of this wide-ranging subject must practice normative discourse. That is, all must study and use relevant social values and obligations in their work. No one would claim that researchers can be ignorant of environmental facts and processes and still provide plausible and useful work. Similarly, policy analysts, indeed, all students of the environment, must not be ignorant of the omnipresent and pervasive role of values and obligations in environmental problems and policies.

The now fashionable concept of sustainable development, as I understand the term (following Schwab and Brower's explication), is significant because of its emphasis on normative considerations. We have seen, in the seven case studies analyzed, that this term is not always used with such normative intent. Environmental scientists, political scientists, and public policy analysts, please take note!

Sustainability and Environmental Justice: A Necessary Connection

Joe Bowersox

In Part 1 I addressed in a very theoretical and rather abstract manner the dilemma of both ethicists and policy decision makers when confronting questions of environmental protection and environmental justice. In summary, I suggested that, given our democratic culture and republican institutions, questions of equity usually lurk beneath policy issues as we seek to maximize voluntary compliance and minimize coercion. Rather than a peripheral and messy issue best avoided, questions of fairness and equity must be central to any policy deliberations, most necessarily environmental policy decisions, which often involve contentious redistributive and regulatory dimensions. I then suggested a deliberative and discursive framework that I believe would sufficiently and successfully incorporate disparate voices and their equity claims in a given policy question.

In this essay I take the principles examined and constructed in my previous piece and apply them to an analysis of the case studies presented in Part 2. Clearly, uncertainty, inconsistencies, and physical, economic, and political constraints normally intervene in the formation of public policies, making them digress from the realm of "oughts" that we normally associate with political theory. Nevertheless, normative theory itself can be a powerful diagnostic tool for evaluating public policy; it can act as an altimeter, if you will, that checks the thinness of the theoretical air surrounding a particular policy mountain or molehill, telling us whether there is enough oxygen present to support political life. Sometimes theory may even suggest that we try an alternative, less precipitous policy pass by which to cross to the other side. If political theory cannot provide such service to public policy, a strong argument can be made that it is indeed irrelevant for modern political life: a mere parlor game for cognoscenti. I refuse to endorse that view.

The case studies of Part 2 demonstrate my point: each in turn evokes some conception of sustainability from which it derives particular maxims or suggestions regarding the "proper" shape of public policies. Yet each in turn has very little to say regarding what sustainability or sustainable development means, and say even less regarding potential normative or contextual arguments against such conceptions. Hence, some theoretical shovel work must be done to expose the foundations of these policy edifices.

Sustainability: Concept and Reference

Before even considering the relevance of questions of justice to the instances and images of sustainability noted in the cases of Part 2, it is worth noting the various definitions of sustainability evident in some of the pieces, though, as Wiener and Percival point out, the concept itself is ambiguous. Nevertheless, I think it is necessary to explore what it means to different people, because, as I suggest in Part 1, behind such ambiguity and diversity of definition may lie not only some tentative areas of agreement, but also a hint of a definite reference for the term. Schwab and Brower use the Brundtland Report's definition: "development that meets the needs of the present without compromising the ability of future generations to meet their own needs" (World Commission on Environment and Development 1987, 8). Campbell-Mohn, on the other hand, defines sustainability as managing "natural systems for the perpetuation of the human species now and in the future," and Laitos suggests this "term of art" requires that "resource decisions . . . be made with a planning horizon that does not sacrifice the future for the present, and . . . defines broadly the beneficiaries of resource policy so as to include a wide range of economic, social, and racial classes." Percival makes a strong argument that within the term sustainable development there is an implicit requirement of accountability. Clearly, even among the small group of writers of Part 2, sustainability means many things to many people, which indeed raises the question whether the term itself has lost most of its relevance in the years after the Brundtland Report.

In addition to the disparity of meaning evident among the authors of Part 2, another vivid example of the tenuousness of the concept's utility is the extremely divergent accounts of sustainability provided by noted ecological economist Herman Daly (1993; Daly and Cobb 1989) and Nobel laureate and economist Robert Solow (1993). Solow defines sustainability as "an obligation to conduct ourselves so that we leave to the future the option or the capacity to be as well off as we are" (181). Because "being well off" is a subjective description, Solow argues that if we leave behind a museum rather than a forest, having invested the profits of the harvesting of the wood in great works of art, that would be okay: one has provided future generations with something of equal value (181, 184–85). Daly, on the other hand, suggests something radically different: in "steady state economics, the economy is one open subsystem in a finite, non-growing and materially closed ecosystem. . . . A steady state economy is one whose throughput remains constant at a level that neither depletes the environment beyond its regenerative capacity nor pollutes it beyond its absorptive capacity" (1993, 56). For both of them, sustainability is an extremely meaningful term—as long as it is defined and utilized according to their individual set of assumptions. As has been clearly demonstrated over the years, neither is willing to recognize the premises or the validity (let alone the legitimacy) of the other's definition. Wiener in turn suggests that most advocates of sustainability tend to target only the private sector, ignoring the public sector due to their own ideological biases.

One would think that as one moved from the apparently subjective realm of economic valuation to a more "biologically" determined conception of sustainability, the room for disagreement and ambiguity might be eliminated. Yet, as Greber and Johnson (1991) noted some years ago, even in the very specific setting of "scientific forestry" there are at least eight different working definitions of sustainability in use by forest biologists, public and private forest managers, and resource economists. Indeed, their definitions of sustainability seem to be driven less by objective measures than by their presumptions of a forest's "primary" function.

Clearly, many might argue that what sustainability means may be dependent on one's political and ethical commitments, not some scientific standard. Thus, one might question whether it is of any value at all to link such an ambiguous concept as sustainability to yet another ambiguous concept such as justice. But, as I suggest later in this essay, I think that our authors *do* point to some basic tenets of what sustainability might look like. Furthermore, if words are to be of any value and not simply meaningless, one must assume that if sustainable environmental policies are possible, so too are unsustainable environmental policies. Hence, it is appropriate to consider the necessary linkage between sustainability and justice, for the same reason that, in Part 1, I suggested that any efficacious environmental policy must at least implicitly address questions of equity. If sustainability is a species of environmental policy, and if all efficacious policies must account for equity, then so too must efficacious sustainable environmental policies.

Now, that argument might be considered merely logical sleight of hand, but I think it useful to consider some historical precedence for the connection as well. In the early "conservationist" days of the modern environmental movement, it was not uncommon for reformers to speak of the conservation of natural resources *and* the conservation of human welfare and dignity in the same sentence: "The principles of conservation thus described—development, preservation, the common good—have a general application which is growing rapidly wider. . . . There is, in fact, no interest of the people to which the principles of conservation do not apply. The conservation point of view is valuable in the education of our people as well as in forestry; it applies to the body politic as well as to the earth and its minerals" (Pinchot 1967, 49–50). In fact, writing at the height of the Progressive Era, Gifford Pinchot made explicit the necessary connections between his new vision of sustainable forest practices and justice: "The conservation issue is a moral issue. When a few men get possession of one of the necessaries of life, either through ownership of a natural resource or through unfair business methods, and use that control to extort undue profits . . . they injure the average man without good reason, and they are guilty of a moral wrong" (110).

Similarly, recent literature looking at sustainability at both the local and global levels reiterates the necessity of developing policies that promote sustainable relations with each other and the environment, as if the former cannot be secured without the latter. For example, in a recent book on sustainable communities, Maser, Beaton, and Smith define sustainability as "the act of one generation saving options by passing them to the next generation, which saves options by passing them to the next, and so on. . . . As such, it will demand a shift in personal consciousness: from being self-centered to being other-centered" (1998, xx–xxi). These authors seem to echo the central theme of the Brundtland Report: given the reality of global interdependence, the interrelationship of ecosystems and biomes, and the transboundary nature of both human and environmental catastrophes, "true" sustainability must endeavor to link human justice and new relations with the Earth in ways that fundamentally challenge the economic, political, and ethical status quo. But no one seems to agree on what that really means, let alone entails.

Approximately one hundred years ago, Gotthold Frege demonstrated that we often get caught up in conceptual distinctions, linguistic hairsplitting, or logical fallacies and fail to see that sometimes we have two different names, or two different concepts that refer to the same thing: the "evening star" and the "morning star" clearly mean two different things, but in reality refer to the same thing, the *planet* Venus (1892, 100). Ultimately, I think this may be the case with sustainability and environmental justice. Although the concepts appear to mean something distinct from each other, they are both simply names for the same thing: a particular set of social and ecological relationships that promote coexistence, or at least prevent capricious exploitation. As I suggest below, this challenges not only the Solows of the world, but also the environmental justice movement in its present condition.

Sustainability, Justice, and a Challenge to Liberalism

In the case studies of Part 2 sustainability is desirable precisely because it evokes concerns for equity that strict efficiency or cost-benefit analysis does not evoke (see Laitos, above). At least two dimensions of equity seem to be present in these authors' visions of sustainability. The most noted one is, of course, intergenerational equity: perhaps the one concept present in sustainability for environmentally oriented as well as the most economistic of writers. If there is any basic agreement on the term, it is something along the lines that a given generation must not impinge upon the capacity of future generations to make their own decisions regarding the utilization of resources for their own happiness and fulfillment. But for Schwab and Brower, Campbell-Mohn, and Lowry, sustainability also evokes concerns for spatial equity: for these authors, there seems to be something in the concept of sustainability that requires something more than the *sic utere tuo ut alienum non laedas* of traditional liberalism: people and their properties are not as unencumbered as we have often assumed, or at least legally recognized (see Campbell-Mohn, Percival, above). The property owner that bulldozes a sand dune on his or her lot, or the mill owner that buys a timber sale on public lands, is in a very real way potentially endangering the private property of a neighbor or inhibiting necessary and valuable ecosystem services. The corporation that sprays its Central American plantations with a harmful pesticide endangers the unborn children of banana harvesters. Such instances suggest that, though sustainability may be a term now bandied about by just about everyone, certain basic components of sustainability posit a serious challenge to liberal conceptions of society, politics, and the rights of the individual or corporation. However, the language of sustainability often employs the hallmark of liberalism: the right of the individual to self-determination and fulfillment without unjustified encroachment or harm by others.

This liberal appeal of sustainability is seen in the first aspect of basic agreement among users of the term. Even for Solow, taking the idea of autonomy seriously means that the present generation has a duty not to constrict or constrain without due justification the capacity of future generations to make their own decisions; destroying the assimilative capacity of a stream or logging the last rain forest without providing at least something in return of equal value and utility is unjust. It is akin to trespass: you have denied the right of a person to the use and enjoyment of his or her property and life. Yet in this case, the person is removed from you in time, not just space. Intergenerational equity seems a logical extension of the liberal notion of negative freedom.

But consider the consequences of the second tenet of sustainability visible in many of these papers: some sort of spatial equity that suggests individuals do not simply have to refrain from "trespassing" on the private rights of another. Rather, individuals may have *positive duties* to act in such a way—even on their own property—as to assure that others' interests or rights in a stable beach or adequate wetlands are not harmed. *Lucas v. South Carolina Coastal Commission* (1992) and *Babbitt v. Sweethome* (1995) demonstrate how contentious sustainability can be in the liberal paradigm: Are we preventing harm or constructing public goods? Are we simply recognizing the rights of others or are we constructing new duties toward others quite foreign to our liberal way of thinking? Are we ready for such positive duties? If Solow and Daly cannot agree on what intergenerational equity might mean, is it likely that we will find agreement on positive duties that so upset liberal political and economic arrangements? Perhaps such duties need further discussion and justification rather than simple assertion.

There are other images of sustainability present in these policy papers that suggest further challenges to liberalism. Consider that Schwab and Brower seem to suggest that one of the major benefits of promoting sustainable development that prevents human-caused hazards is class equity: communities of lower socioeconomic standing are often the least likely to have utilized traditional mitigation techniques and often are the first to suffer the consequences of human-caused disasters. For instance, the African American community without adequate revenues cannot construct a levee to divert the flooding intensified by the levee of the rich white township upstream (Schwab and Brower, above). Rather than simply being the logical outcome of certain autonomous market decisions in which group A acted prudently and group B acted imprudently, Schwab and Brower implicitly suggest that the poor or communities of color suffer more not because of what they failed to do, but because of what the dominant class *did* do. Behind the beneficient hand of the liberal market lies, in fact, the brute force of class and race power.

My point is this: behind the apparently benign concept of sustainability, which Rabe notes has been given lip service in all North American states and provinces, and which Laitos suggests can lead us to recognizing the economic efficiency of valuing ecosystem services, lie some consequences, which, if fully implemented, would severely undermine the liberal state of affairs that characterizes our economy, politics, and culture. It is, actually, a nonliberal vision that I find attractive, and that bears some resemblance to conceptual reforms envisioned by the likes of Walzer (1983) and others: the roots of a perhaps more humane, fulfilling, just, and nonliberal future can be found within the nooks and crannies of liberalism itself, as it faces new social and ecological challenges.[1] The problem is that such consequences are rarely explored, exposed, or argued, as if advocates of sustainability fear to evoke rage against their preferred vision.[2] The authors in the present volume, I would argue, do prove Frege correct: their conceptions overlap and point to a similar challenge to liberalism, yet none dare advocate it directly.

Sustainability and the Quest for Environmental Justice

Chris Foreman (1998) has argued that the environmental justice movement, spawned in the wake of a couple of federal studies done in the early 1980s and popularized by scholars and activists such as Robert Bullard and Lois Gibbs, has misled the public. He further suggests it has ultimately harmed its primary though implicit goal of empowering local communities and doing away with the increasingly illegitimate and bias authority of a faceless, hierarchal technocracy. Citing ample evidence from federal and independent sources, Foreman maintains that the environmental justice movement continues to base its arguments on correlations of discrimination drawn from error-filled research (ch. 2). Although Foreman clearly thinks this is intellectually dishonest, his main concern is that their very important goal (and one to which he is sympathetic) of a much broader social and political transformation is lost behind specious risk claims and tactics that lead simply to a glorification of NIMBY politics (ch. 6). I think that, ultimately, he is correct, and his castigation of the environmental justice movement in its present form should be heard within the sustainability community, for it applies there as well.

This does not mean that I believe environmental justice and sustainability should be dropped either as useful concepts or as political movements. In fact, just the opposite: I believe that the two concepts should be linked—maybe even fused—into a new set of social and environmental relations that promote respect for each other and the nonhuman environment in this generation as well as the next, and in my yard as well as my neighbor's. Speaking of the "rationalizing" desires of reformers in environmental policy that promote sustainability (though he does not use that term) via better accountability, risk assessment, risk reduction, and policy innovation, and the "democratizing" aspirations of environmental justice advocates, Foreman captures the promise of both "competing perspectives": "Environmental justice advocacy is a version of the democratizing critique that gives voice to a range of minority and low-income community concerns, some of which are not fundamentally environmental in the narrow sense. Problems identified by the rationalizing approaches tend to downplay or overlook the social aspirations and popular emotions that underlie the democratizing focus. The conflict between these two reform perspectives is among the most serious challenges facing environmental policy" (1998, 109).

Ultimately, effective environmental policy needs the inclusiveness and community legitimacy promoted by the environmental justice movement *and* the rationalizing practical reforms noted by the essays in Part 2. Environmental justice advocates need the likes of reforms suggested by Schwab and Brower, Laitos, Campbell-Mohn, and Percival to protect their communities from real and potential threats; questions of spatial and intergenerational equity also may redirect the environmental justice movement's tactical NIMBYism. On the other hand, sustainability advocates would benefit from more clearly recognizing the broad social implications of their proposals and find potentially useful allies to fend off the political retrenchments Rabe suggests are always possible.

Most important, I believe that the fusing of the concepts and the movements, with the concomitant explicit verbalization of the principles that lie behind environmental justice and sustainability (inclusiveness, collective legitimacy, accountability, intergenerational, spatial, and economic equity) promote and necessitate the discursive model I outlined in my first essay. As Rabe's essay suggests, the politics of sustainability presently unfolding in the city councils and state legislature of Minnesota resemble that model of openness, value debate, innovation, and concern for legitimacy I advocated. The practice might clearly diverge from the abstract theory, but good theory is at best a roadmap, never as exciting or trying as the journey itself. Throughout North America and the world policymakers and citizens are developing new forums of policy debate, formation, and implementation that bear little resemblance to the traditional models of Dye or Truman; watershed councils, stakeholder negotiations, and consensus-based alternative dispute resolution are emerging as new, effective, and legitimate vehicles for policy change. There is no guarantee that such forums will lead to the vision evoked by sustainability and environmental justice, yet I believe these forums offer a greater opportunity for moving beyond the liberal paradigm (while recognizing its useful legacy) than was imaginable in an earlier era of politics when economic interests (labor and capital) tended to dominate and benefit from the familiar structures of traditional pluralist politics.

Discussion

Audience: Joel used the word "must" quite a lot. "Must" is a very strong word, with positive and negative dimensions. If I want to avoid getting hit by a car, I must look both ways before I cross the street. If I want to win a sweepstakes, I must buy a ticket. But why must I worry about sustainable development? I have no children, so I don't have to worry about future generations, and so why must I worry about the fact that I'm driving a sport utility wagon that wastes resources? If you want to get to the normative side of all these issues, would you consider substituting the word "should" for "must"?

Kassiola: Certainly! The "must" follows from the conclusion that, in my opinion, if one wants to do policy analysis competently, if one wants to have a morally sound environmental policy, if one wants to have an effective social change movement, if one wants to have a world that is both environmentally sustainable and just, then one *must* consider these issues. So you are quite correct in noticing that the "must" is a predicate that follows a very important and complicated precursor. Of course, you could say, I don't want these things, therefore it's not a must, and then I would be forced to engage in a debate about why you don't want those things. The notion that I am using follows a moral obligation kind of must. It is not a commandatory must (as a dean, I never give commands). I would argue it's a moral obligation. That's where the must is derived from. We have a moral obligation to live in a way that produces environmental sustainability and social justice. By the way, my comments were less targeted to sustainable development than they were to normative discourse as a whole, and I am not necessarily endorsing any specific definition of sustainable development. It was the Schwab and Brower essay that at least recognized this, and I know that Joe Bowersox had some problems with that essay, as I did, because they didn't follow through enough, but at least they recognized that if sustainable development is going to have any significance in a normative way for the environmental movement, it has to have more than a narrow, nonnormative definition. When it gets so popular and everybody uses it, you begin to wonder if it has lost its normative bite, and that is a worry. Therefore, I'm not an apologist for sustainable development. What I try to argue is that policy case studies like the ones we are examining, if they want to look at the best use of public lands, or anything else, *must* engage in normative discourse. They did not, and that's what bothered me.

Bowersox: Can I follow up on that? I like what you're saying here, but, being an admirer of Phillipa Foot, I think the must here is a hypothetical, not a categorical, imperative. However, there is one must that I want, and I think it's the one that Joel wants here too. If you're going to have effective policy, it's imperative that the people who are designing the policies raise the value questions, or else you're going to have real problems with justification and compliance. So you at least have to have the debate, and within that debate we can have all sorts of different images, like, Let's let the lights burn because, after all, the millennium is coming and we're all going to be dead then anyhow. We need to have the essential argument.

Audience: I want to ask Joel Kassiola if values are to be dealt with in an autonomous place, all by themselves? I think even in public policy debates that can't happen, because the war raging over public policies is a war of worldviews. A Marxian worldview is radically different from a neoclassical economics worldview. So, in effect, you come down to ask which worldview represents the notion of must we wish to prosecute. Now let me illustrate this with the distinction between *internal relations* and *external relations.* What is my connection with the ladybug? Is the ladybug external to me such that if I squash her it makes no difference to me at all? Or is my identity so caught up, by virtue of the principle of internal relations, that as I kill a ladybug, I suffer? It would seem to me that if you take the *internal relations* tack, which ultimately, I think ecologists, like myself, must take, it's out of this worldview that your policy values will come. The change, therefore, must not be in norms, but initially in one's cosmology. Now that leads me to Joe Bowersox's essay and his wonderful suggestion that sustainability and environmental justice are one and the same thing. I'm not so sure—because it again depends on one's worldview. What cosmology informs one's definition of sustainability? For the classical economist it may be an accumulation of wealth, whereas to an *internal relations* ecologist it is not an accumulation of wealth but the sustaining of biological community. The same thing is true of the notion of justice. What you accept as social justice is contingent on what you take a person to be, and whether you think of justice as individual rights, community stability, or radical redistribution. Can we distinguish between justice and the common good? How do our cosmologies inform our definition of these normative terms?

Kassiola: Let me try to reply. I did not mean to imply that moral or normative judgments should be made in some autonomous way or detached from a notion of empirical reality. Clearly, environmental policymaking needs to be informed by good science, and I would certainly agree with the previous panel that science is an important element of environmental decision making. But that's agreed on, that's socially recognized and, I might add, socially supported. My view about moral argument and policy is not conventional and therefore needed emphasis in my essay. Let me go to your substantive point. I think it's an open question—about how you arrive at normative conclusions. Scientific data and scientific knowledge of the world will be important, but how you assess that knowledge is *normatively* distinct. What I took from your comment is that those normative judgments are going to be derivative from a scientific perspective. If you meant to say that, then I would say it is open to debate and controversy.

Audience: Metaphysics, not science.

Kassiola: Okay, and I wouldn't disagree with that, but what I would try to bring to the people who wrote these case studies is that they're not even thinking about these issues. They're not even addressing them and they are doing environmental policy analysis, and I suspect the policymakers themselves are engaged in this kind of unfortunately detached kind of thinking. I just want to bring them into the arena to worry about normative discourse, and if you want to add, which I did not but I would recognize as well, metaphysical concerns, then we're on the same page.

Bowersox: I think that you're right at one level, that cosmos and ethos—those two conflicting elements—have to be brought into the debate. I think that's actually what I'm advocating. We need a system where we say, Let's get it on: let's talk about how cosmos affects ethos, let's talk about what our values stem from in terms of worldviews. This is part of the problem that we ultimately have in developing effective policies. When it comes down to it, Jesse Helms doesn't see the world the same way that I do. But maybe we're so caught up in one level of disagreement that we don't see that somehow there might be certain things that we can agree on. Why can't we at least have the conversation? For example, there might be something in the various ideas of sustainability in the case studies that might come down to some fundamental, at least contingent, truth claims about the way the world works, that all can agree on. These "truths" might serve as a basis for something else. I think we get so caught up in this conflict of particular concepts, as we had earlier between anthropocentrism and biocentrism, that we fail to recognize the disutility of getting caught up in that sort of conceptual difference. We need to think, Well, how do these things play out? Are there places where anthropocentrism fades into biocentrism? Places where preservation and conservation ultimately end up being the same thing? I'm actually thinking that as we start looking at the way policies are developing, conservation and preservation start looking very similar. It makes me wonder if a lot of these disagreements are useful.

Paehlke: I would like to follow up the first question, reflecting on environmentalists pronouncing "musts." I have often done this myself. But I've come to the conclusion that what we may need to look for in terms of actual policy outcomes are middle-range tools. The "must" is a sort of persuasive device. Governments try to persuade by example, as when Jimmy Carter carried his own suitcase, that kind of thing. Or you can go to the other extreme that involves regulatory rules, that is, that all cars must get so many miles per gallon or they can't be sold or made. Clearly, that's not appropriate either. We need a middle range of economic market-based tools so you make the things that are environmentally costly relatively more expensive, then the debate focuses on How much more expensive? Should it be five times as much or 5 percent more? It's a very big range, and I think you can scientifically and reasonably make calculations about that. You can roughly calculate the cost to health of tobacco per pack. If you add up all the health claims it comes out around $5 or $6 a pack. Then there may be the public action to apply a $5 tax. That's how you arrive at the amount. I think you can do that with regard to any number of other things, and quite clearly, if you're collecting x billion dollars more there, you're going to have to subtract x billion dollars somewhere else. Maybe that's where social justice claims come in.

Bowersox: Okay, we can do that, and I think development of tax incentives and disincentives are a very viable policy tool. We are just starting to realize how beneficial it can be. But then we have to ask the fairness questions. How do we get people to not throw their arms up in disdain, when, for example, in 1992, Ross Perot actually proposed a $.50 per gallon gas tax? Some of us said, All right, this is great, but other people said, This would be unfair.

Paehlke: All taxes are at some level unfair. They need to be unpacked and normatively justified.

Kassiola: I think all policy analysts feel uncomfortable with this kind of discussion, but I think I would drive it to a deeper level with regard to what fairness means in a broad sense and not in a superficial distributive sense. Should different income groups pay the same proportion or the same absolute amount of tax? That's the level that I would like to see explored by policymakers themselves. At least, the analysis must be done so we can have an informed public and policymakers sensitive to the deeper value questions involved.

Taylor: That leads me to a question. I'm interested in the connection between justice and environmental policy. It seems to me that there are two ways we can think about that connection. The first way—and tell me what you think about the distinction, maybe you're not sympathetic with my own reading of how this has been done—is what I would call "wishful thinking." It's the kind of wishful thinking that tempted Marxists once upon a time . . . namely . . . Well, if we only reduce all politics to class politics everything else will work itself out along the way because class conflict causes all other conflicts. . . . Patriarchy will disappear, race politics, ethnonationalism, all this stuff, will disappear if only we do the class politics right. It seems to me that some environmental literature engages in that kind of fantasy: that if we just get our "environmentalism" right all good things will happen, including social justice. I don't think that's particularly true because I think that if we do what we probably ought to do with gas taxes we will be hurting poor people more than rich people and I think what we need to do is to couple our gas taxes with other redistributive policies to find justice. Environmental and social justice need different policies and I would like to see them work together. But there's no necessary correlation between them and that seems to me to be a more useful way of thinking about the relationship between environmental policy and social justice. It also demonstrates, to my mind, how difficult the project really is because you need more than environmental policy, you need your environmental policies working with lots of other different types of policies. And there's no reason to think, given the way a policy is made, that things will always work out the way we like, or expect.

Kassiola: I appreciate your comments. Social policy in a complex society that's this large, with many competing claimants, each with an overall perspective on the world that produces conflict, is always going to be difficult on any level of policymaking. I am well aware that one can engage in what may be viewed as wishful thinking because where one is coming from is so distant from the practice that it's very hard for people to see how existing practice could ever move toward what you're talking about. I think that's where the charge about "wishful thinking" comes in. But I think it's incumbent that we engage in the arguments, complete with normative principle, regardless. The more people who agree to this and engage in it, the more creativity and the more energy we will have to make the connections between the existing practices and our value goals, making thinking increasingly less wishful. Just to take an example: we have this conflict between the environment and jobs. That has been a very important and politically decisive issue. There was once a vote in the Senate to enable the coal miners, who were displaced because of closing down high-sulfur coal mines, to get some retraining money from the federal government. The idea was to start them with 80 percent of their last year's income, for nine months or a year, in order to have a transitional period because they were "environmentally" displaced. Although that bill was defeated, it seems to me to be one of the ways to bring social and environmental justice together. A way to say, All right, we've decided that our society is one in which current job holders are not making an environmentally sustainable living . . . therefore, we have to be concerned about the consequences for those people who are unemployed because of sustainable environmental policy. . . . It would have been a big step in the right direction, but it would have required a different political paradigm to win. It is very difficult, but I don't think we're going to take the more difficult steps until we take the first step, and the first step is recognizing that we need to consider values in this debate. Only then can we begin to resolve value conflicts. I'm not saying there's only one view here about how you implement those values and positions, but the positions must be first stated and then debated.

Bowersox: Socrates talks about what's *good* and what's *pleasurable* and he raises the question of health. Surgery is painful and it hurts to get healthy, but nevertheless we know it's something that is good for us. We've been talking about taxation and such policy, and we've assumed that if we raise taxes in one area we have to lower taxes in another. It seems to me that we need to have conversations about, well, who's going to feel the pain, and is pain necessarily bad? Unnecessary pain on the lower classes may need to be avoided or abated; more pain may need to be the burden of the well-off. Nevertheless, a transition will probably be painful for us all: men and women, rich and poor, white and black. To me, that raises a different vision of what's just and what's fair over and above the redistributive concerns that were raised earlier. These are the questions that I don't hear debated.

Gillroy: But here is where the confrontation between environmentalism and liberalism becomes three-dimensional. To take Joel's point and marry it to another made earlier: it is true that individuals see the world through different cosmologies, different paradigms. Part of this whole process is to utilize normative or moral argument in a way that at least brings people to a self-consciousness of their assumptions, the kind of fundamental assumptions that they're making, perhaps unconsciously, about people, nature, economy, society, and so on. One of the definitions of a paradigm is a series of assumptions you accept without analysis . . . that you just assume to be true without reflection. Maybe it is becoming conscious of our moral assumptions and worldviews that speaks to Joel's approach. The second point is that once you become self-conscious of your "metaphysics" you need to try to figure out some way of speaking across paradigms so that we can have a dialogue about just these kinds of environmental and social justice questions. How are we going to decide on policy winners and losers? How are we going to decipher what the normative and empirical terms of the debate actually mean to the present and future? I think that's one of the things I liked about the Campbell-Mohn essay. She asks private landowners to reconsider their obligations to community. She attempts to make us all transcend our self-interested economic predispositions and consider the normative implications of our motives in terms of duty and obligation to others.

Kassiola: I think that a word we haven't used here is critically important and that's *democracy.* The democratic process can work only when you have an informed public, and that informed public must not just be scientifically informed, which frequently is not the case, but also normatively literate and capable of moral argument rather than mere opinion. This is where we have to start, and we have not yet even begun.

Bowersox: Our present version of liberalism is averse to, and runs from, this sort of normative debate, because it cannot handle it. It can't internalize it because it is more complex than a Rodney King version of the world: "Can't we all just get along?" It's tough hoeing.

Audience: I understand what you mean, but the word liberalism has different meanings. This comes back to the notion of democracy—thank you for invoking it. We need to sustain some notion of a need for community as part of democracy. This is a definition that moves beyond reliance on strong individual rights as central to liberalism. The ecological and the democratic together might be a good worldview to solve both the environmental and social justice issues we have been discussing. *Ecological* in that we are a whole, living together, within rules that produce policy that we all must live with. *Democratic* in that we all must be part of the political process that produces this policy, whether scientific or social. There's the word must again; perhaps I should have said should.

Gillroy: Must and should are good normative words to end on.

Issue 3: A Sustainable Environment as an Instrumental Value?

The Hedgehog, the Fox, and the Environment

Mark Sagoff

The late Isaiah Berlin, in his famous book *The Hedgehog and the Fox,* quotes from the Greek poet Archilochus the phrase, "The fox knows many things but the hedgehog knows one big thing." Berlin interprets these words to define a chasm "between those, on one side, who relate everything to a single central vision, one system less or more coherent or articulate, . . . and, on the other side, those who pursue many ends, often unrelated and even contradictory" (1957, 7). Berlin used this distinction to explain different visions of history. This essay deploys Archilochus' taxonomy to clarify the difference between two opposing approaches we may take to understanding and solving environmental problems.

Those who share the hedgehog view attribute our environmental woes to one overwhelming cause: the expansion of the global economy. Analysts Mathis Wackernagel and William Rees have written that "economic activity on the globe as measured by Gross World Product is growing at four percent a year, which corresponds to a doubling time of about 18 years" (1996, 1). Along with other experts associated with the emerging science of ecological economics, these commentators believe that "the Earth's ecosystems cannot sustain current levels of economic activity." Relentless economic expansion worldwide continually gains momentum: "An irresistible economy seems to be on a collision course with an immovable ecosphere" (1). Only by decreasing or at least containing the size of the global economy can we provide a sustainable environment for future generations.

Environmental foxes, in contrast, believe that our environmental problems arise from many different sources, for example, from poverty as well as from affluence, from economic contraction as well as from economic growth. Rather than prescribing global economic retrenchment as a general solution, foxes recommend management strategies that "combine incentive and regulatory policies, recognize administrative constraints, and are tailored to specific problems" (World Bank 1992, 38). They believe that the world must solve its environmental problems on a case-by-case basis, for example, by developing and deploying cleaner and more efficient technologies. Rather than trying to reduce economic activity, this approach seeks to make it environment-friendly, for example, by requiring industry to use resources ever more efficiently. Environmental foxes do not believe that economic growth automatically will solve environmental problems. But these foxes reject the idea that the expansion of economic activity worldwide must automatically lead to resource depletion and ecological collapse.

In the 1970s, biologist Paul Ehrlich and physicist John Holdren captured the hedgehog perspective in a famous equation that measured environmental impact (I) of a nation or society as the product of its population size (P) multiplied by per capita affluence (A) and the technology (T) of production:

$$I = PAT$$

(1971, 1212–17; Ehrlich and Ehrlich 1990). These three factors, because they are all multipliers, are thought to compound the damage each causes to the natural environment. This equation has theoretical bite because it puts technological advance in the numerator and thus entails that the more technology progresses, the more it amplifies the environmental impact of affluence and population.

Hedgehogs may support with examples their view that technological progress causes more environmental problems than it solves. As technology to cut down trees, harvest fish, extract minerals, construct dams, and so on became more effective, the impact of the economy on forests and the landscape has become all the more profound. Large-scale, high-impact technologies often replace smaller, softer ones. For example, automobiles are replacing bicycles in developing societies such as China. High-yield agriculture brings with it greater reliance on pesticides and irrigation. The depletion of fisheries attests to the baneful effects of advances in technology. For example, when sophisticated methods of harvesting caused oyster populations in the Chesapeake Bay to plummet, the state of Maryland prohibited their use; authorities required that oysterers use only "sustainable" eighteenth-century technologies, such as sailboats and hand tongs.

With examples such as these in mind, hedgehogs reiterate their contention that technology is part of the problem, not part of the solution. "Technological fixes for environmental problems," Ehrlich has written, "often work locally or temporarily but prove unworkable on regional or global scales or over the long term" (Ehrlich et al. 1997, 103). Ecological economist Herman Daly agrees: The "term 'sustainable growth' when applied to the economy is a bad oxymoron" (1993, 267). He adds: "To delude ourselves into thinking that growth is still possible if only we label it 'sustainable' or color it 'green' will just delay the inevitable transition and make it more painful" (267).

The position of hedgehogs, who contend that technological progress must in general compound the environmental woes caused by population and economic growth, contrasts with the analysis offered by foxes, who believe that the world must address environmental problems piecemeal by finding ad hoc technological or political solutions for them. According to this view, technological change can make economic growth "sustainable" because new technologies provide "substitutes for many natural resource–based materials. When shortages do emerge, experience and economic theory suggest that prices will rise, accelerating technological change and substitution" (World Resources Institute et al. 1994, 5).

An environmental fox might argue, for example, that if "captured" forestry and fishing deplete natural resources, tree and fish farming may extend those resources. Roger Sedjo, a respected forestry expert, believes that advances in tree farming, if implemented widely, would allow the world to meet its entire demand for industrial wood using just 200 million acres of plantations—an area equal to only 5 percent of current forest land (1995, 180). In 1993 fish farms produced 22 percent of all food fish consumed in the world and 90 percent of all oysters sold (L. Brown, Hane, and Roodman 1994, 34; L. Brown, Flavin, and Kane 1996, 32). The World Bank reports that aquaculture could provide 40 percent of all fish consumed within the next ten years (Kelleher, Bleakley, and Wells 1995, 4).

The 1992 World Bank *Report on Development and the Environment* takes up the controversy between the hedgehogs and the foxes, although it does not employ those terms. The Bank, a bastion of foxes, correctly characterizes the hedgehog position as insisting "that greater economic activity inevitably hurts the environment." The *Report* continues, "According to this view, as populations and incomes rise, a growing economy will require more inputs (thus depleting the earth's 'sources') and will produce more emissions and wastes (overburdening the earth's 'sinks'). As the scale of economic activity increases, the earth's 'carrying capacity' will be exceeded" (1992, 38).

Against this position, which one associates with ecological economists such as Daly and Ehrlich, the World Bank argues that the environment does not impose physical limits to the growth of the global economy. Rather, societies can "grow" their economies indefinitely as long as they make intelligent policy and technological choices. From this perspective, the world's people can all have refrigerators, for example, provided that each unit uses a refrigerant that poses no environmental hazard. Likewise, people can use automobiles, provided vehicles become more efficient and run on cleaner and renewable forms of energy. If air pollution is a problem, people do not need to drive less but to use cars that run on fuel cells that do not pollute. The 1992 *Report* states: "The key to growing sustainably is not to produce less but to produce differently" (36).

For reasons such as these, environmental foxes contend that economic expansion will not inevitably exceed the carrying capacity of the Earth. They base this thesis on two assumptions, one having to do with our ability to substitute between flows of raw materials, the other with our capacity to manage wastes. According to the first assumption, human ingenuity can always find a way around resource constraints by using resources more efficiently and by finding plentiful substitutes for resources that become scarce. According to the second assumption, the effects of effluents and emissions on the environment can be minimized, mitigated, or otherwise managed. If technological progress finds a way around resource scarcity and if it minimizes wastes and mitigates their effects, the more technology advances, the less economic growth may damage the environment. Environmental foxes differ from hedgehogs, then, because they believe technology (T) often works as a denominator of the I = PAT equation, because it *divides* rather than *multiplies* the environmental effects of population and affluence:

$$I = \frac{PA}{T}.$$

While recognizing that rich societies tend to use technology to tame and domesticate wild environments, economic foxes believe they can also deploy technology to solve environmental problems. The World Bank argues, therefore, that economic growth may be consistent with environmental protection if societies employ improved environmental management systems that combine "clean technologies and management practices . . . to reduce environmental damage per unit of input or output" (1992, 38). As societies become richer, they gain opportunities to create a more healthful and natural environment by adopting cleaner technologies and more environmentally enlightened policies. The World Bank *Report* argues, "This explains why the environmental debate has rightly shifted away from concern about physical limits to . . . concern for incentives for human behavior and policies that can overcome market and policy failures" (10; italics removed).

How Does One Measure the Size of the Economy?

One reason the hedgehog and the fox reach different conclusions is that they define economic growth in completely different terms. The hedgehog measures the size of an economy essentially in relation to its throughput, that is, the sheer amount of matter and energy it consumes and wastes it produces. *"Scale,"* Daly writes, "refers to the physical volume of the throughput, the flow of matter-energy from the environment as low-entropy raw materials, and back to the environment as high-entropy wastes" (1992, 186; Daly and Townsend 1993, 2). Later, he adds: "Throughput is the entropic physical flow of matter-energy from nature's sources, through the economy and back to nature's sinks" (Daly 1993, 326). Thus understood, the scale of an economy can be "measured in absolute physical units" (Daly and Townsend 1993, 2; see also Daly 1992, 185–93).[1] To achieve sustainable development, Daly argues, the economy must stop "at a scale at which the remaining ecosystem (the environment) can continue to function and renew itself year after year" (1993, 270). Economist Herman Daly has deployed this concept of scale "to demonstrate the biophysical impossibility of sustainable growth of people and their things" (Daly and Townsend 1993, 11–12).

The World Commission on Environment and Development, in *Our Common Future* (1987), drew an obvious inference from this conception on economic growth; namely, as economies grow they must logically overwhelm the carrying capacity of the earth: "Many present efforts to guard and maintain human progress, to meet human needs, and to realize human ambitions are simply unsustainable—in both the rich and poor nations. They draw too heavily, too quickly, on already overdrawn environmental resource accounts. . . . They may show profits on the balance sheet of our generation, but our children will inherit the losses" (1987, 8).

This official pronouncement echoes decades of warnings. Paul and Anne Ehrlich, for example, wrote twenty years earlier: "To raise all of the 3.6 billion people of the world of 1970 to the American standard of living would require the extraction of almost thirty billion tons of iron, more than 500 million tons of copper and lead, more than 300 million tons of zinc, about 50 million tons of tin, as well as enormous quantities of other minerals. . . . The needed quantity of iron is theoretically available. . . . Needed quantities of the other materials far exceed *all* known or inferred reserves" (1970, 101).

This approach to economic growth coincides with that taken by classical economists, who, according to commentator Paul Christensen, based their pessimism partly on the doctrine that natural and manufactured resources are complementary, not substitutable. Christensen notes that even Adam Smith divided capital into two complementary kinds: fixed (which included land) and circulating (which included technology and machines). Christensen summarizes: "Manufacturing, it was recognized, was a materials-processing activity that obeys the law of conservation of mass. Manufactured products require material inputs, labor, tools or machines, and the energetic materials required by the laborers and other engines. Thus, an increase in output in manufacturing requires a proportional increase in raw materials. Individual capital goods lack productivity independent of other inputs" (1991, 77).

Accordingly, Daly urges us to "keep the weight, the absolute scale, of the economy from sinking our biospheric ark" (1991, 35). Paul Ehrlich argues that the "most severe problem with the global economic system is its inability to sense that its scale has grown too large for its support systems" (1994, 39). Other ecological economists agree: "Economic growth, which is an increase in quantity, cannot be sustainable indefinitely on a finite planet" (Costanza et al. 1991, 7).

The World Bank, in sharp contrast, measures the size and growth of an economy in economic terms, that is, in terms of gross domestic product (GDP), or the economic value of whatever that economy produces expressed in then-current prices. With that definition, it makes sense to say that a country with a relatively large economy (i.e., GDP), such as Sweden, may damage the environment less than a country, such as Poland, with a much smaller economy.

An example will help us evaluate these methods of measuring the size of the economy. Suppose the global economy, to protect the stratospheric ozone layer, substitutes on a pound-for-pound or gallon-for-gallon basis an ozone-friendly refrigerant for the CFCs that are now being phased out. Let us suppose, moreover, that the prices of the two compounds are the same. For each definition, the economy remains the same size, though one compound damages the environment vastly more than the other. With neither approach would the substitution make the economic "size" or "scale" of the economy "smaller"—unless one defines the scale of the economy simply in terms of the damage it does to the environment. (This would turn the relation between "scale" and "damage" into a tautology.) An environmental fox would claim that this example shows that the size of an economy, however measured, is irrelevant to its environmental impact. In this view, technological progress can both sustain economic activity and save the environment. A hedgehog, in contrast, would say that this sort of substitution distracts us from the real problem, which is that we must have a small economy for a small Earth. Because the Earth and its resources are finite, we cannot expand the economy forever without risking ecological collapse. The very concept of economic growth and of a finite earth shows that growth cannot continue indefinitely on a finite planet.

Because they find an inexorable empirical connection between economic activity and environmental degradation, environmental hedgehogs argue that a problem-by-problem approach, if it does not reduce the scale of the economy, inspires the false hope that we can have our standard-of-living cake and eat it, too. Nobel laureate economist James Meade, for example, writes, "Pollution and the exhaustion of natural resources depend and will depend in the future on the absolute level of total economic activity" (quoted in Ehrlich et al. 1997, 104). Paul Ehrlich and colleagues, citing Meade, have recently denounced those who look to technology to reconcile economic growth and environmental quality—who think T can occur in the denominator. According to Ehrlich and colleagues, those who believe that economic activity can be sustained do a "disservice to the public by promoting once again the dangerous idea that technological fixes will solve the human predicament" (Ehrlich et al. 1997, 104).

The Concept of Natural Capital

To show that the "carrying capacity" of the Earth cannot sustain current (much less greater) levels of economic activity, ecological economists have adopted the concept of "natural capital." More precisely, these hedgehogs distinguish the factors of production into natural capital, human or cultural capital, and manufactured capital (Folke et al., 1994, 4). (The last two comprise *human-made* capital; Costanza and Daly 1992, 37–46.) They further distinguish *natural capital* into three kinds: "(1) non-renewable resources, such as oil and minerals; (2) renewable resources, such as fish, wood, and drinking water; and (3) environmental services, such as maintenance of the quality of the atmosphere" (Berkes and Folke 1994, 129). Ecological economists declare that we have "entered a new era in which the limiting factor in development is no longer manmade capital but remaining natural capital" (Costanza et al. 1991, 8; see also Daly 1994, 28). Economic growth would be limited indeed, because "natural capital was not and cannot be produced by man" (Daly 1994, 30).

If natural capital has become the limiting factor, then the more the economy expands, the sooner it will run out of resources. Thus, the concept of natural capital offers a way to determine the "size" of the economy relative to the surrounding environment. What reason is there to believe, however, that natural capital has become the factor limiting development? Daly answers "that material transformed and tools of transformation are complements, not substitutes." He continues: "Do extra sawmills substitute for diminishing forests? Do more refineries substitute for depleted oil wells? Do larger nets substitute for declining fish populations? On the contrary, the productivity of sawmills, refineries, and fishing nets (manmade capital) will decline with the decline in forests, oil deposits, and fish" (1990, 3).

As these examples suggest, Daly means by "natural capital" nature's free gifts; by "man-made capital" he refers primarily to tools used to acquire and process them. It is clear from these definitions that "the basic relation of man-made and natural capital is one of complementarity, not substitutability" (Daly 1994, 26). The experience of nomadic communities, who exhausted resources in one place and then moved on, supports Daly's argument. Indeed, all hunter-gatherer societies testify to Daly's insight. Nature's free gifts—those that are there for the taking—exist only in limited amounts. The more efficient our tools for capturing and transforming these gifts, the scarcer they become.

Many examples suggest that Daly is right: raw materials and the tools used to capture and transform them are complementary, not substitutable. The collapse of whale populations during the nineteenth century, for instance, demonstrates this thesis: more efficient whaling boats quickly depleted whale stocks. Game disappeared from Kansas, moreover, when pioneers introduced better rifles: more bullets resulted in fewer turkeys and in the virtual disappearance of prairie hens.

The problem with this argument, however, is that no environmental fox, mainstream economist, or anyone else would disagree with it. It seems irrelevant, in other words, to the thesis foxes do assert about the substitutivity of resources. When economists like those at the World Bank speak of substitutability, they do not refer to the relation between raw materials and tools used to capture and process them. Rather, they assert that resources are substitutable one for another so "that a satisfactory substitute can always be found for the role of any one of them" (Barnett and Morse 1963, 11; quoted in Ehrlich 1989). Daly acknowledges that the substitutivity that matters to mainstream economists has to do with the use of one resource flow in place of another, not with the relation between man-made and natural capital. "Orthodox growth economics," Daly writes, "recognizes that all particular resources are limited, but does not recognize any general scarcity of all resources. The orthodox dogma is that technology can always substitute new resources for old, without limit" (1979, 71).

Environmental foxes who believe in the substitutability of resources, in other words, have Thomas Edison in mind, not Captain Ahab. When Edison invented a lightbulb that burned electricity and when synthetic lubricants substituted petroleum for whale oil, whales ceased to be resources and became aesthetic marvels and cultural icons—which they remain to this day. Similarly, pioneers like Pa in *Little House on the Prairie* relied on tools of capture (rifles) to increase the supply of resources (prairie hens). Frank Perdue, in contrast, substitutes a plentiful resource (cheap feed) for a scarce one (wild stocks). The poultry industry in the United States produces 278 million turkeys a year without firing a shot. As a result, we may protect wild turkey populations as an aesthetic, moral, cultural, and religious heritage for future generations, but not because we think our great-grandchildren will depend on hunting them for Thanksgiving dinner.

Technological advances not only allow more plentiful resources to do the work of scarcer ones, but they also decrease the resource input to production per unit of economic output. The progress from candles to carbon filament to tungsten incandescent lamps, for example, decreased the energy required for and the cost of a unit of household lighting by many times. Compact fluorescent lights are four times as efficient as today's incandescent bulbs and last ten to twenty times as long (Ausubel 1996, 164–70). Comparable energy savings are available in other appliances; for example, refrigerators sold in 1993 were 23 percent more efficient than those sold in 1990 and 65 percent more efficient than those sold in 1980, saving consumers billions in electric bills ("Appliance" 1995).

Many hedgehogs point out that increasing fuel efficiency cannot solve energy problems over the long run, because sources of energy are limited. The economist Nicholas Georgescu-Roegen, for example, acknowledges that in some cases, "it may be that the same service can be provided by a design that requires less matter or energy" (1979, 98). He points out, however, that even if we go far in the direction of greater efficiency and resource substitutivity, there is a limit. Energy is always required to make one substance do the work of another—to restructure plentiful carbon molecules, for example, to make nanotubes stronger than steel. Where is the energy to come from? "The question that confronts us today," as Georgescu-Roegen has written, "is whether we are going to discover new sources of energy that can be safely used" (98).

Environmental hedgehogs and foxes answer this question in opposing ways. They argue that solar and other renewable forms of energy will become broadly available and can be safely used. Amory Lovins, among others, describes commercially available technologies that can "support present or greatly expanded worldwide economic activity while stabilizing global climate—and saving money." He adds that "even very large expansions in population and industrial activity need not be energy-constrained" (1991, 95; see also Johnsson et al. 1993).

Environmental hedgehogs, in contrast, argue that those who agree with this statement by Lovins "believe in perpetual motion machines" and "act as if the laws of nature did not exist" (Ehrlich 1994, 41). The second law of thermodynamics, they argue, indicates that energy is bound to run out. The availability of inexpensive and environmentally benign forms of energy, for example, solar or geothermal power, would not solve problems but create them, because the size of the economy would grow in response, further threatening ecosystems. Ehrlich and colleagues point out that "no way of mobilizing energy is free of environmentally damaging side effects, and the uses to which energy from any source is put usually have negative environmental side effects as well. Bulldozers that ran on hydrogen generated by solar power could still destroy wetlands and old-growth forests" (Ehrlich et al., 1997, 101). Daly writes: "Even the advent of commercially viable fusion energy . . . not only would fail to eliminate pollution constraints on economic growth, it would actually hasten human-kind's progress to the brink of ecological disaster" (Daly and Townsend 1993, 9).

The problem with this answer, however, is that it verges on defining economic growth or activity in a way that ties it logically to environmental damage. How do Daly and Ehrlich know that each form of energy, even solar or fusion, if it powers economic growth, entails ecological disaster? To be sure, bulldozers that run on solar or fusion power may well convert old-growth forests to ski resorts, wetlands to golf courses, and so on. Many of us believe, moreover, that old-growth forests are intrinsically better than ski resorts and, similarly, wetlands than golf courses. We may agree that by replacing nature with commerce we disgrace our moral, spiritual, and cultural principles. Where, however, lies the certain disaster as far as human welfare is concerned?

Human beings have been replacing magnificent natural environments with tawdry commercial ones for centuries, and we owe much of our economic success to this practice. People flock to golf courses and ski resorts. Absent a clear showing that the economy is running out of a resource for which no substitute can be found, predictions of ecological disaster merely claim that economic activity or "scale" will cause ecological depletion or collapse, the very thesis that is to be proven. Engineers informed about emerging energy technologies, such as Lovins, tout the capacity of solar and other "renewables" to "grow" the economy and still save the earth. One must weigh their technical expertise and empirical knowledge against the a priori belief asserted not by engineers but primarily by ecologists and ecological economists that the second law of thermodynamics dashes all such hopes.

Coping Capacity, Not Carrying Capacity

Environmental hedgehogs, however, offer a strong empirical and historical argument to show that even if technological solutions to environmental problems can be found, societies will not adopt them. The way a society behaves will depend largely on its ability to set and attain environmental goals through political deliberation and action rather than by market exchange. We must drive nonpolluting cars, for example, to prevent climate change, but the market alone is unlikely to bring them to us. How well we protect nature requires political, not just technological, innovation.

Taking up this truth, many environmental hedgehogs argue that even if technological paths to sustainable economic growth are conceivable, we lack the political will to find and follow them. Ehrlich and colleagues write, "Above all, one must recognize that what is technically and economically feasible is often sociopolitically impossible" (1997, 103). They quote demographer Nathan Keyfitz, who comments, "If we have one piece of empirically backed knowledge, it is that bad policies are widespread and persistent" (103).

Of course, foxes would agree that bad policies are widespread and persistent. Energy consultant Michael Brower, for example, writes, "At heart, the major obstacles standing in the way [of a renewable-energy economy] are not technical in nature but concern the laws, regulations, incentives, public attitudes, and other factors that make up the energy market" (1990, 26). The question then arises how societies can substitute better policies and incentive structures for worse ones. The environmental fox suggests that this must be done on a policy-by-policy, case-by-case, problem-by-problem basis. What kind of strategy does the hedgehog recommend? Apparently, the hedgehog would favor policies that would greatly slow and indeed reverse economic growth to decrease affluence and delay technological change. Are policies of this kind, presumably aimed at increasing global poverty or at least reducing wealth, more likely to be enacted than the kinds of problem-by-problem "fixes" environmental foxes propose?

Environmentalists meet stiff opposition from industry and other groups whenever they seek to require cleaner or more efficient technologies. It is hard, for example, to pass laws even to get the worst gas-guzzling cars off the road. Yet the hedgehog alternative—to require worldwide economic retrenchment—

hardly appears more politically popular. It is one thing to get people into nonpolluting, efficient cars; it is another to try to take away their vehicles altogether.

Nevertheless, environmental hedgehogs make a good point when they change the focus from *carrying capacity* to *coping capacity.* Even if the planet sets no physical limits to economic growth, they may say, there are plenty of social and political limits. We lack the management skills, the political organization, and the cooperative goodwill to make and to follow the policies necessary to sustain economic growth in a divided world. If hedgehogs focus on coping capacity, however, it seems they would then agree with the World Bank that the environmental debate has "rightly shifted away from concern about physical limits to . . . concern for incentives for human behavior and policies that can overcome market and policy failures" (World Book, 1992, 178). They may be less sanguine than the Bank, however, that the world can avoid these failures.

The Distinction between Nature and the Environment

Wackernagel and Rees (1996), among environmental hedgehogs, seem most open to the idea that technological advance can support global economic expansion while reducing society's demand on nature, provided that we change our incentive structures to favor less polluting and resource-intensive economic activity—in short, that we deploy policies to overcome market and policy failures. Wackernagel and Rees answer the question of whether technology is friend or foe as would the World Bank, by saying it depends. They write, "If new technology is to reduce our Ecological Footprint, it must be accompanied by policy measures to ensure that efficiency gains are not directed to alternative forms of consumption" (25).

When Wackernagel and Rees speak of the demand society puts on nature, they apparently include everything from the use of forests to absorb carbon dioxide to the development of a beautiful mountainside for a ski resort. These aspects of our "ecological footprint," however, may concern us for quite different reasons. The buildup of CO_2 in the atmosphere can be tied causally to all kind of untoward consequences. The construction of discos, Ho-Jos, go-gos, and casinos in pristine and magnificent wilderness, in contrast, may offend us less because of its untoward consequences for human welfare—actually, people enjoy and are willing to pay for these entertainments—than as a matter of principle. Many of us believe on moral, aesthetic, and religious grounds that the oldest and most magnificent aspects and areas of nature should be protected for their own sake rather than simply or primarily because they contribute to our well-being. Indeed, human welfare more often provides a reason to transform nature, for example, to create farms, housing, highways, and so on, than to maintain its natural integrity and history.

Anyone who looks out a window can see that the intrusion of economic activity on the landscape has become more and more apparent. Much of this intrusion can strike us as ugly, indeed wrong, for aesthetic and spiritual reasons, even though it might contribute on balance to human well-being. If the lands where Pa hunted prairie hens become intensively managed grain-producing systems supporting factory farms, for example, the whole operation may disgust us, whereas the old pioneer economy strikes us as romantic. But life was a lot harder for pioneers than it is for us. We have transformed nature so that it does not kill us as soon as it killed them. It is not true that we transform nature always at our peril. Actually, as the pioneers knew, nature is itself full of perils. It will not feed, clothe, shelter, or sustain us. Human beings are tool-making animals, and this is how and why we survive.

Even if the current agribusiness arrangement could be self-sustaining—all the chicken manure recycled to fertilizer and the whole thing running on solar energy—we may think of it as less appealing than the natural landscape it replaces. If so, we probably ascribe intrinsic value to nature, as we should, while condemning on aesthetic grounds what is artificial and commercial. We can and should distinguish those aspects of our ecological footprint that have bad consequences from those that are intrinsically unappealing, for example, because they are ugly, commercial, tacky, and disgusting, whereas nature is beautiful and sublime.

The distinction between nature and the environment helps divide those aspects of economic activity we may condemn on intrinsic grounds or as a matter of principle and those that concern us on instrumental grounds because of their consequences. The values nature inspires tend to be religious, moral, and aesthetic. Cultural ideals, judgments, and principles impel us to appreciate and to protect nature for its intrinsic qualities and not only for the long-term benefits it may offer humankind. Our basic normative approach to nature, therefore, is not necessarily instrumental or even prudential. It is also founded in religious experience, moral reflection, and aesthetic judgment. We value nature, in other words, for its history, its beauty, and its inherent dignity, which is to say, for what it means to us as distinct from what it does for us.

The environment, in contrast to nature, comprises just those aspects of nature that are useful and that we therefore value for welfare-related reasons. The environment is what nature becomes when we cease to value it as an object of cultural, religious, and aesthetic affection and regard it wholly as a prop for our well-being: a collection of useful materials (natural capital) and as a sink for wastes.

Many of us regret the extent to which the economy has transformed the natural world—turning dells into delis, arcadias into arcades, and paradises into parking lots. We may even share a sense of guilt about what human beings have done to the beauty and majesty of the natural world. We might even think we ought to be—and will be—punished for our sins. The threat that the seas will rise up against us has slipped from a religious fear to a scientific prognostication. The prospect of rising sea levels, nevertheless, may express our traditional spiritual insecurity before God for destroying Creation as much as our recognition of physical limits to economic growth.

Paul Ehrlich, Herman Daly, Mathis Wackernagel and William Rees, and other ecological economists correctly point out that the more lean the economy grows, the meaner to nature it may become. If houses can sustain themselves entirely on passive solar energy, what will stop us from building second, third, and fourth "vacation" homes? If cars get 800 miles per gallon, we may pave over everywhere and spend all our time in sustainable traffic jams. Many ecological economists point out that after we meet our basic needs, further production and consumption of material goods seems to be pointless, even if technology makes it possible. The moral objections to the transformation of nature for commercial purposes may be more important than the prudential objections. We may ask, then: What will stop market forces—the commercial juggernaut that transforms into commercial sprawl all that is distinctive, meaningful, and expressive in nature—if technology removes the physical limits to economic growth?

The hedgehogs are certainly right in seeing economic activity as a scourge on nature, in the aesthetic sense, whatever its relation to the environment conceived as natural capital. There seems no doubt that the growth of the global economy will continue to transform the natural world. If cars become part of the hydrogen economy, completely nonpolluting and fueled by solar energy, society may simply build that many more highways, parking lots, and commercial strips. There would be that much less of the natural world to value for its own sake. Economic growth, even though technology may make it sustainable, after a point is no longer desirable. This does not mean that the global economy, as it grows, must overwhelm the carrying capacity of the Earth. But it does suggest that much of what we care about for cultural and religious reasons, the historical and aesthetic integrity of the natural world, is, indeed, in crisis and under attack.

Why Not Foxy Hedgehogs?

Bryan Norton

We are here to discuss the concept of sustainability, its meaning and its usefulness in environmental policy discussions. I cannot disagree with my colleagues on this program, who have mostly—and, I think, correctly—found the concept to be ambiguous, vacuous, or otiose. I can find little in common in the various references to sustainability in the background papers from Part 2, and of today's commentators, at least Mark Sagoff and Robert Taylor have recommended abandoning the term as useless. If we take current usage as the determining factor in assessing the value of the term, it is difficult to disagree with their assessment. I find myself agreeing with the evidence presented, however, and yet I do not accept the conclusion that we should give up on the term altogether, which may suggest that I have a somewhat different view of the term's value and role in policy discussions. My goal is to suggest, very briefly, a strategy that I think might lead to the rehabilitation of the term, and to the recognition of a somewhat different role for the term sustainability in our discussions of how to protect the environment.

I believe that whether to embrace or give up a high-visibility, low-content term such as sustainability depends as much on one's theory of language as on substantive environmental commitments. So my first task is to sketch a few assumptions about the nature of language and linguistic change, assumptions I hope will help to make sense of my attempts to rehabilitate this problematic term. I hope this approach suggests a possible strategy for developing managerially and politically useful normative terms with scientific content. Then, I discuss Mark Sagoff's useful distinction between environmental "foxes," who see environmental problems as a collection of specific, unrelated assaults on our aesthetic senses, and environmental "hedgehogs," who seek to reduce or reinterpret every environmental disagreement as one more manifestation of the relentless overexploitation of a tragically finite planet by too many humans. My goal in these remarks is to show that the either/or choice Mark offers us is both unnecessarily unattractive and nowhere near as clear-cut and all-or-nothing as he makes it sound. Finally, building on the approach to sustainability proposed by Jan Laitos (this volume) and my assumptions about how language works in situations such as this, I discuss what role a rehabilitated concept of sustainability might play in discussions of environmental policy.

Some Assumptions about Language

As noted above, I think our reaction to the term sustainability and the related phrase, sustainable development, is probably based as much on our view of language and terminology in science and public debate as on our substantive and particular views. Understandings of the nature of language range from essentialist views, which see language and its categories as reflective of natural realities and kinds, to more conventionalist views that language and its categories are bequeathed us by our culture. I lean toward the conventional end of this continuum, and I emphasize also that, being conventional, language is highly dynamic and subject to change, especially under changing conditions. Once we reject the static view of language, the question arises whether a term such as sustainability, however suspect in current discourse, might acquire usefulness through creative redefinition, affording us more precision in the terms we use and thus improving our public discourse on important topics. If my colleagues are right that, on the basis of current usage, the term sustainable is meaningless, one option—besides that of expunging it from our vocabulary—is to follow the advice of the philosopher W. V. O. Quine, who once allowed that, of all terms, it is most seemly to assign meanings to the meaningless ones.

As a preface to our discussion, then, I wish simply to state, and explain very briefly, two assumptions I make about language and the task of developing a useful and perspicuous language for environmental policy discussions. My goal in listing these assumptions is to make my own viewpoint clear and to frame our discussion by describing an attitude toward language that makes sense of my optimistic attempts to rehabilitate the term sustainable. My two assumptions, with the briefest of explanations, are:

1. Language is a dynamic and flexible tool and serves many purposes. Linguistic usage is continually changing both as a result of chaotic and "noisy" elements in speech and society, and also as a result of conscious suggestions and attempts, as in science and philosophy, to clarify or make particular terms more precise for particular purposes. In choosing or altering terminology, it is important to recognize that language is often used for purposes other than providing unambiguous descriptive terms that can be useful in communicating scientific results. Other uses include communicating complex ideas and theories to the general public; serving as a shared language for specialists from different fields of study; and increasing people's motivations to act in certain ways. We should consider, therefore, the possibility that the term sustainable will be use-

ful for purposes other than providing precise descriptive measures of the state of our environment.

2. If we are reflective about language and our use of it in particular and important situations, it is sometimes possible to improve on the language codified in everyday speech, making it more useful for purposes such as scientific description and political communication. Improving on terms often involves explicit or implicit "conventions" that link the term to a related body of knowledge or theory, either scientific, political, moral, or whatever. It is this relationship to acceptable theory that can make the term sustainable nonvacuous and perhaps more useful. But this means we cannot substantially improve the use of the concept until there emerges a more satisfactory theory of environmental management and, especially, of the moral relationships that obtain across generations.

Given that I operate on these assumptions about language and our use of it in public discourse, I pay far less attention to the undeniable problems that beset the term sustainability in current discourse, and more attention to how the term might be useful if it were embedded in a nonideological theory of environmental management. Accordingly, most of my work on sustainability is designed not to learn how the term is currently used, but rather how the term *might*—with suitable understandings and stipulations—function to improve public discourse and thus improve the likelihood that democratic societies will, through participation and political action, enact better environmental policies. When unqualified and unexplained, this goal may seem question-begging: it may sound as though I think I know the correct policy and am proposing language that will help me to achieve my selfishly formulated goals. I think I can avoid this charge by arguing not on the basis of the usefulness of a given term, but rather on the basis that there are certain problems and gaps in the current discourse about environmental policies, and that changes in available terminology might help us to fill those gaps. In this way, I argue for improved terminology on the more neutral premise that there is a problem of inadequate dialogue, communication, and understanding in the discussion of environmental policy, and that it is at least reasonable to attempt to salvage the term by building it into a general theory of environmental obligation.

Perhaps my point will be clarified by noting the current confusion about environmental policy goals, how to characterize them, and how to relate environmental *values* to physical *models* that are constructed to track the changing environment. I characterize this difficulty with current discourse as follows: in environmental policy discussions there are, to my knowledge, no terms that clearly relate changes in the physical environment with associated social values. Consider an analogy: in political discussions of the economy, economists can invoke their theoretical constructs and at the same time link described changes using those constructs simply by discussing the "growth"—or lack thereof—of the economic sector of the society. We are all aware of the theoretical and practical problems involved in treating economists' constructs as perfect indicators of aggregated individual welfare in the society, yet it is undeniable that the availability of terms that readily link economists' models to widely held social values helps economists bring their analyses to bear on policy discussions. These terms relate to measures that can be generated by economists' professional tools *and these terms are interpreted by the public and policymakers as having an unquestioned link as a key social value.* Again, I too have problems with specific growth indicators such as GDP, but my point is that such terms can play a very important role in public discourse.

I believe that environmentalists often do less well in their attempts to influence public opinion and policy because they lack such terms to link their descriptive statements about measurable or predictable changes in the physical environment, statements that make sense within their descriptive models, to identifiable social values. As a result, ecologists who express concern that a given policy is reducing the "resilience" or the "diversity" of a particular system have only begun the process of explaining *why the public and policymakers should care about the changes they describe.* Normative sciences such as economics and public health usually employ general terms—such as "growth," "health," and "disease"—and also more specific "indicator" terms—such as GDP, an individual's body temperature, and infection rates for a communicable illness—as useful indicators of impacts on social values. Given this, it seems reasonable that the normative science of environmental management should have some indicator terms that similarly link bodies of descriptive knowledge with social goals.

Accordingly, I see the creation of terms such as sustainability, ecological integrity, and ecosystem health as nascent attempts to bridge the gap between environmental science and environmental policy discussions. Obviously, whatever my philosophy of language, I will not be able to explain and justify the use of all, or even any, of these terms today. I would like to convince you, however, to separate two types of questions that, when confused, make it harder to see why someone such as I might like to save and improve on these concepts rather than expunge them from discourse. One set of questions, which has to do with particular terms—sustainability, integrity, and so on—and here the issue is whether these specific terms can be made precise and whether they are related to important social values, is a worthwhile one. But whatever position one takes on the particular terms, it is also important to address the more abstract question of whether we need descriptive normative "bridge" terms as a means to link scientific models to important social values. Once these two sets of questions are separated, it makes sense to argue that, given the need for better communication among scientists, decision makers, and the public regarding scientific models and environmental goals, it is worthwhile to make a conscious effort to give some specific terms more substantive meaning by linking them to a specific theory of intergenerational obligations.

Thus, I am not arguing for the term as it is currently used, but rather am advocating improvement of the term for functional reasons, as a response to a particular kind of failure of communication about environmental policy. Given these clarifications, it may now be possible to respond to Sagoff's choices between foxes and hedgehogs.

Should We Be Hedgehogs or Foxes?

Sagoff apparently is convinced that one must be *either* a hedgehog *or* a fox. He would have us decide between the single-minded carrying capacity model, which reduces all environmental problems to one of growth in the scale of human activities, or joining him in having no theory at all, basking with him in haphazard many-mindedness. I share Mark's concern with reductionism of all kinds, and prefer with him some form of pluralism. But Mark refuses even to consider that there might be some integrative theory that would make sense of varied environmental goals, relating them to each other and placing them in a coherent framework of understanding. He implies that we must choose between hedgehogs and hodgepodge. I, on the other hand, fear that, if we have no theory at all relating environmental change to human values, we will have no basis for deciding on general priorities or even of knowing which questions to ask first. Environmental policy with no theory at all will remain ad hoc and driven by interest-group wrangling, and there will be no vision of an environmentally better world—no general understanding of what environmentalists are *for.*

I think it is unfortunate that Mark chose as his only example of hedgehogs the ideologically loaded theory of carrying capacity applied at the global level. He equates having a general understanding of what we are about in seeking an improved environment with accepting an ecologically reductionistic and highly questionable unifying theory. That makes it pretty attractive—but, I think, too easy—to embrace foxiness. If we consider some different types of theories of sustainable development, it may be possible to imagine a theoretically guided approach to the big picture in environmental problems, without going all the way to the sort of physical reductionism embodied in the I=PAT formula.

For example, take the general structure suggested by Jan Laitos, who adopts a general theory of environmental value: that the public lands (and perhaps other environmental assets) should be managed to maximize economic efficiency, measured in broadly market terms. But he adds to this general theory an important "overlay," which he calls "sustainable development" and describes as tempering the search for efficiency with a concern for "two variables not otherwise considered by the efficiency criterion: equity and time." He sees this idea as overlying efficiency considerations with the ideas of social justice through application of a "time calculus (the future should not be sacrificed for the present)." Although Laitos's essay left me wanting to hear much more about how this "time calculus" would work, in what ways it would modify efficiency calculations, and a host of other issues, it does seem to me to provide a case of what might be called a "foxy hedgehog" approach. It starts with a basic theory (à la the hedgehog), but ends by complicating the theory with cross-cutting and limiting purposes, so as to provide some basis for a discussion of constraints and priorities that might significantly change the overarching, single-minded theory and satisfy some of the concerns of the foxes among us.

In short, it seems that Sagoff is pushing us into making an either/or distinction in a situation in which there are many alternatives and many ways to link theory—including moral theory, economic theory, scientific theory, and democratic theory—in an attempt to make sense of, and relate to each other, the myriad environmental problems we face. Worse, he has assumed that the only way to be a hedgehog is to ignore the concerns of foxes, and force every moral concern into a theory that rests more on imagination and a priori reasoning than on a careful observation and analysis of the facts of particular environmental problems. Surely there is a lot of theoretical ground between I=PAT and the chaos of basing policy on individual aesthetic tastes.

Hunting for a Hedgehoggy Fox

I think that one reason the term sustainability has fared so badly in environmental discourse is because most participants in the debate—especially the hedgehogs from economics *and* the hedgehogs from ecology and ecological economics—have assumed that the term should be given purely descriptive content. Ecologists and ecological economists, as Mark explained, have tried to link the term to a description of the comparative scale of human economies with their ecological and resource context, resulting in an equation, I=PAT, that describes a physical relationship between human populations and their resource base. Welfare and resource economists, on the other hand, have reduced the concept of sustainability to a relationship among the opportunities for consumption (income levels) across generations. Neither of these reductionisms leaves any room for voluntaristic, moral commitments on the part of communities to protect and sustain certain features of their environment. As I have argued elsewhere, these reductionistic interpretations of sustainability commit what the philosopher of language J. L. Austin (1962) called the "descriptivist" fallacy. Austin argued that many terms, especially terms in ethical discourse, function more as "performatives," in that, when uttered under appropriate circumstances, these terms can enact a commitment rather than describe a state of the world. He used the examples of saying "I do" during a marriage ceremony and saying "I hereby christen you the *Queen Elizabeth*" while striking the ship's prow with champagne. In the context of an ecosystem management project or a watershed management project, analogously, communities might articulate long-term social values and commit themselves to protect certain features of their environment or landscape for the future as a community-performative act.

I noted above that I am drawn to Jan Laitos's general approach to the idea of sustainability. He adopts the general theory that good environmental policy makes choices about public goods on the basis of efficiency, but he then qualifies this hedgehoggy goal by placing it within a larger context in which there is a commitment to choose policies that are also sensitive to future outcomes and impacts on future generations and designed to increase intragenerational equity. This foxy hedgehog's approach, in other words, takes a general theory of environmental value and qualifies it with other considerations based in other theories. Although drawn toward this more hybrid approach, I differ with Laitos on two counts. First—and this may be only an omission on his part—he does not make clear the extent to which different communities might choose somewhat different mixes of these variables, nor does he seem to recognize that democratic process and community involvement might be involved in articulating broad goals that can decide questions of which long-term impacts are considered unacceptable in a given community. Second, he seems to treat equity, both inter- and intragenerational, as a matter of descriptive "variables," and I doubt such questions can be resolved simply by quantifying descriptive variables.

Although I agree with Laitos that sustainability should provide us with a broad theoretical orientation toward the future, and that this should be accomplished by piecing together elements of economic and moral theory, I would not use efficiency as the basic theory. I believe efficiency calculations—especially if they embody discounting across time based in "time preference" of individuals—are inherently short-sighted in their outcomes. This short-sightedness will either overwhelm or be at war with the morally based concerns of intergenerational equity. I would start from a slightly different direction in finding a hybrid system, looking for a multicriteria system to govern their interactions. I see problems of choosing sustainable policies as involving problems of temporal and spatial *scale,* and I would introduce a theoretical approach for arraying impacts over multiple scales of space and time prior to introducing any moral discourse. Here, I am thinking of something like hierarchy theory of C. S. Holling's (1995) "panarchical" model. Then, with a broad method for arraying impacts across time, I would attempt to match particular environmental "problems" with impacts at different scales. But this second step cannot be purely descriptive, because our very notion of a problem depends on what values we are committed to. Like the fox Sagoff, I am a pluralist about values and place weight on the process whereby communities come to identify, articulate, and commit to the protection of certain features of their environment. We cannot—short of reliance on a priori ethics—know what a society or community should save or "sustain," until we know what is valued by members of that society or community. And determining what is valued by a *community* (as opposed to aggregating preferences of individuals) requires a public process and a political act. This act, in turn, requires public deliberation, articulation of values, and an ongoing, participatory process.

Personally, I favor the starting point of the adaptive managers, such as Holling, Walters, Lee, and others, who adopt a multiscalar model of local ecosystems, expect management to be an iterative, experimental struggle toward better policies, and argue that sustainability should be worked out within a consciously reflective process of examining both causal relationships between human societies and their natural environments and also of considering and reconsidering environmental goals and values. On this reading, the concept of sustainability begins to look more like an empty vessel, empty until social values—especially commitments to the future based on an articulation of what we value most dearly today—are articulated and added to the empty vessel. But values such as this—I have called them "community-constitutive" values elsewhere—should not be thought of as the aggregation of static individual preferences, but as expressions of the emergence of the concern of one generation for the next in the context of a Burkean community: one made up of those who have lived in the past, those presently living, and those who will live in the future. These values, understood as expressions that emerge from an ongoing process of community reflection and self-definition, can then guide broad decisions about social values, environmental goals, and policy priorities within a community.

Because I view these communities as locally based and do not think that the concept of sustainability commits us to more than a general moral concern for the future of our community and to a process of value articulation and constant reexamination, I suspect my theory is pluralistic in a more fundamental sense than is Laitos's overlay of equity considerations on an efficiency-based model. In this sense, my approach may qualify me as that of a "hedgehoggy fox": I see problems of environmental policy as emerging from local, community-based concerns, as does the fox, but I also hope to develop a theory of intergenerational obligations that captures the common, cross-community concerns that are embodied in a moral concern for the future, which adds into the mix the hedgehoggy idea of a common ideal of sustainability that will be expressed more specifically in many different communities. This bottom-up, case-based approach to building a theory of sustainability seems to me to hold considerable promise for clarifying the nature of, and perhaps even guiding, environmental management.

Discussion

Sagoff: I'm much happier with my essay now, if only because it led to that distinguished and extremely helpful analysis. Largely, I agree with Bryan, but I disagree in many details. I'll mention only one.

When you talk about normative science, you immediately think of medicine. The thing with medicine is that we know what the goal of medicine is, namely, to restore human health and to relieve pain. Science ceases to be normative as soon as we no longer agree on its purpose. Agronomy might be another normative science: we're trying to make a lot of crops cheaply and regularly. You can name sciences that have particular purposes and you can call them normative sciences, but what that hides is that we just happen to agree entirely on what the purpose is. Well, in that sense, I think economics is not—decidedly not—a normative science, because nobody knows what we want to do with it. When it is a matter of GDP, of expanding GDP, maybe you could argue that the models help us do that in agreement. But those that criticize economists say, Well, GDP is not a measure of welfare or well-being; these two have nothing to do with each other. Nobody really knows what to do with GDP. And that's an indication that we really have not agreed on what we want the "science" of economics to do. There are a number of possibilities: we want low inflation and high employment and we seem to have that, but no economist or economic modeler could ever have predicted or explained why. Everybody is baffled by it. Now we have low inflation and full employment without a theory to realize these two conditions at the same time. . . . This loathsome, egregious resource or environmental economics that presupposes allocative efficiency—that is, giving people more and more of the things they want to buy—it is very difficult to find anybody who thinks that comes to anything but the tautology of "willingness to pay" and "welfare" defined in terms of each other. So you maximize willingness to pay for things that people want to pay for. Why? Because that's what gives them more welfare. How do you know? Well, welfare is what they're willing to pay for. You come to a hopeless and banal tautology. Nonnormative sciences, sciences that really don't have a purpose, bury their emptiness under tautology. So I think that our real problem in making environmental management a normative science is mainly what we want to do with the environment. If that is game management, you could see it as a normative science: you can have all these critters out there and you can blow their heads off or whatever—that's the point of game management. It becomes a normative science. If you're against the norm, then you're against the science. It is terrific! So I haven't the foggiest idea how we would have a normative science of environmental management as we don't have a clue as to what we want to do or what the purpose of the environment or of an ecosystem ultimately is. And without that, we don't even know what environment or ecosystems are; that is, we don't have a sense of them.

It is the same with sustainability. . . . Is it allocative efficiency, layered over with the pale cast of equity and future? I think that all indications are that future generations, come what may, will be much better off in economic terms than we were from our great-grandparents. And not because we preserved the environment at all but because we have completely changed it, making it places like Lewisburg, Pennsylvania. There's not a native species in sight; there are only buildings, and you get your dinner from the cafeteria. That's what has made us better off: technology has made us better off. That's the difference, not environmental protection. And we will find future generations will be much better off because technology is still in its infancy. The growth in genetic technology, in information technology is still before us—I need not tell you we are just putting our toes in those waters. . . . So it would seem to me that the current growth in wealth resulting from technology will continue: future generations will be far richer, far more able to consume. To say that we ought to save resources for them is like our forebears saying that they ought to have saved coal or whales. It doesn't seem a plausible argument to worry about future generations given the march of technology, which is the decisive factor in the production of wealth. . . . But we do need to have the kind of participatory political institutions that will enable questioning. However, what we find—and this is my disagreement with Bryan—is that these damn normative sciences undercut the basis of democratic deliberation, discussion, and problem solving, because the scientists in the progressive spirit say, You can't let the people talk. We know because we have this model . . . we are going to do the efficiency and the allocation, and we are going to have the willingness to pay, and so on. And at that point you just have to sit down in the pew and just let the scientists do their thing until they get exhausted—as I now am, so I thank you.

Norton: It does seem to me that once you add the sort of aggressive technological optimism, as Mark has given us here, then I think you are starting to shift the ground. But I'm not really inclined to take on Mark's technological optimism, given that he has defined it quite narrowly as maintaining levels of consumption. What I would argue is that the pursuit of technological options and its unconstrained consumerism is reducing the options, or in a broad sense the opportunities, for the freedom of people in the future. Perhaps, if they have no nature to worship, they have less, in a sense, than we do, even if they can consume more. So what I've been working on over the past couple of years is a theory that tries to break apart welfare, in the sense of an ability to consume, from maintenance of options for the future. Here, options are not defined as purely economic welfare values but rather as some sense of freedom and opportunity and perhaps spirituality. Then, I would think that we would be moving toward a theory that could admit to the technological optimism argument in that narrow sense, and still argue that we do have a broad sense of obligation to the future and that obligation would be to maintain the range of choices that they have. That would fit very well with what Mark said.

Audience: Am I right that both of you are agreed that ultimately value is located in the human community, and that all considerations, including noneconomic, aesthetic, and religious ones, are human considerations?

Sagoff: Human beings care! We care deeply about nature and we care about nature for its own sake. We care about nature because we think it is created by God. We care about nature because of its intrinsic, glorious, aesthetic characteristics. We care about nature for any number of reasons that are quite independent of human well-being. So we have to distinguish between human values and what those human values are about. If you're asking, Do I take into account only human values? I say, Of course. There are no other values. The fox and the hedgehog can't speak. It is the sheerest kind of anthropomorphism—the sort of thing that Teddy Roosevelt would have laughed at—to think that we could impute preferences or values to animals. The only values that there are, and the only values that count, are human values. But I would also say that as human beings, the content of our environmental values is about nature, not about our own welfare. So we care about nature in itself, but it is *we* who care. If there are other beings that care about nature, I would be glad to bring them into our democratic political process. But I wouldn't do it as a guardian relationship because I don't think any two people would agree as to what the interests or the values of the animals were, outside of our own interpretations.

Audience: I think I understand what I would suspect is a deep bifurcation between humans and nonhumans. Humans impose their own meanings on the "other" in an attempt to comprehend it.

Sagoff: Well, every time we try to comprehend we are imposing. If you knew a person who could comprehend, then have him come up here and tell us X. I will find another person who could comprehend from a different religious background and he'll tell us Y. What I want are those two people to enter a political process and agree on a common environmental policy.

Kassiola: Bryan's and your last comments, Mark, are something I want to focus on. How is it that communities make choices and how does this relate to the hedgehog and the fox? If value pluralism is adopted and there is no rational way to choose among value conflicts because you don't have an overarching principle to resolve incompatible differences, then it seems to me that communities have to make those choices individually. And different communities can make different choices, at different times. And that gets us to the democratic process. In a lot of the environmental literature this point is not emphasized. How do we prepare communities to make these choices? How do we relate the conflicts on the policy level to communities making choices and recognize that different choices, made by different communities, might be appropriate for each? I'd like you to elaborate on that.

Norton: It does seem to me very important that we reframe the debate and the discussion so that diverse viewpoints are expressed, explained, developed, and defended in the political process. I don't accept that pluralism leads to relativism because I believe much of this disagreement can be resolved with respect to people moving closer to each other in terms of what they think is important. But even when that doesn't happen, what I'm impressed with is how often you can actually find a policy that will satisfy quite a range of pluralistic value positions. What I would emphasize is the importance of *process.* The last thing we need is another "elicitation of preferences," as if individuals lived inside boxes sticking their preferences out from under a crack to an economist who calculates and aggregates them and tells us what we want. The community has to make a real effort to listen to all of its components and to create a process where win-win situations are likely to be taken advantage of, where disagreements are reduced to the extent possible, and where fair adjudication takes place when there are irreconcilable differences.

Sagoff: It's like regulatory negotiations. I can think of dozens of cases, such as watershed committees. The Quincy Library Group, where loggers and die-hard environmentalists and others took over when the Forest Service wouldn't do anything and developed their own management plan. A deliberative community that works out a consensus on the basis of reasons is rational. Somebody who elicits preferences, well, people don't have preferences, they have reasons, they have attitudes, they have beliefs. But someone who elicits preferences and then tries to calculate aggregates is not rational, but silly. The rational way to do things is to ask people what they think and to get them to defend that on the basis of reason. You get a representative group of all the different sides, a problem to solve, and accountability and responsibility. That's democracy and that's rationality. Al Gore's reinventing regulation program did this. It is trying to devolve environmental problems: What should we do with this forest? What should we do with this watershed? What should we do with this wetland, this estuary? Giving these responsibilities to boards of trustees, to committees, to diverse representative groups, who then become responsible for solving the problem and finding a purpose or being able to balance the various purposes that these different "environments" might serve.

Kassiola: Let me add just one more point about potential conflict between the political definition of community and the ecological definition. One concern that I thought of is that "politically," the smaller unit might be best . . . the substate level, neighborhood, or even watershed may be ecocentric. But when you get to the ecological requirements of nature, especially because of impacts that might be global, you really don't have the level of institutions available to make decisions that are ecocentric. This raises the problem of how the "process" supported by both of you would work.

Norton: Well, it's a good point, and I think my theoretical response is that the big shift that we need to make in our worldview is toward a different relationship between the present and the future that gives us a more multilayered conception. I believe that this is what Leopold meant by "thinking like a mountain," which is only metaphorical and obviously has to be developed. Hierarchy theory from ecology seems to give you some kind of tools for seeing how small and local processes relate to larger ones in a descriptive model. Now, what we don't have yet and what I think is very important is a better understanding of how individual and small-community values relate to these larger values. I'm not sure if Mark would say, Well, these larger values are just irrelevant. I don't think he'd ultimately say that. And if not, we would both be struggling to develop some sense of layers of human community. I've actually been experimenting with an idea that our values are in an important sense scaled as we experience them: we have *consumptive* individual values, but they have a very short time horizon and a relatively local individual impact; then we have *community* values at a different layer, as we see ourselves as members of society. And if we extend the community to include the *ecological community*, we may, in some sense, move into a form of biocentrism or ecocentrism of some sort. But I don't think anyone has gotten that worked out yet.

Taylor: We've got to be able to distinguish among local, less local, not local, and global problems. There are lots of different types of environmental work to be done. And some problems will need to be addressed in a big forum. Some can be dealt with much more locally. There's not one big policy blueprint for all.

Audience: Back in the eighteenth century, the Vietnamese developed a technique for turning swamp into productive land for the cultivation of rice. And even after the devastation of war they've now moved to number two in the world in the export of rice. That's an application of technology. Number one in the export of rice in the world is the United States. We do this through the heavy subsidization of the price of rice. But there are generations living right now that don't have very much, while we have a lot.

Sagoff: Let me address that. The price of rice is now the lowest it has ever been in history. The reason why we have those subsidies is because rice is so immensely inexpensive; of course, the Japanese do even more subsidization. Where something is a total glut—where you can't get rid of it because there's so much of it—you have to subsidize it. Now, the reason people starve has nothing whatever to do with the amount of rice in the world or the ability of people to produce it, which is pretty much infinite at this point. It has to do with the fact that in Rwanda they're killing you because of who you are. You look: where there's famine, you will find injustice, you will find racism, sexism, and cruelty. And for a person such as Paul Ehrlich to stand up and say, Well, you know, there's a famine because there's not enough food, that deflects the attention from where it should be, which is focused on man's inhumanity to man. It deflects it from the civil wars, racism, and so forth, to his schtick, which is somehow: We don't have enough resources. So people die in the millions because Garrett Hardin says, We can't produce for everyone. Of course we can produce for everyone! We have enormous overproductive capacity. We didn't know what to do with our potato surplus last year. They all rotted because people couldn't even pay the transportation costs to get them to where they were needed.

Audience: I have a problem with what you're saying. What is Monsanto's role in this? What does justice mean here . . . is it a matter of democracy and decentralization, letting people produce their own crops rather than having to buy their food from corporate farms? Does the environmental community see this as an issue?

Sagoff: I'll tell you something about Monsanto that's important. It is very easy to make transgenic crops. That is to say, your agricultural extension can do it. And there are lots of advantages to transgenic crops. Monsanto and Novartis are going to divide agriculture between them. Already 40 percent of our soybeans are transgenic "roundup" crops. Everybody predicts that it is going to be 80 or 90 percent in the near future. Monsanto is going to integrate these farms into a transgenic economy of scale, and we don't like it. But how is it that they are getting away with it? Why isn't it that the local extension agency or your state government and/or the small producers can't be part of the transgenic revolution? I'll tell you why . . . because to get a transgenic canola, you have to spend $200 million field-testing it. Why? Because Monsanto was able to use the safety concerns cynically and self-consciously raised by environmentalists to build up an entry fee to developing transgenic crops, which are the crops of the future, so that no small operator could do it. The same with pharmaceuticals. We have to accept in this society the idea that there are only a few big pharmaceutical companies. . . . Why are there so few? Because it takes $200 million to bring one of those products to market. Because the safety of drugs is so important that we are willing to accept that high hurdle to market entry. Monsanto has therefore used the environmentalists and their safety concern to protect markets for things that are otherwise utterly and completely safe. Tomatoes are not going to eat up Chicago! But corporations exploit that fear. Their lobbyists dwell on it to get the same kind of field-testing requirements for their transgenic crops as for pharmaceuticals. That means small-time operators can't compete with them. Safety concerns also lead to the fiasco of patenting genes. Safety has nothing to do with patenting, but we focused on the patents! We are locked in this horrendous situation largely because environmentalists have been unable to make the proper moral arguments to create more environmentally sound circumstances. Maybe it is not safety. Maybe we have to move past instrumental economic or health values and engage in real moral argument about policy ends. Otherwise, we just play into the hands of corporations, I'm sorry to say.

Issue 4: A Sustainable Environment as an Intrinsic Value?

Sustainability: Restricting the Policy Debate

John Martin Gillroy

Oscar Wilde argued that the definition of a cynic is a person who knows the "price of everything and the value of nothing." I would add that a positivist or empiricist is one who understands the measurements of everything but the integrity of nothing. My argument here is against the skeptical empiricist who would limit our discussions of environment and policy to instrumental value considerations alone. We need to start with a few definitions and distinctions.

First, the distinction between policy ends and means: the ends of a policy are the goals, but more, the determinants of the policy and the standards by which we will judge accountability and success. The ends of the policy represent our core values and are the reasons why we act collectively. The means of the policy is the way in which we achieve, or implement, our ends. Here we consider the different alternative routes to the same objective and make judgments about the "proper" and "reasonable" way to get there. Both the means and ends of policy require standards of evaluation.

We also need to distinguish between cost-benefit methods and cost-effectiveness analysis. Cost-benefit is a process based on the principle of Kaldor efficiency that judges both the ends and the means of a policy in terms of its net monetary benefit (e.g., both the environmental standard itself and the permit process that implements it). On the other hand, cost-effectiveness analysis expands the value conversation by leaving the ends of policy to be judged by principles other than economic efficiency, while it saves Kaldor considerations for the means to those ends. For example, setting ambient air quality standards by health requirements, then implementing a market trading permit system to implement them is an example of cost-effectiveness (Gillroy and Wade 1992, 5–7).

Third, we should distinguish between instrumental and intrinsic value. Some would call me a Cartesian dualist for doing this, but I will anyway. Intrinsic value is value independent of use, whereas instrumental value is use value, whether it is a tree for lumber, an animal for its fur, or a person for his or her slave labor.

Last, I want to define "modern positivism" as the belief that no intrinsic value exists. Positivism traditionally divides fact from value and concentrates on the superiority and independence of the former from the latter, in effect ending the conversation about value by denying that it exists. Natural scientists argue that their study and analysis is "value-free" and based on fact and empirical observation independent of normative evaluation. Although this traditional positivism remains in only a few circles, its modern version admits that values are important but argues that any normative content in one's analysis can also be based on empirical foundations related to observed behavior, human preference, and instrumental use value. These are assumed to have real physical properties and to be tangible, and therefore measurable, dimensions of human agency.

From these distinctions, let me state my argument, namely, that sustainability has a clear and concise "empiricist" definition that works only to support the continued instrumental use of the environment as a means to our policy ends. This definition becomes confusing and innocuous when it is used to evaluate the intrinsic, nonpositivist, nonefficiency properties of humans or the natural world. The confusion and lack of effect in these cases come from the fact that sustainability cannot evaluate these dimensions of humanity or nature, and so it seems ineffective, fuzzy, and useless, or "all things to all people."

The use of sustainability to "represent" other than economic values is both cynical and positivist. It is cynical because it has contempt for what are the most important qualities of living things and also because it pretends to do what it cannot. It is positivist because it co-ops noneconomic values into economic language so that they might be "measured."

The Limits of Sustainability

We meet to discuss the state of humanity's relationship to the natural environment and the methods and concepts that we take into the next century as the moral basis of our rights, the foundation of our definition of social justice, and the standards for our interdependence of spirit with nature.

Concepts such as rights and social justice imply moral values that inform the democratic debates that shape our policy lives. Knowing that we have rights implies the knowledge that humans have intrinsic value independent of their use by others. But this implication does not end the democratic debate, as some would argue, but enriches it. Understanding ourselves as having both use and humanity gives us two levels of trade-offs instead of one and a richer moral discourse about who we are and how the rights of the individual are to be weighed against the integrity of the community. This is not controversial and is part and parcel of our constitutional republic. In this sense, we are all dualists in that we acknowledge both our own social utility and our essential capacities as human moral agents.

However, instead of enriching this democratic debate even more by adding the instrumental and intrinsic value of nature to the instrumental and intrinsic value of humanity in policy trade-offs, we are met with cynicism: first, by the counterargument that there is no such thing as intrinsic value for nature, and second, that the search for it is some totalitarian plan to undermine democratic institutions that are represented by instrumental value alone. First, I fail to understand how a dialogue entirely in terms of our, or nature's, instrumental value is inherently democratic. It was the disagreement about whether individuals were property or persons that generated policy debate in the 1850s, not the assumption that all humans were properly measured in instrumental terms alone. (Would a civil war have been fought over that?)

Second, why would one assume that adding a category of value eliminates all other categories? Even if the Cambridge Dictionary of Philosophy (1995, 829–30) is correct and instrumental and complementary value come out of, and relate to, intrinsic value, this does not preclude a democratic process deciding that intrinsic value needs to be traded away for instrumental. We have already traded the destruction of species for human progress. The point is to fully evaluate what is at stake in the decision, and for these purposes, the simple fact is that the inclusion of intrinsic value enlarges, enriches, and makes democratic debate *more* complex, not less.

The same cannot be said for the concept of sustainability. Although proponents of its use argue that with it they are enlarging the terms of policy debate, this is not the case. Sustainability is simply another in a long line of definitions for efficient use of the environment that began in the colonial period with "maximum" use, continued in the age of environmental statutes with "optimal" use, and now, on the verge of a new millennium, focuses on "sustainable" use (Gillroy 2000, ch. 3; Andrews 1999).

The concept of sustainability limits the political conversation to matters of efficiency, even though it claims more. Worse, it improperly evaluates the non-efficiency dimensions of environmental questions by extracting from capacity and integrity only those factors that can be empirically quantified and processed. Sustainability, from this perspective, essentially deals with instrumental values and does not make the democratic discussion more complex, but leaves us with efficiency as the primary principle of policymaking and preference for use as the primary guide to environmental policy. This also makes efficiency the principle for both the definition of sustainable policy ends and the selection of sustainable means to those ends, making cost-benefit methods essential to decision making.

An example here is how sustainability treats future generations, which are an essential component of its definition. Sustainability is a positivist doctrine. It focuses on observable human action, physical development, and the resource use of nature, conditioning this use only with a concern for future human use. But how does it integrate intergenerational justice? It does so not by analyzing the common intrinsic capacities of human beings but through the preferences of present generations for environmental resources. Future humans have no physical presence, observable actions, present needs, or manifest preferences, and their inclusion in policy deliberation can be done only by considering the present preferences of real people for future savings. Therefore, sustainability as a positivist doctrine depends on the present generation's preferences to dictate the future's resource base, and argues for conservation measured by the current expected utility of any resource to the future, rather than what we owe future generations in terms of our common humanity or our capacity to flourish as moral agents. This makes sustainability an instrumentalist doctrine, not just in terms of humanity's use of nature but in terms of our use of the future and their resource base as means to our ends.

Sustainability is also a means doctrine that recognizes the environment as a materials balance of resources-to-goods-to-waste and that concentrates on the implementation of human policy so that this transformation process conserves over time based on present preference. Here, sustainability could be a sound principle to set standards for policy means, as it implements ends based on other principles in an efficient manner. However, when sustainability is used to measure both ends and means it becomes a kind of cost-benefit analysis with a smaller discount rate in consideration of future resource availability. This comes out clearly with an examination of the field of ecological economics.

Ecological Economics

According to Costanza (et al. 1997), ecological economics "accepts much of the neoclassical [economic] theory regarding allocation" (1997, 80). This bases the field firmly on the foundation of a Kaldor definition of efficient allocations based on "individual preference" and the "ability of the individual to pay" for the goods he or she wants (80). The difference between ecological economics and neoclassical economics, according to these authors, is that two other "goals" are added to efficiency: "sustainable scale" and "fair distribution" (79). Specifically, we need to "first, establish the ecological limits of sustainable scale, and establish policies that assure that the throughput of the economy stays within these limits. Second, establish a fair and just distribution of resources using a system of property rights and transfers. . . . Third, once scale and distribution problems are solved, market-based mechanisms can be used to allocate resources efficiently" (83). With these changes, the claim is that the policy generated by ecological economics will protect ecosystem health and lead to "natural capital preservation and restoration" (86) and a shift in the "burden of proof . . . by requiring the demonstration of safety by the emitter before [natural systems] use" (194).

At first glance, by integrating these new elements it would seem that we have enlarged the policy debate. It might seem that the trade-off structure is more complex, by letting nonefficiency principles (sustainability and fairness) define policy ends while we limit efficiency to the definition of means alone, creating a cost-effectiveness analysis for environmental policy. But there is less here than meets the eye.

This is only cost-effectiveness analysis if sustainability and fairness are based on a principle, or moral value concept, other than instrumental efficiency, for if they are not, sustainability and fairness reduce to efficient allocation and ecological economics becomes neoclassical economics with a low discount rate.

For sustainability and fairness to be other than efficiency, they must contain a principle that is able to value the nonuse dimensions of humanity and nature. To value other than instrumentally, sustainability and fairness must relate to intrinsic qualities of humanity and/or nature in order to define ends in terms distinct from efficient means. This might require language related to the rights of individuals, their capacities as moral agents, and the duties to themselves and others generated by these capacities, or a concern for the cultural integrity of community or the functional integrity of nature. But the language of ecological economics uses no such terminology.

In discussing "ecosystem health," for example, ecological economics speaks of "harvested sustainability" and "optimal throughput" of goods into and out of the materials balance. "Ecosystem/ecological services" (Costanza, et al. 1997, 81–92) are defined and even "trusteeship" is referred to in terms of "willingness to pay" (82). "Fair distribution" is put in terms of property rights, as these are a necessary perquisite for market/resource trade (83), the way Ronald Coase defines them.

Ecological economics claims a pecking order of values. First comes sustainable scale, then fair distribution, then efficient allocation. Others may claim that this makes ecological economics undemocratic, but not I: I would welcome such a complexity of trade-offs and policy considerations. All policy is a process of debate where trump principles are decided on through discourse. No, the question is whether there is a pecking order of values here at all, or just one value in three masks.

In the end, ecological economics declares that as we move into the twenty-first century, "we need a different criterion for what is 'optimal' " (Costanza, et al. 1997, 140). This quote tells the story. What we have here is sustainability, fair distribution, and efficient allocation all being reduced to economic terms for the purposes of redefining optimal use of the environment so that it protects present resources for future people's use. Within ecological economics, it allows efficiency to determine both the means and ends of policy, and therefore it is not cost-effectiveness analysis as first assumed, but cost-benefit analysis with a lower discount rate.

Wiener and Percival

A perfect example of modern positivism among the essays in this volume is Jonathan Wiener's contribution. Here, like the ecological economist, his entire worldview is so dominated by the market model and its inherent separation of policy and morality that he uses a single vocabulary of Kaldor microeconomics to describe both market failure and what he calls "government failure." He insists that government decisions can be as counterproductive for the environment and the concept of sustainability as market failures, which is not in itself an unreasonable point, but he fails to perceive that the state should be judged on its own terms and not those of the market. Like a person who applies human moral categories, such as rights and interests, to animals, Wiener judges government decision making not in terms of its unique public duties to ecological or human integrity but in terms of its being too strong or weak in terms of such market measures as private property rights, instrumental values, and responsiveness to preferences for efficiently balancing "countervailing risks."

Specifically, when he speaks of government failure to protect rights, Wiener is not talking of a right to a clean environment or to protection from risks that are beyond the control of the individual, but the state's failure to "create exclusive property rights." When he blames government for passing laws that push pollution from one medium to the other, he fails to appreciate that this failure comes from the predisposition of policymakers to see the environment through the eyes of the market as a series of species for extraction and media for sinks, rather than as an integrated ecological whole with value independent of our use that might be considered in making policy choices (Gillroy 2000, ch. 3). When he talks of a government "free-rider" problem, he assumes that nature is a private rather than a collective good. His solution to government failure is "liability," a market solution.

Throughout his argument Wiener essentially judges failure entirely in market terms, no matter its origin in markets or government. Essentially, he blames government for its failure to do what a market would do if it were able to operate, and therefore he creates government failure as a subcategory of market failure, with the same principles and assumptions as if these must apply to both settings. Alternative normative principles and assumptions that might define the state as an independent entity with distinct duties and ends, and a concern for intrinsic value, does not enter his analysis. Overall, he seems to assume, at a subconscious level, that sustainable policy must be human- rather than nature-centered and his preoccupation remains that of the good modern positivist, with "the basic inability of the state to underpin market activity."

Meanwhile, Percival, in his search for global accountability, also assumes the principles and assumptions of the market, but with a distinctly different and more optimistic cast. Specifically, he is in search of a way to take government and its duties, responsibilities, and ends seriously on their own terms, and is willing to consider that these definitions may require nonmarket vocabulary. Beginning with the premise that sustainability is a "malleable" concept, he finds that "the popularity of the concept may stem from its very malleability" and moves immediately to acknowledge that the current debate about sustainable futures "emphasiz[es] the importance of market strategies." This acknowledgment, however, makes all the difference, for it is a removal of the author from the condition of true believer in blind modern positivism to the recognition that one might have to free oneself of market terminology to properly evaluate and define environmental accountability.

Therefore, though he does examine "property rights," "tradeable allowances," and "liability" as means to accountability, as does Wiener, Percival also suggests that citizen participation and information provision are unique functions of government that define duties within the public sector. He examines efficiency as a conventional point of departure for market models of sustainable development, but he also focuses on fairness and equality as alternative principles that might redefine accountability and sustainable development. In the final section of his essay, he argues for "expanding" the reach of law globally so that it is better able to control market forces that both reduce humanity and nature to instrumental values and narrow accountability to considerations of economic costs and benefits alone. After quoting Kofi Annan that "the spread of markets far outpaces the ability of societies and their political systems to adjust to them," Percival ends his argument with the suggestion that the secretary-general's words "reflect a growing appreciation that additional institutional mechanisms for holding multinational businesses accountable will be required in an era of unprecedented global capital flows and trade liberalization." This search for new principles and concepts is an acknowledgment of the limitations of sustainability and a step away from modern positivism.

Laitos and Lowry

The specific failure of the modern positivist definition of sustainability becomes most clear in the essays of Laitos and Lowry. Here, sustainability as the new standard for positivism supports efficiency principles in policy and law and works well defining efficient ends for cost-benefit evaluation. It fails when it is used as a principle in support of intrinsic value or preservation ends. Laitos and Lowry demonstrate that sustainability works only to narrow the policy conversation to matters of the optimal and instrumental use of nature.

Laitos examines public land policy with an eye to establishing an efficient allocation of nature between what he calls "preservation uses" and traditional "resource uses." One of his measurements for preservation is recreation, which is defined as the "price of a recreation visitor's day" or the "number of visitor days" (Table 1). This argument works, as these are economic measures of a basically instrumental and economic nature, which is what instrumental positivism can best evaluate. However, when he tries to measure preservation uses independent of recreation he is reduced to consideration of "monetary value per acre" and "imputed market value of services" (Table 2), which say nothing about whole systems, functional integrity, or uniqueness of any natural system subject to restoration or management. Like all instrumental analysis, his preservation use category extracts what can be measured in economic terms and ignores the rest because the principle of efficiency cannot evaluate it.

Overall, Laitos argues all "uses" in terms of quantitative monetary factors, willingness to pay, and preferences, with efficiency as a core principle. He is then able to "overlay on the efficiency model the notion of sustainable development" with greater ease and effect than any of the other case study authors. He concludes, as a natural extension of his cost-benefit methods, that it would be useful to factor into public land decisions a "balancing of economic and environmental and social equity interests in light of intergenerational sustainability."

Overall, he argues that sustainability can "improve on the current multiple-use paradigm and be a useful adjunct to an efficiency criterion." But this is like saying that sustainable efficiency can be mapped onto optimal efficiency in a cost-benefit argument where economic factors determine both the ends and means of policy. The policy conversation is not enlarged but restricted.

Lowry examines national park policy in Canada and the United States from within an institutional conflict between those supporting traditional multiple-use management and the new trend toward ecosystem management and preservation of whole natural systems. In a series of cases from both countries he consistently demonstrates that multiple use is more easily measured by sustainability analysis, whereas the preservation of whole ecosystems that are of no recreational or other use to humanity is very hard for either nation to analyze or support as "sustainable" policy. He finds that it is more difficult, within the concept of sustainability, for agencies to plan and implement preservation than use and that "unsystematic" growth of the parks in both countries has foiled the long-term plans to integrate and restore whole ecosystems like Yellowstone and Banff.

He finds that there is a tendency to "provide immediate benefits" rather than benefits for "future generations" and that "long-term planning" never really replaces "crisis" management. Even though Lowry wants sustainability to expand the conversation about national parks with concerns for preservation as well as use, the concept is not up to the task. Even with a new management scheme, we are restricted in what sets policy ends because we are restricted in how we value ourselves and nature.

What works well for Laitos fails for Lowry, and for the same reason, namely, that sustainability is a form of optimal efficiency that grafts well onto a narrowly focused economic analysis of the environment utilizing cost-benefit analysis but cannot expand the policy debate by bringing in noninstrumental, nonefficiency values or principles to define policy ends within a cost-effectiveness analysis.

Conclusion

If we continue within the conventional market paradigm and refuse to consider a new vocabulary of public responsibility and intrinsic value, modern positivism will define sustainability as a new category of optimal efficiency and it will remain the dominant principle of twenty-first–century environmental policy. Then the public debate will be restricted by the extent of the market's power to evaluate and its strict separation of politics and morality, which is marked by the following characteristics:

—Argument limited to the evolution of instrumental value
—Argument limited to valuation of parts with use, not wholes with function
—Argument implying the goal of subsistence, not flourishing
—Argument assuming use without justification (default drive)
Therefore, the survey of alternatives is not critical
Thresholds must be specified and benefit goes to more rather than less use, "guarantee[ing] that the natural capacity for environmental self-renewal is fully utilized" (Goodin 1976, 176)
—Argument supporting the market as the primary allocation and distribution mechanism

If we wish to expand the policy debate and move from cost-benefit to cost-effectiveness analysis so that we take account not only of efficient allocation but other environmental values, then we need policy argument to be able to take proper account of:

—The whole, not sectors
—Nature-centered as well as human-centered policy options
—The flourishing as well as the subsistence integrity of natural systems
—The assumption of nonuse where use needs justification
—Nature as a collective as well as a private good

If we are to move to an alternative methodology, perhaps ecosystem management, we will not be successful unless we also expand the value base on which this methodology is built. This will require that, in addition to thinking about nature as instrumental to our means, we must also consider it as having value independent of us. This expansion of moral argument in environmental policymaking is not for the purpose of ending discussion but of enlarging it to take account of the complexities that surround us and that are at stake in our democratic policy choices.

Instead of the positivist with a hammer to whom everything looks like a nail, we might envision a policymaker with a more diverse toolbox able to evaluate the ends and means of policy on different scales and with appropriate concepts, depending on the characteristics of the issue (Gillroy 2000, ch. 1). This policymaker might be able to judge the integrity as well as measure the economic worth of human and natural systems and make decisions with the *value* of everything as well as its *price* in mind.

Comments on Sustainability

Bob Pepperman Taylor

Reading through the case studies in Part 2 of this volume, I'm struck by three things. First, there is no real consistency from essay to essay in the use of the term "sustainability" or related phrases such as "sustainable development." Second, the idea of sustainability appears to do very little conceptual work for any of these authors. Finally, whatever work sustainability does do is much more modest than we might expect or even hope. In the end, it seems to me that sustainability suffers from many of the same problems that "intrinsic value" does, and we should be wary of using the term for similar reasons. Let me consider each of these issues in turn.

Campbell-Mohn says very little about these goals and principles of sustainability other than to suggest that the primary aim of sustainable development is "to manage natural systems for the perpetuation of the human species now and in the future." She also suggests that a precept of sustainable development is "avoiding pollution at the beginning, in the extraction phase, through managing natural systems," although it isn't clear how this goal grows from or reflects her previously stated understanding of sustainability.

Some of the other authors give a little more attention to the meaning of sustainability, but the conceptualizations are very general and not consistent throughout. Rabe, Laitos, Percival, and Schwab and Brower draw on the classic statement from the Brundtland Commission, which defines sustainable development as "development that meets the needs of the present without compromising the ability of future generations to meet their own needs." In addition, as Laitos quotes, it is a "process in which the exploitation of resources [and] the direction of investments . . . are all in harmony and enhance both current and future potential to meet human needs and aspirations" (World

Commission on Environment and Development 1987, 43). But Laitos also notes that sustainable development has become "a term of art within federal agencies, which employ it in various programs as if its content and scope were accepted as a norm." The reality, he points out, is that there "is no consensus about the meaning of the term, and no new laws have been enacted that further define it." Schwab and Brower actually begin their essay with a much more literal understanding of sustainability than the Brundtland definition they eventually turn to, suggesting it is the idea of ecological viability: "Our growth-acclimated society has exceeded the Earth's capacity to sustain us while we continue to follow a growth-at-all costs lifestyle." Lowry recognizes that "sustainable development can take many forms" but also suggests that it has fundamentally to do with meeting the needs of the present generation without harming the options for future generations. Last, Wiener seems to have a market definition of sustainability as efficiency.

The meaning of sustainability in these arguments thus runs from undefined, to a very general principle about the rights of future generations, to what Locke calls "enough, and as good" as that enjoyed by contemporary generations (*Second Treatise*, par. 27), to a notion of ecological limits beyond which environmental catastrophe lurks.

This leads to my second observation: sustainability is not really doing much moral or conceptual work in any of these essays. Let's begin with Rabe's discussion of Minnesota's innovations in environmental policy. Rabe concludes that these policies suggest "that it is indeed possible to begin to respond to the sustainable development challenges set forth by the Brundtland Commission." This may be true, but it is also true that the reader has learned very little here about the notion of sustainable development to help her or him evaluate the claim's accuracy. Rabe notes a "shift toward sustainability" in the 1990s in Minnesota, but the policy developments he then traces could easily, perhaps best, be described as something like *reasonable and successful environmental policy*, or *good and aggressive environmental policy*. This is all well and good, of course, but it isn't clear why the word sustainability is required to make the point.

We find a similar quality in some of the other papers. Campbell-Mohn argues that communities can rightly regulate private property to promote public goods such as environmental protection. Although I think this claim is certainly true, it succeeds or fails regardless of Campbell-Mohn's comments about sustainability. Lowry's discussion of how political institutions in Canada and the United States make it difficult to pursue a coherent management policy owes very little to the notion of sustainability and a great deal to a fairly conventional (and very important) analysis of the political pressures on public administrators. Schwab and Brower argue that we should do a better job of controlling growth and development and mitigating the effects of various hazards. As convincing as this point is, it seems profoundly disconnected from the apocalyptic tone of the opening passages, in which they convey a notion of sustainability as basic ecological viability. Or consider the disconnection between the following passages. In the first, the authors suggest that "sustainable development has emerged as a paradigm with the potential to give human beings the perspective and the power they need to rediscover our proper niche in the Earth's panoply." The idea here seems to be that sustainability promises a transformation of the human relationship to nature, a discovery of an entirely new way of living. In the second passage, however, the authors argue that the notion of sustainability, when combined with a proper understanding of "natural hazard mitigation," allows for a more humble outcome: "Neither principle necessarily proposes a 'no growth' policy for communities to become less vulnerable and more sustainable. Rather, these concepts advocate for the safety of future population rise through conscientiously controlled growth and development." This would seem to owe a lot to notions of administrative effectiveness but very little to "our proper niche in the Earth's panoply."

Jan Laitos appears to value the utility of market efficiency above most other possible values, but he considers the notion of sustainability toward the end of his article in order to introduce an element of intergenerational equity to his evaluative structure for thinking about the uses of public lands. It is immediately clear, however, that he is uncomfortable with such nonutilitarian considerations because they are hard to define and measure, don't necessarily promote efficiency, and seem to mean different things to different people. To the degree that sustainability is thought of as intergenerational equality, it is a coherent enough notion, although as Laitos recognizes, it would require a great deal of analysis to explain exactly what intergenerational equality could possibly mean. At least partly because of this, Laitos seems very ambivalent about the idea, and it is clear that the most important message of his essay concerns his interpretation of how market efficiency requires a greater emphasis on preservation than on "multiple use."

Another way of looking at this point is to suggest that all the values being defended in these essays can easily be expressed in more conventional, old-fashioned language. A number of the authors—Rabe, Lowry, Schwab and Brower—promote a kind of administrative rationality and coherence; Campbell-Mohn defends a notion of local public goods; Laitos and Wiener advocate efficiency and, to a lesser degree, equality. Now, I am not one to think there is any magic in words themselves, and if we decide to call any or all of these values sustainability, so be it. But it isn't clear to me what the benefit of doing this would be, especially given the long-standing traditions of using other words and languages to express the same ideas. What does the idea of sustainability add to the arguments themselves? In actuality, it seems that less rather than more clarity is gained by employing a term that some of these authors themselves recognize is ambiguous and even politicized. If the presumption is that we need some new value such as sustainability to guide us in our struggles with contemporary environmental problems, the presumption seems clearly to be false. These authors themselves, sometimes seemingly contrary to their own intentions, offer trenchant analyses of environmental problems by drawing primarily on very conventional moral tools. In fact, the force behind their theses owes much more to these conventional ideals than to some new values captured by the word sustainable.

It isn't surprising that this should be the case, because even granting a "coherence" to the idea of sustainability, the notion is of limited moral usefulness. This is most easily seen when sustainability is thought of explicitly as a kind of ecological or environmental viability, as it is at the beginning of Schwab and Brower's essay. If the bottom line of our environmental evaluations is ecological viability, then there is clearly a huge range of options available to us in pursuing environmental policy. In my home state of Vermont, for example, we have, over the course of the past century, cut down the forests and then regrown them. Both the deforested and the forested Vermont have supported human society to a greater or lesser degree. My suspicion, however, is that few advocates of sustainability would be happy with a policy promoting the devastation of the forests like that experienced in Vermont in the past century. The reasons to oppose such policies, however, would have to be more rigorous than arguments about ecological viability. Rather, they would have to defend a very specific understanding of the particular kind of life and environment that is not only necessary but is also desirable for human beings (which, of course, may include a strong appreciation of, respect for, and humility before the natural world). What is necessary may be a part of the good, but it is certainly not the whole of it.

Even if we think of sustainability as providing for a kind of intergenerational equality, it is difficult to avoid the idea of sustainability as a kind of environmental necessity. If we think of this equality in more robust terms than a minimal equality in chances for living, it becomes very difficult to understand exactly what such equality would mean. What our generation does, produces, invents, creates, and becomes not only has an impact on natural resources; it also sets the parameters for the possible character(s) of future generations in sometimes predictable but primarily unpredictable ways. The invention of modern democratic institutions altered the fundamental conditions of existence for those who inherited them, as has the invention of the computer and the electric guitar. What would it mean for us to be equal with people living under radically different social, technological, and political conditions from us, conditions that it is impossible for us to imagine and understand? This is a huge problem, especially if we consider that there is no good reason to believe that those future generations will value the same things we do.

Beyond the minimal requirement of bequeathing a viable environment to those who follow us, sustainability tells us very little about, say, why we should leave future generations a more heavily forested, less polluted, better preserved natural environment. Why we should do these things owes much more to our conceptions of the good, and our commitment to transmitting this good as best we can to our descendants, than to a notion of sustainability. By focusing on sustainability, we are always tempted to make too much of arguments about viability and not nearly enough about arguments for the good.

Some conception of human good beyond survival informs all the essays in Part 2, but the reliance on sustainability obfuscates rather than clarifies these notions. For example, we can certainly muddle along with our national parks despite the incoherence of our current political approach to managing them. The power of Lowry's analysis is his claim that we will seriously fail to achieve our (rightly held and justifiable) goals for our national parks if we don't find a way of making our policies and implementation more coherent and cohesive. Arguments about sustainability seem like poor candidates for replacing these more important and difficult arguments about the full range of desirable environmental policy. In this sense, sustainability appears to be less powerful or useful than we would wish or require for our environmental deliberations.

It is sustainability's weakness, however, that also accounts for its attractiveness. Even though sustainability directs us to think about environmental necessities and viabilities, and thus tells us little about what policies to pursue once these minimal requirements are cared for, the seductiveness of the idea lies precisely in its appeal to necessity. If a policy is required in order to maintain sustainability, that policy would seem to be beyond the reach of principled opposition. In Part 1 of this book, I suggested that environmental ethicists have been drawn to the idea of "intrinsic value" to silence environmental debate, to develop a value that would, in Eugene Hargrove's words, "trump instrumental values" and thereby silence all possible objections to preservationist environmental policies. Although there are other ways of thinking about the idea of intrinsic value (remember the example I use of Mark Sagoff's discussion of *Charlotte's Web*), on the whole, as I argued in Part 1, the concept has evolved from the desire in the environmental ethics literature to put an end to debate about environmental policy rather than to democratically inform that debate. Sustainability appeals to us, at least in part, for the same reason, for it would appear to be an absolutely conclusive moral argument. It lends an air of necessity to our advocacy. If we are unable to make an argument about what we must of necessity do, we are forced into the position of having to argue what we should do. The latter, sadly, is always harder to defend than the former.

The essays in this volume, however, show us that the more difficult project is the one we must attend to. If my argument above is correct, these authors are defending values and ideals that cannot be fully understood or explained by appealing to necessity. When environmental viability is at stake, by all means we must recognize and respond to this. I certainly would not want to underestimate the contribution of environmental science in teaching us about truly unsustainable human behaviors and institutions. But despite the practical problems such unsustainability presents, these are not the morally difficult dilemmas we face, and environmentalists discredit their own arguments when they try to reduce all environmental problems and choices to matters of necessity. Even with matters as important as those addressed by the authors of these essays, it is clear that that is not the primary or fundamental problem they each face. To try to reduce these problems to sustainability is to cloud rather than illuminate our judgment.

One of the consequences of appealing to sustainability is parallel to a consequence of appealing to intrinsic value: it seems to allow the certain knowledge of experts to replace the uncertain evaluations of citizens. Schwab and Brower's discussion of "competing values" is instructive here. They point out that different constituencies may view the ethical imperatives of hazard mitigation differently. They also point out that values can compete with one another, such as when we are sometimes forced to balance environmental protection against public safety. They suggest that planners can devise strategies for minimizing these conflicts, but they also recognize that "scholars" disagree on the proper way to understand the "ethic of conservation and protection." There is no reason to think that the idea of sustainability does anything to resolve the value conflicts that remain after consultation with experts (and remain they most certainly will), or that it will allow us to discover the proper ethical balance among values promoted by different constituencies or different experts.

At the end of the day, judgments will have to be made when we are confronted by environmental problems and dilemmas. Despite the implication that the vaguely scientific notion of sustainability can direct us in making these judgments, there is no such magic bullet to be discovered for all those decisions that concern matters above and beyond environmental necessities. Because there is no reason to think that lawyers or planners or public administrators have any special access to knowledge of the correct way to make these judgments, it makes practical sense to minimize our appeals to sustainability. To do otherwise is to confuse the issue by appearing to promise that normative decisions can grow from the scientific or technical knowledge of experts rather than the judgments of citizens.

Discussion

Norton: I agree with John that the ecological economists do retain essentially an economic perception of value, but I don't see why he's so quick to give them the term sustainability, as they would also say that they include all of those values, the nonuse values, or passive-use values, or whatever you want to call them. It seems to me that throughout the debate, more in environmental ethics than in economics, there is a tendency to narrow human values down to the bare essentials of subsistence. Subsistence may also include conspicuous consumption but doesn't include spiritual values or anything like that. So you first collapse the concept of human value, and then say it's missing something. Well, there's something perverse about that. It's like chopping off someone's arm, then blaming the person for not being a whole human being. What you want to do is to force people to see that when we talk about human values it can't be reduced to those mere "ciphers," as Leopold used to say. Why give the economists the term sustainability?

Gillroy: If I am giving the economists the term sustainability, it may be because it is defined by their principles and can evaluate nature and humanity in only economic terms. Sustainability does certain things well and other things not so well. Sustainability cannot protect an old-growth forest. Saying that sustainability has evaluative power that can protect ancient forest, independent of our use, doesn't help that old-growth forest if in fact it can't produce adequate policy. Sustainable policy, from within, for example, the Multiple Use and Sustained Yield Act, ends up evaluating a tree, any tree, ancient or not, as board feet of lumber. We are deceiving ourselves to think otherwise. Making sustainability represent what it cannot properly evaluate is not only dishonest, but it inhibits the development of policy paradigms that might successfully create law to protect old-growth ecosystems. Ultimately, I'd like to see sustainability represent one alternative principle for policy choice while others were developed and applied as alternative moral arguments. Then we could have a discussion about how different principles would affect policymaking. Would they produce different policies for wolves in Yellowstone, or old-growth forests, or global warming? How would each trade values? How would humanity and nature be worse off or better off from each alternative? I want to challenge people to think of what is a concrete, well-formed alternative argument here that has its roots, its foundation in different values, in different principles than efficiency and sustainability. This is what I have tried to do in my latest book, *Justice and Nature.*

Kassiola: Schwab and Brower emphasize the "transformational aspect" of the principle of sustainability. Would it be your suggestion that they should not use sustainable development if that's what they want, as it requires more, in an evaluative or moral sense, of sustainable development than it is capable of giving?

Gillroy: I think that's a good case, Joel. Schwab and Brower acknowledge that there is a moral discourse that has to take place, and that there is more than use value that has to be considered, and that humanity creates hazard, or creates the concept of hazard. My problem is that they stop there. Why not carry the analysis through? Because, if you have a worldview or ideology that uses certain terminology and certain ways of organizing the world, it evaluates only what it can normatively grasp, and it takes into consideration only what it can properly value, and what it can't take in, you can't fully analyze with it. I think that was the problem with Schwab and Brower. Sustainability can't properly evaluate "transformational value" because this would require more than an efficiency standard. This limited the analysis. Alternatives are hard, because basically we've had economic language separating political and moral debate and dominating policy analysis for so long. In policy schools we feel a responsibility to teach only the language of efficiency and markets to our students, because it's the only language they're going to need when they get into the job market. This attitude creates, of course, a self-fulfilling prophecy. We don't expect policy, or law students for that matter, to really understand normative discourse or the formulation of moral argument. What could Hume or Kant possibly tell them about policy? But by adopting a concept such as sustainability, you adopt the meanings and the limited moral range of that concept, making it necessary to move past the language of economics to get at intrinsic valuations of anything, human or natural.

Bowersox: I guess I'm like Bryan here; you don't want to concede that.

There is a group of writers like Herman Daly who argue that sustainability is not about economics but about morality. That is a sort of argument that makes sustainability in some conceptions very subversive.

Gillroy: Yes! But sustainability is about the morality of efficiency. To assume otherwise would place expectations on it to evaluate more than prices, costs, and welfare benefit. That's what I'm worried about. If we want to do more, we have to be conscious of the fact that we need a new language of environmental values. We can't just fall back on "willingness to pay" and "throughput" and "materials balance" language. We need something else to compete with that language. And I'm worried that the fuzziness of sustainability and its ill repute may come from the fact that we think it's capable of doing more than economic evaluation, and then we are disappointed with the policy results, that is, when the Forest Service continues to cut old growth. It is not "preservation" and "least cost" rolled into one. It's making environmentalists happy, when it really shouldn't, because it's not really doing any work for them. I think we have to do what Joel talked about earlier. We need to engage in the debate about principles and produce alternative policy arguments if we want policy to fit our principles.

Taylor: I'm all for the use of subversive words. I heard on National Public Radio a story about the gay bars that are opening in Beijing. The language that the gay community is using is the language of "comrade." They are calling everyone their comrade, which I think is lovely, and wonderfully subversive in the face of the Communist Party, as it should be. And to the degree that we can use words like this rhetorically to make points, that's fine, that's good. But I see sustainability being a smokescreen for lots of people who want to create things that I don't like. I don't know if the word can be saved and used subversively or not; once again, I'm completely pragmatic on this. If it can, use it; if it can't, let's do something else. I don't read the economics literature, I read the environmental history and environmental ethics literature. In this literature, there's some incredibly rich material that's not using sustainability language but using the conservation tradition. One example is the wonderful book *Common Lands, Common People* by Richard Judd about how conservation was not just an elite movement, how it did grow up from the experiences of the farmers of New England. That's the kind of thing I look at and say, Gee, we've got to draw on those traditions and build them up. They're there, it's not as if they're not there to be discovered and worked with. *Landscape and Memory* by Simon Schama is another example. It shows the dangers, potential, and terrors of nature thought in the West, and is a great lesson to us in that sense.

Norton: One more comment along these lines. It seems to me that it is really not necessary to go into a tug of war over sustainability with the economists. It seems to me, in fact, that economic sustainability, or whatever you want to call it, is, I assume, based on some kind of obligation that we not impoverish the future. Seems like a fairly natural, and necessary, condition. But I don't see it as anything close to sufficient. And when people talk about sustainable communities it seems to me that in that usage there is no effort to reduce it to just not impoverishing the future. So admittedly, the word may be in a tug of war, but I would advocate not fighting against the economists' usage. Instead, say, Okay that's part of it, but there's a lot of other stuff that's part of it too. And as long as you have a broad concept of human value, especially a more communitarian one than the economists can even imagine, then I think the term starts to take on some kind of legitimacy.

Gillroy: If that is true, it is not happening. Are you thinking of Michael Walzer's communitarianism?

Norton: Edmund Burke is good enough.

Taylor: The world is a strange and wondrous place. . . .

Gillroy: We'll go with Burke. I'm amenable. I think that's possible, but I don't think that's what is being done. I would say we're not wrestling enough with the economists over the term if we really want the term to mean something more than efficient, optimal maximization of use value. We ought to wrestle more with them, force more language into the debate, and maybe make this rearranged pecking order between a sound ecology and a sound economy, that the ecological economists talk about, real, as they cannot. Then we can have an argument about old-growth forest: keep it, or cut it down! We need more terminology, not less, and more paradigms, not fewer. Everything doesn't reduce to one moral principle or normative, evaluative idea. In that way I'm "antihedgehog." There is a normative complexity out there that we need to decipher to do environmental policy well. And we can't really decipher that with this one market-based model. Maybe sustainability started out without that model. It was meant to replace it; it was meant to at least compete with it. Maybe it got co-opted along the way. Maybe it got pulled in, and the fairness and scale parts of it were toned down for the efficiency allocation parts of it, which really are its identity.

Kassiola: I think that the last two comments, emphasizing the reductionism of the economists' use of sustainability and the idea that you can expand on that and show how it is reductionist, overlook an important point. We are talking about being taken over by another discipline. There is something special when economists get a hold of a concept and it gets into the mainstream institutions and gets enacted by the policymakers. If you substitute another social science, or if sociologists did this, or psychologists did it, the impact on society and policymaking would be very different. And therefore, there is something politically and socially significant in the economists doing what you're suggesting: taking over the concept, reducing it, and institutionalizing it. And going up against that is a very different kind of battle than if you are just getting involved in a debate between political scientists and sociologists, or philosophers and sociologists. Sustainability has now been institutionalized from the U.N. on down. And now it becomes an aspect of the political landscape, with all the powers behind that. That puts the points that John made in a different context, and we need to look at that under these conditions.

Gillroy: Yes, I think that's true.

Audience: Is it too much to ask to define sustainability as we want it, and then try to convince others that this is the proper way to think of it?

Gillroy: There are two levels here. On the personal level, it can mean anything you want it to mean. On a policy level, I worry about it, because policy is made based on what people assume concepts mean: how people interpret statutes, how they write words into regulations, how they define them is very important in policy circles. And to the degree that sustainability, for whatever reason, has an efficiency-based definition, it is going to produce efficiency-based policy. That's what I worry about. A dialogue that includes alternative normative models is not happening.

Taylor: There's a big fight in my home state of Vermont right now about cutting timber. A hundred years ago, Vermont was cut over, over 90 percent cut over. Now it's 90 percent forested again. The fight that's being fought at the moment is not about whether to cut it all over again; that's not politically on the table. It's about what *percentage* we are going to cut all over again. And it will probably be a relatively modest percentage, but meaningful. That's where the fight is going on. The word sustainability doesn't do anything for me in addressing this particular fight, which is a real live fight. I don't see sustainability, as a normative concept, doing that much work. The fight is over what kind of landscape we want to live in, what kind of place we want to live in, what our understanding of that place is, what kind of economy we want to build in it, and those very specific things are what the arguments are about now. I don't see any clarification coming from reducing it to fights about sustainability. I think it would just muddy the waters. I have my view of who should win, and I hope we do. But actually, I think the situation from the perspective of democratically fighting it out is not too bad right now in my home state. Lots of parties are involved, lots of words printed, a lot of public meetings, a lot of anger.

Gillroy: But the questions you ask, Bob, will be answered based on people's worldviews, or paradigms, which will combine distinct values and assumptions and make them policy. One of those paradigms will be based on sustainability and that begs the question: Can sustainability help with Vermont forest policy or old-growth forest policy? Bob Paehlke said earlier that we should probably just stop logging our temperate rain forests. Right now the old growth of the Tongass Forest in Alaska is being clear-cut at a rather fast pace. When the Clinton administration tried to slow this process down, the Alaska delegation just went ballistic. I'd like to hear a sustainability argument about an old-growth forest, and I'd like to hear, for lack of a better word, a preservation argument, or an intrinsic value argument, or an ecosystem integrity or biodiversity argument about that same forest. And then we could have a "Vermont-type" democratic debate about which of these ought to triumph in the end, or ought to be the trump card in a new statute or new Forest Service regulations.

Bowersox: I agree with John on this point. We tend to utilize concepts like sustainability when we're afraid of really getting down to what we mean, and the idea is to make discourse more explicit. If when saying sustainability we really mean "respect for ecosystems," we should be talking about respect for ecosystems. On that you're right, and that's the type of language that we're not hearing.

Gillroy: Maybe in Vermont?

Taylor: No, we don't have any old-growth forests, we cut them down in the last century. Too many good oak trees. The fight is really much more about what Vermont means to me, what my town means to me, what the landscape of this town means to me, what the visual landscape is and its importance, what is important about how people make their living here, what's the importance of how we do business here, what's the importance of the kind of landscape we are going to give to the next generation. . . . I think that's the real stuff in Vermont.

Gillroy: But does sustainability help answer these questions?

Taylor: I don't hear much talk about sustainability on this particular issue. Let me give you another example. I live across the street from a hundred-acre pond, which is a very beautiful pond, was natural, half natural, then it was dammed up; it's twice the size it was. It was dammed in the early 1960s as a water reservoir. Lots of wildlife: there are bobcat, in the backside there are bear and moose that come around, a lot of waterfowl and wading birds—it is pretty nice. There is a new semipublic group that has taken control of the pond and surrounding land. And they're a park service, and their job is both to conserve nature and to make it accessible to human recreation. You realize soon that these don't always go hand in hand, and there are fights. But the fights have been very, very particular. How close to a neighbor's house can they build a walkway? How close to the bird's breeding ground can they build a walkway? Should there be mountain bikes? Should there be motorized boats? For example, do we have a walkway where the bitterns live? You know, bitterns are a neat bird. And the park service says, Well, they'll just have to find some place else to live. And we say, Well, around here there is no place else for them to live, they have to live here. I suppose it is about sustainability for the bittern, but even though the language has never come up, it's very clear what we're fighting about: the rights of families with their kids to walk around the site versus protecting a place for the bittern. I don't see any clarification coming from the term sustainability. But you know, maybe it will. I think everyone knows what we're fighting about. I don't think there's any lack of clarity about what is at stake.

Gillroy: But it may be the case that the bittern, like the rest of us, need alternative arguments based on the nonuse or intrinsic value of nature in order to have any change of winning such an argument, or even competing with those that promote the "efficient" or sustainable recreational use of nature's resources, as any argument based on sustainability plays into their hands. Is the bittern a species for use, or a valuable component of a greater natural system that is the pond? I think, ultimately, we can be fair to the bittern only if we reach a new level of clarity about what is at stake, and this is possible only if we unpack this question based on distinct and competing moral arguments with distinct and competing policy ends. Sustainability cannot do it all!

Conclusion: Democratic Competence, Accountability, and Education in the Twenty-first Century

I think the more democracy, the better, on these types of issues. The more points of access for the public to use, the better.—Bob Pepperman Taylor

The democratic process, for elected leaders, is frightening, even if they claim to be democrats with a small "d." Somehow, we have to . . . educate a generation of elected leaders not to be frightened by the messiness and the risk of open discussion.—Joel Kassiola

Let me overstate it at the beginning, which is that I think democracy is an overrated concept.—Susan Buck

As we conclude this volume it may be useful to consider the points on the intellectual map to which the conversation has taken us. We began with a brief discussion of some of the historical and philosophical origins of the moral austerity in environmental decision making. From that conversation we tentatively concluded that several philosophical and practical obstacles will confront environmental law and policy in the twenty-first century, including (but not limited to) questions of the role of science in policy debate, the implications for social justice of environmental management, the place of pragmatism and instrumentalism in environmental policy, and the dilemma of considering intrinsic values in policy formation. In Part 1, our eight discussants paired off to address these obstacles abstractly and preliminarily, drawing different conclusions about the normative implications of the four areas of concern to environmental policy. In Part 2, seven case studies specifically written for this volume examined concrete, contemporary environmental policies with regard to their utilization of, or relationship to, the broad notions of sustainability and sustainable development popularized in the 1980s by the Brundtland Commission's *Our Common Future* (WCED 1987) and the work of various environmental policy analysts and economists (e.g., Daly 1980, 1996; Solow 1992, 1993). Those case studies covered such diverse topics as management of North American national parks, factory emissions controls, and natural hazard mitigation. In Part 3, our discussants returned to their individual topics in light of the case studies, analyzing them for their sensitivity to the normative and empirical ramifications of the concept of sustainability, as well as their overall potential performance. With these case studies as a point of departure, the discussants met again for a series of workshops and a final roundtable to examine the current state of environmental policy. The essays presented in Part 3 illustrate some unique reflections on the possible paths of policy evolution over the next few years. However, because the essays for this book are all original and the drafts of each were circulated to the discussants, it may not be surprising that those reflections elicit related concerns—particularly with regard to fundamental questions on the nature of democracy, its utility and consequences for policymaking, the relationship between public and private values and policy formation, and the effects of efficiency concerns in a democratic polity. These concerns became the subject of the concluding roundtable, and the pages that follow rely on the thoughts and comments emerging from the roundtable in the discussants' own words. After briefly examining several questions regarding the role of democracy in environmental policy, we conclude with some final thoughts on the prospects and perils facing an austere environmental policy and law as it enters the new millennium.

Sustainability and Democratic Competence

As the quotes that begin this conclusion indicate, several of our discussants believe that the primary challenge facing citizens, policymakers, and scholars in the area of future environmental policy is reconciling substantive and procedural democratic claims with the scientific and bureaucratic requirements of policy formation. In one sense, as suggested by John Martin Gillroy, this challenge is simply a species of a much broader genus of problems facing political governance since the Enlightenment: "Whenever I think about democracy, I think about Rousseau and his distinction between the general will and the will of all. It always occurred to me that consent as the 'will of all' was very important to him, but he was worried that people would consent to the wrong things and he depended on the 'general will' as a corrective mechanism."

From fifth-century Athens and Republican Rome to the emergence of democratic movements on the crumbling frontiers of the ancient regimes of seventeenth-century Europe, a central concern has been the capacity of mass publics—often comprising uneducated, untrained, or inattentive citizens—to make rational decisions for the collective good of the polity. The fear has been that citizens may fall prey to their individual interests or the caprice of persuasive demagogues rather than focus on the common good. Filmer attacked self-rule on such accounts, and Madison and other Federalists severely limited the role that men "in leathern aprons" or without "sense and property" (see Wood 1969, ch. 12) would play in the young American Republic for fear of their incompetence and potential democratic excesses. Even Marx, often hailed as the spokesperson of rule by the masses, could not help but question—in fact, disparage—the capacity and reason of the typical inattentive and easily misled human being. Describing the use of the *Lumpenproletariat* by the reactionary Provisional Government bent on defeating the French Revolutionists of 1848, Marx echoes Plato's conclusions[1] about the average citizen: "[They were] thoroughly malleable, as capable of the most heroic deeds and the most exalted sacrifices as of the basest banditry and the foulest corruption. The Provisional Government paid them one franc 50 centimes a day, that is, it bought them. . . . And so the Paris proletariat was confronted with an army, drawn from its own midst, of 24,000 young, strong, foolhardy men" (1969, 219–20).

In terms of humanity's relationship to nature, concerns about the scientific competence and political sense of the citizenry has been a recurrent theme in environmental disputes since the early conservation movement. Pinchot (1947) advocated scientific management of the nation's resources by a bureaucratic elite precisely because he and his Progressive allies trusted neither the intent of the timber barons nor the transcendentalist musings of Muir and his fellow Preservationists—both were equally self-centered and dangerous to the "Public Good." In the "First Wave" of contemporary environmental literature of the 1960s and 1970s, such wariness also prevailed. Hardin's (1968) classic statement of the "tragedy of the commons" derided the incapacity of liberal democratic regimes to restrain individual appetites and suggested that a much more coercive "politics" was necessary. William Ophuls went so far as to suggest that in light of the level of environmental degradation and the inability of individuals to understand the emerging biophysical realities of scarcity, "democracy as we know it [can] not conceivably survive" (1977, 152). The comments of Hardin and Ophuls caused much ire and served as fodder for a generation of scholars rebuking environmental "doomsayers" for their elitist and apparent antidemocratic leanings, but they were simply the most blunt of a host of writers (e.g., Heilbroner 1980; Catton 1980; Ehrlich and Ehrlich 1990) expressing doubt about a democratic culture's ability to respond to environmental dilemmas. As Bob Pepperman Taylor notes, for many of us this apparent rejection of democratic procedures and ideals in favor of environmental management by scientists and other elites served as a catalyst for our work: "I don't think that democracy is an overrated idea. . . . An awful lot of what really drew me into the environmental ethics literature to begin with was precisely an antidemocratic attitude that I found so appalling. And that attitude was that professional scholars were going to decide what we should believe and then tell us."

Most discussants agreed with Taylor's desire to reinvigorate democracy, yet what democracy might mean and to what extent democratic outcomes should be the basis for environmental management is still vague. Most readers who have experienced intractable public debates over environmental conflicts no doubt can sympathize with Buck's quote at the beginning of this conclusion. Sometimes democracy *may* be overrated; as numerous studies have demonstrated, the average citizen may significantly *overestimate* the risks posed by a nuclear power plant, a carcinogen, or a pesticide while seriously *underestimating* risks associated with air pollution, the single-occupant vehicle, or local tap water (C. Foreman 1998, 68–84; Percival et al. 1992, 607). If citizens are likely to make decisions or empower others to make decisions on their behalf based on inaccurate information or false perceptions, one may conclude quite logically that democratic input may not lead to more effective or reasonable[2] policy.

This is indeed the concern that has made many environmental managers and policymakers wary of reconfiguring policy processes to make them more democratic and participatory and, purportedly, legitimate in the eyes of the average citizen. It is historically ironic that in the early twenty-first century a distrustful yet easily manipulated populace is demanding greater say in policymaking at a time when environmental problems and other arenas of government involvement are increasingly complex, technologically dependent, and with more dire consequences when subjected to incompetent management. The democracies of the late eighteenth and nineteenth centuries—experiencing more mundane tasks of state building, infrastructure development, and expansion with much smaller populations—were probably more capable of (and practically less threatened by) entertaining broad-based democratic governance. Hence, in the early twenty-first century we appear to be more philosophically committed to democratic processes at the same time that such philosophical commitments run head-long into the practical demands of competent governance.

Nevertheless, our discussants questioned the notion that citizens are scientifically incompetent or lack access to competent science and thus may hinder more than they help the process of determining a fully informed policy strategy. Indeed, as Bob Pepperman Taylor points out, an open and contentious process where scientific presumptions are challenged can make for more realistic, cost-effective, and legitimate outcomes:

> We have a superfund site in the middle of town. It's an old coal tar site. The EPA came in and said, "Listen, there is only one thing to do. You've got to build these huge earthworks. You've got to take this little backwater where this stuff is and bring in about 10 square kilometers of dirt, put it on top of it, and ventilate it. It's going to cost 50 million bucks, and it's going to take a lot of time to build, and then it's going to cost millions of dollars each year to maintain." All of a sudden, all the lefties in Burlington [Vermont] were running from cover and saying, "This is crazy. This stuff is inert. It's not polluting." . . . Before you know it, we had all of our scientists saying that the best thing was to leave it alone. . . . The EPA had a process of community response to their own proposals, and has generated a very interesting three-year conversation plan that I think has been excellent. I want more of those points of input. The more the better.

As the recent history of the Emergency Planning Community Right to Know Act (42 U.S.C. §§11001–11050), with its provision for an annual Toxic Release Inventory (§ 313) and its citizen suit provision (§ 326), demonstrates, citizen access, information, and monitoring can be an extremely effective management component (Percival et al. 1992, 622–28; C. Foreman 1998, 126–30; Bryant 1995). Indeed, state and federal regulatory agencies increasingly rely on citizen participation and input—even promoting programs that train citizens to do basic, competent, and useful scientific monitoring of water bodies, soil, and airsheds. Thus, though there may be truth to the argument that citizens lack the technical knowledge to participate in environmental decision making, it is clearly not an unequivocal conclusion derived from unquestionable empirical evidence. Citizens can and do often intervene and *correct* the scientific and technical information utilized by politicians, policymakers, and managers (Press 1994, 100–103).

Who Is Represented? Who Is Accountable?

Even if the technological competence of citizens is not a concern, other shortcomings of citizen participation and democratic institutions were noted that are both more justified and more problematic. For instance, some discussants questioned whether participatory structures such as stakeholder negotiations, watershed councils, or even public hearings were sufficiently representative of the communities they aspire to reflect. Others argued that the predominance of certain local (and potentially parochial) interests can outweigh the needs and interests of broader, but more diffuse, constituencies. Of major concern was the potential disparity in political power resulting from varied economic power and the influence of statewide and national interest groups. An exchange among Robert Paehlke, Mark Sagoff, John Martin Gillroy, and Susan Buck captures some dimensions of this concern.

> *Paehlke:* The notion that the ideal decision-making apparatus is small-scale, transparent, democratic decision making works most of the time in Vermont, where you've got a . . . fairly diverse economy and a town meeting day, and a two-century habit, [compared to] . . . a small town on Vancouver Island, where there's one employer and no chance of any other. And it's trees that they're after. I see these debates happen all the time. It's between three or four tourist operators and the schoolteacher who are for environmental protection and 642 people who are, regrettably, against. The lesson of that is that maybe one needs . . . to start with economic power first.
>
> *Sagoff:* But often you do get a lot of representation in a small group. There was a famous example called the Quincy Library group in upper California in the Sierras that put together a management plan that involved Sierra Pacific Industries and local environmentalists. There you have a real balance of economic interests. . . . The extractive industries in the West now are less than 2 percent of the entire [economy], so [there is] tremendous diversity. And of course, I agree that you have to start with diversity of interests, but we have a lot more of that in many places than we did before, so we can do a lot more with democracy. The problems occur when local groups ask national environmental organizations to send representatives and they won't. They don't because they don't want to be involved in these types of good faith negotiations. . . . They're not going to negotiate because they can't compromise. If they compromise, they'll lose their urban supporters. . . . So you have these national organizations actually turning against their local chapters, which are trying to save their communities. And it's this kind of dynamic that we can now study and see where we are on the spectrum that goes all the way from the old-time timber barons [controlling everything] to a pure Athenian democracy where everyone has an equal amount of property in there for an equal voice.
>
> *Gillroy:* Is equality of social position a condition of good environmental dialogue? In order to talk about trees, must poverty and equality and other social justice issues be already taken care of?
>
> *Sagoff:* You have to have a diverse number of interests that can hold

their own in debate. So it is not equality. But obviously, if there's only one employer in a community then you can't have that. In that case, government has to come up with some sort of science to balance the economic power. But where political power is well distributed and where the courts will look out for discreet insular minorities, then you can have good stakeholding negotiations—and those will be democratic.

Gillroy: How many places is that?

Sagoff: I think in wealthy societies like the United States and Canada—outside of certain company towns and certain old-time throwback pockets—it's pretty ubiquitous: in an area like this [central Pennsylvania], for example, and certainly in Vermont, where these issues have to be faced. But the national environmental groups need to get involved with local problems. Everyone who studies this—Don Snow's (1992) book on the national conservation movement and so forth—always points out a tremendous tension now between the national groups and local chapters, both in chasing for dollars and also in the kinds of responsibility and accountability that they will undertake. The thing about stakeholder negotiations is that those who make the decision must be accountable for it. And it's that kind of accountability that does not work all the way up to Washington, whether you're talking about the bureaucratic agency or a national industrial or environmental group.

Buck: To what extent do you think regional differences have a substantial impact here? I was listening to you and thinking that, in North Carolina, one of the real serious agricultural problems that we're facing is that there's nothing you can produce on small farms that makes as much money as tobacco. And small farms are going under. They're going under to conglomerates. . . . And we also have a large number of black family farms that are going under as well. I don't know what the general level of education is in North Carolina, but when I was going to graduate school in Virginia in the early 1980s the average education for adults in the state was eighth grade for men, sixth grade for women. And that included northern Virginia. I'm listening to you and thinking, I wonder if balanced stakeholder negotiations can happen in North Carolina, Alabama, Mississippi? So my question is, How much of this "democratic" capacity is social, cultural, and regional?

Like Buck, others are also less convinced than Sagoff of the economic and political diversity of most environmental policymaking bodies, and hence questioned their ability and legitimacy in democratically fulfilling their policymaking duties. Again, Kassiola and Paehlke set the tone.

Kassiola: There's a problem with the stakeholder analysis and that is representation of the general public. . . . [The question is] . . . whether or not there is someone at the table who is speaking not from the stakeholders' point of view but for the overarching public interest. And I think for people who have seen the joys of special interest liberalism and seen the problems of that, the stakeholder analysis is problematic. From the Rousseauian point of view, we would need much more education of the public. The Quincy Library group did not hold public meetings in the sense that we would have public meetings. And it's the public institutions that probably need rehabilitation or some kind of improvement. That starts with the leaders. It starts with campaigning. It starts with public information. Can the public understand the issues? Do they get enough facts? Do they understand the choices involved? That also gets to the media representation. If the media represent the old-growth forests in a certain way, the public get their information only as the media represent it. So even if you want to have a referendum—if you want to do anything to [ascertain] the public will—it's more likely to be a mirror of the way the media represent it, and that's problematic from a democratic point of view.

Paehlke: Problems are—more often than we think—on a fairly large scale. . . . You've got to get a jurisdiction of decision making big enough to capture both the costs and the benefits. Sometimes that's even international—there are lots of examples of that. This is just a caution on localism. Clearly, where localism works better and you get a higher proportion of the population who have a sense of the issue and concern about it, you are more likely to get a participatory reality. . . .

There are [other] dimensions of democratic practice in contemporary society that are clearly worrisome and about which people might want to think. For example, the diversity of access to media as a way of communicating is declining—other than the Internet, and that's maybe the one exception. Especially in Canada, we have had more and more mergers of newspapers. . . . I think that's one issue that democracy itself could address: assuring community access time on radio and TV. Perhaps allowing one to have a radio station with a fairly low amount of capitalization . . . so you [don't] need $28 million to have a media outlet. . . . The other [concern] is the ability only a few have to transform dollars into power within the electoral process. . . . Obviously, it's a problem because the people who have to make the rectification are those people who got elected under the old system and are beholden to it. It's a vicious circle that doesn't ever seem to break. And it's the same in many systems. But again, there are ways to change the rules. You cap the maximum amount that can be put in, or you give a tax break up to $100 for donating money, but not above that. . . . We love democracy, but we also have to think about how it might be more democratic.

Democratic Alternatives: Substance over Process

A major focus of discussion became the ways in which traditional liberal democratic and pluralist notions of politics and policymaking could be either overcome or augmented. Like many analysts in environmental policy and environmental political theory, our discussants moved to further define the weaknesses of the present system and outline alternatives. In the process, the fundamental commitment of environmentalists to democratic processes was questioned, given the popularity of strategies, such as litigation, that appear to undermine or circumvent more participatory processes. An exchange between Bowersox and Gillroy exemplifies this concern.

> *Bowersox:* What we are talking about here is an image of democracy that puts to shame the image of our political processes that we have today. We are clearly having a vision of democracy different from the pluralism that we have now. We want something that is more substantive. . . . We want more of a system that allows for valuational debate—and in one sense, we are kind of all liberals because we are afraid to put forth one value as overarching; we want the conversation to continue. But there's a lot that we find, within our present political and policy systems, that is incomplete. . . . One of the other conclusions that we must accept is that if we are going to espouse democracy, we have to be willing to lose. Oftentimes as environmentalists, [I think we have] . . . a real fear of losing. . . . That's why we have Joseph Sax (see Sax 1972) saying, Go to the courts because we can't win with the population. We are going to have to be much more honest about this: if we're going to espouse democracy, sometimes we're going to lose. That's a consequence.
>
> *Gillroy:* What if the loss is a species? Or the last stand of old-growth forest?
>
> *Bowersox:* Well, humans make a lot of mistakes . . . that's a part of the human condition. We don't want to lose those [biological] pieces, but in the long run, isn't democracy the best way to save those pieces?

Bryan Norton put forward one image of what the new system might look like and argued for a broader conceptualization of democracy.

Norton: On the question of democracy, I generally agree with the line of thought that says that at some point environmentalists have to decide whether they're going to be democrats first or environmentalists first. And eventually they have to make their choice. I have made my choice and I'm going to be a democrat first; and that means that you do have to lose once in a while.

On the other hand, I think that we may be using an unnecessarily simplistic conception of how democracy works and how it might work better. The model that I find very intriguing is based on stakeholder negotiations. . . . I know that there are many problems with the stakeholder [model], and it does seem to already assume that the communities are fragmented and so forth. But what I think can work is a situation where you do have people around a table, a much smaller group than the whole community, where each person in the group accepts responsibility for his or her constituency.[3] In successful water management projects, for example, what happens is you take these people who volunteer and you put them right in the middle. And what happens is that they come into a small group of people—the advisory committee, or whatever it is called—and as members of that advisory committee they learn to be friends with people that they didn't think they could ever be friends with. They start to develop some kind of trust as a group. For example, they learn that the person who represents the timber company actually has children and that he or she cares about those children. You learn things like that, and you start to develop trust within that small group. Then a transformation occurs!

Once the group develops some trust among themselves, then you start seeing that you might find some win-win situations, you might find something acceptable. So then a plan starts to emerge. As a plan starts to emerge, that's the crucial democratic point. You do not want that advisory group to suddenly become a dictatorial group. It is at that crucial point when those people have to go back to their constituents and say, I've been talking to these people, they're not as bad as we thought they were, we're starting to make some progress. And we're starting to look at a proposal that would look something like this. . . . Every one of those members of that advisory group has the same problem: they have to bring their people along. Because environmental problems are complex and technical, you cannot expect that they'll be dealt with on the evening news. You cannot expect that people who are working two jobs and taking care of aged parents and four children are going to be on top of the details of these negotiations. What you can hope is that they trust their representative and that their representative can say, Listen, I think this can work. I know that there are problems with it, we might lose a few jobs, or, We may have to give up some areas for cutting that we'd like to keep, but it's better than anything else we're going to get, so let's give it a try.

What I'm saying here is that democracy in this sense is a sort of three-level thing: we've got the general population, a very small number of whom are really committed; we've got the decision makers; and what you try to do is set up in the middle between them a group that's willing to take some risks and hopefully has enough personal capital to be able to go back and reeducate the people who trust them. Ultimately, we may have to have a referendum or something like that, but what's really important is that we not think democracy is limited to one-time referendums. If that's what democracy is about, we're lost.

Norton's vision of an alternative to existing decision-making processes may seem like simply more of the same: at one level, stakeholder forums, watershed councils, and other similar forms of dispute resolution and decision making appear as nothing more than new renditions of representative democracy. Indeed, some argue that such councils and forums are ultimately a step backward—back toward more localism, less accountability, and more secrecy in decision making—because such forums are usually quite restricted in size and may be populated with self-selected or administratively appointed individuals, thereby lacking authority and legitimacy within their purported constituencies (see Thomas 1995, 120–123). This remains a problem to be addressed and studied further. Nevertheless, for the discussants, democracy becomes less a system recognized by its procedural elements and more defined by its substantive characteristics.

Sagoff: Democracy happens when people of often diverse and opposing views, beliefs, and interests agree to become responsible for solving environmental problems and become accountable for the solution over the long run, to their constituents, to each other, and to the community that has to suffer the consequences. What is the alternative? The alternative [is] having Congress pass a law that means nothing, delegating its responsibility to an agency like the U.S. Forest Service. . . . Congress can't make and won't make and will never make the necessary political trade-offs, nor could it on a case-by-case basis, where it's different every time. So it empowers the agency to do this—whether it's the EPA, or the Forest Service, or BLM, or whatever. But the agency doesn't have the political legitimacy with the people to make the political trade-offs. So what does the agency do? It needs to have some sort of legitimacy but it has none. These are not elected officials. The agency has a "science" . . . [and the] science habitually and regularly tells the agency that whatever increases the agency's budget is right. . . . So the alternative to democracy is having the agency making decisions in the name of a science that does very little but expand the agency's [power]. That's the alternative to democracy. Democracy depends on the ability of people who are concerned: they don't have to be local, but they are the ones who are concerned with the problem and represent all the constituencies of society and who are willing to sit down and . . . become responsible and accountable for a solution to that problem by working collaboratively together. . . . Whether [the outcome is] good for the environment or not depends on your theory of what's good for the environment. But the real outcome is that we survive as a community and we survive as a democracy. And we do not resort to violence. Everyone is willing to lose occasionally. When everyone is willing to lose occasionally, everybody will win much more often.

Bowersox: So democracy becomes an iterative process. It doesn't stop after developing one good; it doesn't stop after one dissent or one conflict. It is dialogic process—it is an epistemological tool for achieving . . . contingent decisions that are constantly revisited and tested. And one of the benefits of democracy is that you come back and you can fight another day. We're going to lose and we're going to lose some things. But the alternative perhaps is to win the battle but to lose the war.

The discussants seem to suggest some procedural innovations to democratic processes that transfer greater decision-making authority to local bodies (with appropriate input and authority for bodies representing broader constituencies and interests) and maximize openness, citizen input, and accountability. Although concerns were raised about what these innovations may look like (in terms of intermediate structures, their selection, representativeness, and legitimacy), most agreed that environmentalists too often have been distrusting of democratic decision making, even when it is clear that environmental causes have flourished when championed by grassroots, participatory, and democratic movements. Most agreed that democracy was necessary to assure a diversity of interests and the participation of groups and individuals representing differing levels of economic and political power. However, there was no agreement on what the appropriate distribution or number of representatives is, nor whether such distribution exists or potentially exists in various regions of North America.

Democracy and Education

These evocations of democracy minimized speculation on procedure and focused more on the substance and outcomes of democracy: for many, democracy was not simply a means of adjudicating and resolving differences between groups and individuals pursuing their own self-interest (the essence of liberal pluralism), but also an epistemological process in which more voices, more data, and more actors promote the adoption of richer, more accurate, and finally more legitimate public policies. Implicitly, panelists articulated the notion that democratizing and rationalizing the policy process are not exclusive but, in fact, necessarily interdependent tasks (C. Foreman 1998, 130–36). Indeed, what was articulated in a manner reminiscent of John Dewey was the fundamental relationship between democracy and education.

Some of the discussants view democratic processes as a *form* of education: stakeholder negotiations, advisory committees, watershed councils, and public meetings can be effective means for interjecting new information and pertinent perceptions into decision-making processes and educating citizens and decision makers alike. However, Joel Kassiola made it clear that as environmental policymakers, managers, and citizens grapple with questions of sustainability, they must also consider what educational requirements and curricula would be necessary or prudent for promoting democracy, environmental protection, and social justice simultaneously.

> *Kassiola:* I think when you look at our data about the numbers of Americans who are functionally illiterate and who are scientifically illiterate we're not starting with a good base. . . . We can make improvements, and we need to demonstrate to people the impact of environmental issues on their lives. . . . Dewey made a really important point about democracy and I think that it's really instructive here. This is the "shoe pinches" argument, and I think it's one to consider. No matter how informed the salesperson is, when you go to buy a pair of shoes, the salesperson needs to ask you if the shoe pinches. And we need to understand how decisions that are made by elites, even if they are accountable, affect the public, and how the public can get back to the elites. . . . We need to have more access between the leaders and the people, and I go back to those horrible data about the American people with the greatest number of years of formal education being so illiterate and insensitive to these issues. One of the things that came through today is the difficulty in getting the public to understand moral discourse and normative discourse. We have a very serious problem when the average person will be dismissive in a morally

> relativistic way about making moral judgments. We have a huge educational problem there.
>
> But I do not think that it is scholars working in the various fields . . . represented up here . . . [who are best suited to] voice and make a contribution to . . . the ideas that we would like elementary education teachers to communicate about the environment in the third grade and in the fourth grade, and what . . . we want teachers in high school to communicate. I don't have the answers. . . . I don't know enough about curriculum building in K to 12, but there are specialists and they need to get in contact with the environmental specialists. But I think that before we try to build that bridge, we need to have a message that we agree on. Would it be, You should try to teach the fourth-graders sustainable development? What would that message be? . . . Do we have infinite resources so that we could live [with] 12 billion people in fifty years, or don't [we] have enough resources? . . . One of the things that I'm sure that you all know is that curriculum building is politics by another name, and that's why it is so contentious. . . . We need to deal with that too. So I introduce this point as a goal that we haven't even recognized. . . . What is a journal like *Environmental Ethics* . . . doing about public education? What are they doing about addressing what the public knows about our environment? Does the public understand the limits of the neoclassical economic model? Do they even understand what the neoclassical economic model is? Many of our colleges have been at fault in not relating to the public, and then we berate [the public] because they don't know anything.

Few environmental writers—with rare and notable exceptions, such as David Orr (1994) and C. A. Bowers (1995, 1997)—have addressed the relationship among environmental protection, democratic institutions, and education systems. Although there are journals directed toward disseminating curriculum among education professionals (e.g., *Journal of Environmental Education*), their focus is primarily, and perhaps necessarily, limited to those immediate demands of teachers in the field. Hence, the connections among general requirements of curricular areas such as the natural sciences, social studies, and civics, the desire for more ecologically oriented units and clusters, and the political realities of achieving consensus and support from school boards, state oversight agencies, and federal officials have rarely been addressed systematically or strategically. Joel Kassiola perceptively noted that one of the main reasons the connections between education, environmental policy, and democracy might be overlooked in the literature and research is the nature and presumptions of the academy itself: "Like most things on campus, the college of education tends to be isolated from the arts and science faculty. And I think this is just one example among many where it's not a good thing for that to happen."

Democracy and Public Values

Kassiola thus suggests that in addition to greater attention to the connection among ethics, public values, and environmental policy, scholars and activists must address the divisive nature of the academy. Connections must be made between those who address or explore the ecological, social, political, and philosophical origins of our environmental dilemmas and those who educate future generations of active citizens. Yet it also seems clear that Kassiola is correct in suggesting that another major reason for a lack of attention to education in environmental policy is the political nature of curricular reform itself. One simply has to look at the social and political fallout resulting from conflict over sex education, evolution, and school prayer. Any attempt to introduce environmentally oriented curricula shares a common—even fundamental—trait with those areas of very visible public conflict on education issues: the desire for the curriculum and the outcry against its adoption are primarily defined by conflicting value systems. How does one develop interconnected educational and political institutions that promote sustainability (however defined) when there will always be a portion of the population uncaring about the environment? Susan Buck expresses this concern.

> *Buck:* The people that I come into contact with outside of the university on a day-to-day basis know very little about a lot of things that I think are really exciting and interesting and important, and they don't have sterile lives. They have interesting, full lives and they have lots of things to occupy them, and worrying about the environmental and sustainability and old-growth forests are simply things that they are not concerned about. And I have some real problems with our deciding what they should be concerned about.

Reflecting on Buck's statement and her concern about regional differences in education and economy, John Martin Gillroy noted similar concerns.

> *Gillroy:* We do seem to speak as if there are universals to this democratic formula. . . . If you relate that to Joel Kassiola's earlier point about people's distrust of value argument, it brings up an interesting question: Is what people have to be convinced of factual material, or is it really about values

and principles, and about moral argument? Many of us have been in front of audiences, to talk about, for example, environmental facility siting . . . we have lots of facts with us, none of which the audience wants to hear. What they want to hear about is why this decision was made, why they are not a part of it, and whether they are being exploited or not. They didn't want to hear the facts, they want to hear the value arguments about what is at stake and how the policy addresses their autonomy as active citizens.[4] They want their "rights" acknowledged and their participation encouraged, and that's a harder argument to make, it seems to me, than the strict factual matters of, for example, a low-level radioactive waste site. I'm wondering if the regionalism you're talking about affects value arguments as easily as it affects factual arguments, or are these value arguments universally hard to make?

The Tyranny of Democracy? Remaining Questions

Building on this concern, Joe Bowersox drew a nonenvironmental comparison, which sparked a lively and telling debate regarding majoritarian processes, the protection of minority opinions, and the delicate nature of political legitimacy in democratic states. The ensuing conversation regarding values, religion, and democratic dispositions illustrates the hope panelists felt for achieving sustainable policies through democratic means as well as their concerns and frustrations regarding our ability to change public perceptions and values.

Bowersox: I think there's a problem here and I think that it gets back to some of John Gillroy's concerns about democracy. Let's go back to the South in the 1960s or the 1950s—a perfect example of some of the things that we fear. When faced with the white Southerners' treatment of African Americans, do we simply say, Well, that's regional, and let them continue? We made the decision then to say no. This is why we have the federal government; this is why we have a Supreme Court. But I wonder if there was another alternative available for us—for those of us who have concerns about democratic governance—to maybe come to the same conclusion that was morally justifiable but maybe didn't involve the same sort of mechanism.

Gillroy: This is the essential democratic question, though, isn't it? It's one thing to say we can wait for a point in time when everyone has the personal epiphany that the environment counts, and when we all have the collective personal epiphany, then we all start asking our representatives to do the right thing.

Taylor: The heart of the civil rights movement was not that [Southern states] were making bad decisions. The problem was that they didn't have democracy.

Gillroy: But what do you do in those circumstances?

Taylor: That is a separate set of issues. You do what you have to do to make the decision democratic.

Gillroy: But the situation was democratic in the sense that the set of [Southern] governors who were standing in the doorways were responding to what they thought their constituents wanted.

Taylor: Huge numbers of people were disenfranchised.

Gillroy: Exactly! But under those circumstances, how do you make it democratic in a democratic manner?

Taylor: You might not do it in a democratic manner.

Gillroy: Okay, there you go. I just wanted you to say that. . . .

Taylor: That's not the problem that we're facing in terms of the environment. Except in rare cases where local communities lack the preconditions of a healthy democratic decision-making process, democracy functions well. . . . Of course, the degree [to which} we can encourage those preconditions is a good thing.

Gillroy: But do we encourage those conditions or do we *create* those conditions?

Kassiola: Who's the "we"?

Gillroy: Presumably, in this instance, and in Joe's example, the federal government, in the same way it did or tried to do in the civil rights case. Eisenhower sent the 101st Airborne to Little Rock, and doors opened. . . .

Taylor: I would love to see more discussion about redistributive politics as a prerequisite to democracy rather than as something that we ought to talk about democratically, but I'm in the minority on that—a very small minority on that. I don't think they are parallel situations.

Gillroy: But this is a critical question! It's hard to talk about from a liberal standpoint. But the plain fact of the matter is that there are times—because of whatever injustice or whatever situation—that the democratic processes are not working or are not representative. That could be true about an environmental question as easily as it could be true about a social justice question. Under those circumstances, whose responsibility is it to bring the ground-level conditions up to the point where these small groups can work? Where people can meet at a table with some basic level of equality and basic personal autonomy? That is a question that environmentalism has ignored for a long time. All our talk of democracy has this as a prerequisite. Have we been assuming that all come to the table with a certain amount of equal baseline power when this is not the case?

Taylor: On the contrary: I think environmentalism was preoccupied with that question for a long time and I think that is the reason that environmentalism generated some of the most astoundingly antidemocratic literature in the contemporary canon. William Ophuls's book (1977; see also 1997) is absolutely amazing, and Paul Ehrlich's (1970) recommendations. . . . I think the fact that we are having the conversation about the connection between democracy and environmentalism shows a tremendous growth and change in . . . the environmental movement, and I think it's a terrific development.

Audience: All this talk about epiphanies is scary. Remember where this came from: on the road to Damascus. It was a religious conversion that happened. Are we asking or hoping that people have a religious conversion so that they suddenly become rabid environmentalists?

Kassiola: What is the difference between paradigm change and a religious conversion? Actually . . . a worldview change is deeply profound and on the level of what many people might mean by a religious conversion. But it doesn't have to be religiously based. I don't know if we'd get agreement from the group that a worldview change from the "business as usual" is necessary. [But] . . . I'm not comfortable with the religion language and I would prefer other metaphors. But such change is deeply profound and rare and difficult. . . . But not impossible.

Buck: I think the mechanism is less likely to be analogous to a religious conversion. I think that happens, but I think it is less likely. What you're more likely to see is . . . what sociologists call cognitive dissonance: where you think you know the way the world works but the weight of evidence [against that view] begins accumulating. You think you can explain it away just so long, and then you realize, I had it wrong. And I think that's much more likely to happen with a shift in thinking in environmentalism rather than with this "epiphany" type of conversion. I worry about the sustainability of these conversions. Sometimes people convert back.

Gillroy: I do not mean the term in a religious way. I am thinking, as Joel is, of a paradigm shift. Then the question becomes: How can politics and justice facilitate this paradigm shift? Do we wait for moral enlightenment, or provide the basis for it? What is the job of government? What is the job of politics? Is the job of politics to accelerate that change but wait for it, as through education? Is the job of politics to inform that process with law? Or is it not to wait for "epiphany" at all but to simply make the changes that capacitate individuals to transcend fear, uncertainty, and their self-interest for greater appreciation of humanity in themselves and ecological value in nature? I wonder about this. . . .

Buck: Isn't it politics that provides the process, and leadership that provides the impetus for change? That's one of the things that worries me about environmental policy. A couple of years ago I did some work in Great Britain looking at biodiversity implementation. . . . What really, really troubled me was that every single place I went where implementation was going well it was because there was an individual there with both the interest and the power to make change happen. And the places where it was just limping along—just to keep Westminster happy—lacked that individual. That's a very troubling policy insight, because it means that there's no way to structure the institutions.

Audience: But why is it that you have to wait for this individual to emerge? Why isn't there a notion of individuals emerging with training . . . developed through precisely the educational process you are hoping to develop?

Buck: Perhaps we should go back a step and look at the people who were leaders and see what made them that way.

Audience: Community organizing has got this whole idea of leadership development down—they've been doing it for a while. You look at who is emerging. Give them a little more training. This is not a mysterious process.

Buck: It needs to be somebody with the power. It needs to be somebody who is on the "council." In England it needs to be somebody with the title—I mean somebody who has literally got the title.

Audience: But that too is an iterative process. People are often given power by virtue of having assumed leadership or demonstrated leadership. I guess all I'm saying is that I'm not sure that the model is linear as a way of describing it. I do a lot of popular economics education and this is also an area where people have values and principles. Many of them, by the way, *are* religiously informed. [They have a vague feeling] of fairness and they want the world to be fair, but they are deeply cynical. They believe that the economy is not fair. And these are not people who would describe themselves in any way as Marxists or socialists or whatever, but they're quite cynical about power relations and they generally feel that they are on the losing end of power relations. Very often, what I do is provide facts that run counter to their cynical beliefs about things, or run counter to [the facts] . . . they wished existed. So they'll say, Well, we can't raise the minimum wage because if we raise it there will be all this unemployment. And I then say, Well, actually, we've done some studies about raising the minimum wage and it turns out that it's not all that true that if you raise the minimum wage there's going to be unemployment, and they seize on these facts. Part of it is that I play a role as a legitimizer . . . [of] certain ideas that they want to believe already. Joel Kassiola's point that we're very uncomfortable with moral argument and that people don't know how to do moral argument—they may not know how to do moral argument but they do have moral hungers and moral beliefs. And often what they get from a process like this is permission, whether that permission comes in the form of an authority or comes in the form of empirical data that they can then marshal, or whether it comes in the form of learning that there are others who share the same values and beliefs. So then you get an "epiphany," or an "ah-ha!" of surprise, or a paradigm shift, as you provide certain kinds of tools that people need in order to *enact* beliefs that they already have. . . .

Taylor: We've all had similar experiences. . . . I certainly found similar patterns of behavior in people. . . . I have a lot of respect for the moral inclinations and impulses of people I know outside of the academy. In fact, I sometimes wonder if I have the same level of respect for the moral impulses of people inside the academy. That doesn't make me an anti-intellectual or skeptical about the need to do scholarly work, but my inclination is much like yours.

Gillroy: Is what we want to tell policymakers, then, that they should go and found an environmental leadership and community organization school? Is that what we want funding for? Have we turned into hedgehogs? Democratic hedgehogs? That's a different kind of animal.

As a conclusion to the evening, the discussants spoke of agreement and what recommendations could be made to policymakers.

Bowersox: Can we agree on anything? Actually, yes, I think we can agree on some things. It *is* clear to a lot of you that there is more work to be done. There is something coming out of the conversation today that I find attractive, and I would like to hear more about the next steps. Somehow, there is something about democratic processes that we think is desirable. But it's not just something about procedure, there is also something substantive about democracy that . . . has been overlooked, in terms of a lot of environmental policy, and that is [the substantive] democratic methods available to address some of these value questions that we face.

Taylor: I was just going to say the same thing. I thought that the one thing that seems to strike a chord among many of the discussants was the idea that the more open and democratic the process of environmental policymaking, the better. One of the things that we want, that we could agree on, is that we want to make institutions as open as we can. I think the more democracy, the better, on these types of issues. The more points of access for the public to use, the better.

Kassiola: I tried to think of some quick and brief phrases. I think one is that "Economics Is Not Enough" . . . another is that "Moral Philosophy Counts." With sustainable development, we learn that this is just a technique that needs moral principle to make it work. For policymakers, we agree that business as usual is not sufficient, and that somehow, as scholars, we need to get other scholars in the policymaking process to communicate this to the policymakers. They need to be open to doing things differently and thinking differently. The democratic process, for elected leaders, is frightening. Somehow, we have to train or educate a generation of elected leaders not to be frightened by the messiness and the risk of open discussion. This is a point that hits home. From where I sit, even in our own institutions of higher learning, we need faculty and administration to look at openness and democratic discussion as well. We don't really train people for the requirements of genuine dialogue. I think we need to do that.

The conversation observed above should demonstrate to readers the sense of trepidation and frustration as well as hope with which the discussants consider the future. Readers may have been surprised with the general dismay with which the discussants approached the case studies in sustainability (presented in Part 2) in their responses in Part 3. For many activists, policymakers, and scholars, sustainability and sustainable development hold promise for reforming the existing market-oriented, growth-presuming, resource-consuming, wilderness-destroying social and political paradigm. Indeed, some of the discussants retain a modicum of optimism that the term can be given more meaning and serve as a starting point for bringing together science, economics, politics, and social values. Nevertheless, most of the discussants saw in these case studies the same skepticism (if not hostility) toward normative argument, and hence tacit reliance on the market, that was observed generally in public policy and law in this volume's introduction. At best, the discussants echo Joel Kassiola's words above, noting that presently, sustainable development is a technique in search of a moral principle. At worst, the discussants claim sustainable development is simply "the market" by another name.

Furthermore, as children of the Enlightenment, the discussants express little hope that a single worldview or paradigm will descend on North America (let alone the world), uniting all in disposition and purpose toward saving the Earth and abandoning the market. A collective "road to Damascus" experience certainly is not foreseen, and although individual "epiphanies" may be more possible, there is no guarantee that the conversions will last, nor that one's revelation will lead to the same commitments as one's neighbor's. In essence, our discussants, like Locke, Madison, and Adams, assume reason and intuition are fallible. This is a fundamental starting point for all of them.

As such, some of our discussants apparently *assume* a democratic polity, and in fact a democratic value system. Even though Buck *does* (*over*)state her belief that democracy is an overrated concept, she, like many of the discussants, sees no other viable alternative political and social system. This is important and not trivial, for it is indeed one logical outcome of the Enlightenment skepticism explored in the introduction. *If* there is no unquestionable, apodictic, and universally recognized political or moral authority, *then* one has to assume that rival claims to such are subject to dispute, and *by themselves* not enough to legitimate or justify a political, social, or moral order. They can claim legitimacy only to the extent they gain voluntary assent.

Thus, although most of our discussants express dismay with consumer culture and a citizenry more attuned to their latest media-fueled desires than the health of a watershed or the complexity of an ecosystem, most are reluctant to suggest that the populace is wrong, or that the values that may be expressed by the population at large are per se illegitimate. Rather than blame the ethical choices of individuals, some of our discussants primarily point to structural problems that prevent dissemination of relevant environmental information and education, skew political and social discourse toward particular economic interests, inhibit political participation, and prevent the expression of alternative values.

Their suggested reforms are admittedly sketchy and incomplete, in part due to the nature of the preceding conversation itself. Most seem to favor a more open decision-making process that allows citizens (with or without specific scientific expertise or information) greater access to and greater oversight of elected and unelected officials and the information they use to determine policy priorities and strategies. Acknowledging the importance of expertise, discretion, and leadership among administrators and resource managers, there is a general consensus among the discussants that, unlike Ophuls (1977, 1997), Heilbroner (1980), and many other early political ecologists, an autonomous, authoritative, and ecologically oriented state apparatus is neither likely nor desirable.[5] Most suggest that it is just as likely that professionals will suffer from information constraints and bias as that the citizenry do, and that concerned individuals and activists can play an important role in overcoming these problems (Torgerson 1999, ch. 4; Majone 1989; Dryzek 1993; Bowersox 2000; Percival, this volume). Nevertheless, our discussants remain ambivalent as to the likely structure of this more open and democratic process, and simply mention various possible scenarios ranging from the extreme of town meeting democracy to stakeholder negotiation.

Yet democracy does not simply happen, and neither does *better* democracy. Some analysts do suggest that positive information effects, value discussion, and policy development occur in processes such as stakeholder negotiations and collaborative decision making (see Kearney et al. 1999; Busenberg 1999; Shannon and Antypas 1997). Nevertheless, most of our discussants do not share Mark Sagoff's assertion that such nontraditional processes will be necessarily and sufficiently representative or free from coercion by well-organized and/or well-funded interests. Thus, suggestions are made to gird citizens and future activists with sufficient economic power and knowledge to counteract these potential disparities in influence. Some of our discussants even hint at radical transformation of our educational system and economic redistribution as prerequisites to a truly democratic and green politics (see also Dryzek 1996, 1997; Baxter 1999).

Some might consider these tentative reforms by a few of our discussants to be politically naïve with regard to the nature of power (see, e.g., J. Connelly and Smith 1999, 112–17; Lukes 1974). John Martin Gillroy suggests as much during the exchange on desegregation. Similarly, Doug Torgerson notes that there is no guarantee that the results of green democracy "will not be stillborn" (1999, 160), and Robert Goodin scoffs at the notion that such commitments to a political *procedure* will necessarily produce environmentally *substantive* ends (1992, 168). Nevertheless, Bowersox, Norton, Sagoff, and Taylor do recognize that there is no guarantee of green outcomes, and do suggest that they are fearful of utilizing nondemocratic means to achieve green ends. Perhaps they are reluctant to have environmental democracy take on the worst aspects of the ancien régime (see also Torgerson 1999, 168).

Is this naïve? Ultimately, the reader must decide. However, some of our discussants may be making two claims. The first claim is that *any* political response to environmental degradation that will require sacrifice *must* maximize its reliance on democracy to achieve legitimacy and any hope for implementation. Second, these discussants may be asserting that there is something about these democratizing elements themselves that, while appealing to fundamental liberal principles of fairness, autonomy, participation, and openness, *in fact* undermine contemporary liberal pluralism (see Dryzek 1997; Torgerson 1999).

This is, in fact, the fundamental distinction between some of our discussants' conception of democracy and the contemporary liberal democratic paradigm that has its roots in the Enlightenment skepticism and concomitant market assumptions we examined in the introduction. Although they differ in the degree of their endorsement, some of our discussants focus on the structural shortcomings of the existing paradigm and thus challenge the fundamental assumptions of contemporary political and economic life. Classical liberalism assumed the existential autonomy of the individual: our religious beliefs, philosophical commitments, economic transactions, even our very constrained "public lives" were radically independent from those of our neighbors. Perhaps itself more of a normative ideal than an accurate description of the human condition (see e.g., Kingdon 1999, ch. 3; Arendt 1963, 137), this radical autonomy and its epistemological distrust of fundamental truth claims promotes settling valuational conflicts by preference aggregation. Hence, autonomy and epistemological incommensurability foreshadowed liberalism's reliance on market mechanisms for noneconomic as well as economic decision making (see introduction). In the conversation above, some of our discussants fundamentally challenge this predisposition; others articulate a clearer defense of individual autonomy.

Rather than assuming that one's private beliefs and actions are conceptually and functionally separate from the social, economic, and political context in which one lives, some of the discussants implicitly question this "public/private" distinction and instead view the social, political, and economic commitments of individuals as growing out of and influenced by this broader context. Rather than radically autonomous, individuals are thus in large part creatures of their surroundings: what they hold to be their "own" beliefs, values, wants, and needs are in fact artifacts of their social (and even ecological) environment.

Although not articulated by all of our discussants, this understanding of the human condition allows some of them certain advantages in their pursuit of a more ethically principled and more ecologically sound environmental policy. It suggests that their desires to transform education, the modes of political participation, and the distribution of economic power may indeed lead to the emergence of a new, environmentally oriented, principled paradigm over time, and that this can be done without resort to violence and unacceptable forms of coercion. Although education and resource redistribution are indeed forms of coercion, the democratic and inclusive methods of decision making discussed may allow implementation of these means with greater legitimacy. Yet, tellingly, these few discussants do not specify the extent to which government ought to enforce such means.

Hence, this nonliberal strategy poses particular risks for some. First, such a strategy could fall prey to claims of cynical manipulation and social engineering (see Huber 1999). By assuming a nonautonomous and malleable human nature, they tempt the use of unprincipled methods to achieve principled ends (Torgerson 1999, 168). Clearly, most of our discussants are not suggesting this level of malleability, and others, like Buck and Gillroy, more clearly maintain a defense of the individual. But some believe that by changing particular social conditions, a new consensus may be constructed. In fact, the methods of this manufacture of consent are akin to the ones many green critics of liberalism claim are used pervasively by economic elites to maintain their hegemony. If this is indeed so, one may wonder whether greens are simply naïve with regard to their ability to outcompete these existing and entrenched interests at their own game.

Furthermore, by potentially targeting an even more fundamental liberal commitment than the market—that of autonomy and individuality—some of our discussants risk marginalization of their argument by contemporary society. Part of the appeal of democracy to greens is its familiarity and "givenness" within contemporary society. Compared to other social movements that have periodically challenged the liberal paradigm in North America, environmentalism has been able to tap into powerful strains of Anglo-American political thought, such as rights rhetoric (e.g., Nash 1989), populism (e.g., Cable and Cable 1995), and rugged individualism (e.g., D. Foreman 1991). It is one thing to critique the market as unfair or undemocratic: as Paine (1953 [1776]), Jefferson (1785, 1813, 1816), and Thoreau (1966 [1849]) demonstrate, this can be done readily from within the liberal tradition. It is quite another thing to challenge the contemporary seat of *all* value in our cultural and political ethos: the individual and his or her autonomy.

Hence, we return to the beginning of this volume—to the fundamental realization that public policy in general and environmental policy in particular depend on metapolicy: the ideas, values, and principles on which any policy is erected. In these final pages our discussants have sketched out elements of new metapolicies on which to construct an ecologically sound society in the twenty-first century. Are these metapolicies sufficient to justify and legitimate the transformation of contemporary social, political, and economic institutions, or do they need additional attention to potential criticisms? Again, that is for the reader to contemplate. Nevertheless, what our discussants *have* done in the preceding pages is articulate a series of first premises that must be debated and tested before any policy can be fashioned. By doing so, they have exemplified the type of exploration and questioning that must become routine if we are to transcend the moral austerity of markets and have truly principled environmental policy for the new century.

Notes

Introduction

1 The concept of a dominant argument is related to the degree of persuasiveness one argument has in comparison with the others available in the policy debate. Also, our focus on policy as argument instead of policy as power comes from our contention that power itself is built on argument and that even an authoritarian state needs to be persuasive to both the general population and to those who support the center of power. This is not to discount theories of class or hierarchy, but to attempt an understanding of what lies at their foundation.

2 Through election, violence, and so on. One might argue that the "Republican Revolution" of 1994 was an attempt to cause the demise of a dominant line of argument as to the role of the state. It could also be argued that it was an attempt to revive a long-time argument about states rights that was waning.

3 Competitive principles, but not presently the core principle of the dominant policy argument.

4 Majone (1989, 145) describes the metapolicy as a series of concentric circles with the core on the inside. The model presented here is our own design.

5 We find political ecology useful in this context for it emphasizes their focus on politicoeconomic structures as they respond to propositions posed by ecology: the study of closed ecosystems responding to biophysical laws.

6 See also Ophuls and Boyan (1992, ch. 4).

7 Kassiola is actually much more self-conscious about this strategy and is one of the very few writers who utilizes the "limits to growth" debate to do some fundamental questioning of the place of economic and scientific thinking in politics.

8 Public lecture, fall 1989.

9 For a full development of this argument see Bowersox (1995, ch. 1).

10 Actually, some apparently do take just this view. See Hardin's bison hunter definition of scarcity and morality (1968, 1245).

11 According to Locke, these relations, as well as our ideas of the moral actions and duties that arise from them, are available to us through the use of our rational faculties. See *Essay* 1.3.13.

12 Tully stresses this teleological dimension of Locke's motivation. See 2T, 26, 134, 159, 165; Tully (1980, 43–50, 98). This is quite different from the Locke portrayed by Ophuls or McPherson. The same problem is confronted by those who read Adam Smith's *Wealth of Nations* without first reading his *Theory of Moral Sentiments* and assume markets without morals.

13 The inclusive interpretation has been most forcefully advocated by Tully, who argues that everyone in Locke's theory "has a claim right not to be excluded and to demand that others make room for them, correlative with their positive duty to do so" (1980, 71). In this interpretation, we appropriate from the common itself conditionally, not exclusively. We come to have property in the right to conditionally use the common for support (128–29; see Locke 2T, 29). This position is in sharp contrast to that of McPherson (1962).

14 We leave aside the controversy surrounding McPherson's claim that labor is a truly alienable property (1962, 211–14).

15 Here we follow both Tully and Rapaczynski, who, in contrast to McPherson (1962, 212), see Locke saying human productivity does not mean more goods are available even for the propertyless, but rather that each needs less property to fulfill his or her wants and desires, leaving more for later appropriators. See Tully (1980, 148–49); Rapaczynski (1987, 207–12).

16 Quoted in Rapaczynski (1987, 210); emphasis added. Interestingly, Tully's later work (1993) affirms Rapaczynski's focus on productivity, and reads Locke's arguments in chapter 5 of the *Second Treatise* as justifying English colonization of North America over the "less productive" utilization of the land by either the indigenous inhabitants or the French voyageurs (Tully, 1993, 155–66).

17 For the connection between rational industry and one's own moral fulfillment and happiness, see also *Essay* II.21.60.

18 One need only consider the political and ethical controversies surrounding the right to privacy and the U.S. Supreme Court's decision in *Roe v. Wade:* 410 U.S. 113 (1973).

19 Although we do see this tendency in much of the existing literature, the contributors to this volume, as well as writers such as John Dryzek, Douglas Torgerson, John Barry, Tim Hayward, and Avner de-Shalit, seem much more sensitive to the dynamics of this logic.

20 The coherence and tenacity of this framework is amazing! Although disagreeing with the desirability of continued accumulation, Ophuls and others argue that, because of assumption 1, coming resource scarcity will require greater reliance on the coercive apparatus of the state.

21 Two general exceptions to this rule who seem to implicitly recognize Alper's warning are Kassiola (1990) and Alan S. Miller (1991, 80–85).

22 A similar strategy can be detected in environmental ethics. Although not using political authority, these writers resort to such various systems of implicit coercion as deontological rights theories and holist theories dependent on scientific analogies to quantum physics that produce "law-like" responses. See Callicott (1989, ch. 9).

23 Parts of this section were adapted from my review of Ophuls 1997, which appeared in the *American Political Science Review* (Gillroy 1991: 948–49).

Buck, Science as a Substitute for Moral Principle

1 Although ably served by its personal and committee staff, the Congressional Research Service and the Office of Technology Assessment, Congress also relies heavily on information and analyses provided by the bureaucracy.
2 There is a bit of chicken-and-egg quality to a discussion of discretion and expertise. The agencies are given discretion partly because they have expertise, but one of the reasons they have developed expertise is that they must exercise discretion.

Kassiola, Why Environmental Thought and Action Must Include Considerations of Social Justice

1 For a discussion of global injustice and the environment, see, Athanasiou (1996). Also, for excellent worldwide data indicating the extreme nature of global inequality, see the *World Resources* series, including 1996.
2 I would add to this question *even if the concept of "social justice" is contested,* as the antisubstantive moralist and proceduralist Rawls would emphasize.
3 See Lane (1998).

Norton, Sustainability: Descriptive or Performative?

1 I credit Alan Holland for convincing me how difficult it would be, conceptually, to maintain the idea of "capital"—as in "natural capital"—and at the same time to provide a sharp and operational notion of strong sustainability.

Sagoff, Are Environmental Values All Instrumental?

1 The author gratefully acknowledges support from the National Science Foundation. The opinions presented are those of the author alone.

Gillroy, A Practical Concept of Nature's Intrinsic Value

1 This word is used in Norton's essay and in the Wilderness Act itself. See 16 U.S.C.A. §1131 (c).
2 These cases concern the Clean Air Act (*Chevron v. NRDC* (467 U.S. 837, 1984)), which states that the agency must have only a permissible construction of the statute; the Endangered Species Act (*Palila v. Hawaii* (649 F. Supp. 1070. 1986)), which argues that an agency can move before harm is encountered; the National Environmental Policy Act (*Vermont Yankee v. NRDC* (435 U.S. 519, 1978)), which contends that the agency decides on what substance environmental impact statement procedures will have; and both the Multiple Use–Sustained Yield Act and the National Forest Management Act (*Robertson v. Methow Valley* (490 U.S. 332, 1989)), which establish that the agency decides what constitutes reasonable procedures. Finally, in *Babbitt v. Sweet Home* (115 S.Ct. 2407, 1995), the justices allowed a preservationist construction of the law that tied potential harm to species to the potential adjustment of their natural systems even through private logging operations.

Campbell-Mohn, Sustainability in the United States: Legal Tools and Initiatives

1 For instance, sustainability is the cornerstone of the National Forest Management Act (NFMA; 16 U.S.C. §§ 1600–14; see Gale and Corday 1991, 31). It requires the U.S. Forest Service to prepare long-term land and resource management plans (16 U.S.C. § 1604), and mandates that the Forest Service limit timber sales to the quantity that can be removed annually in perpetuity on a sustained-yield basis (16 U.S.C. § 1611[a]). In addition, the Multiple-Use Sustained-Yield Act (MUSY; 16 U.S.C. §§ 528–531) requires the Forest Service to manage for uses such as recreation, range, timber, watershed, and wildlife and fish purposes (16 U.S.C. § 528). See also the Marine Mammal Protection Act (MMPA; 16 U.S.C. § 1361[2]); and the Fishery Conservation and Managment Act (16 U.S.C. § 1801[b][4]). Some of the pollution abatement statutes also incorporate elements of the sustainability objective. For example, see the Federal Water Pollution Control (Clean Water) Act (33 U.S.C. § 1281[g][5]).

2 This concept is developed in the literature on the zero-infinity dilemma, whereby some threshold of ecological tolerance is breeched, resulting in irreversible environmental consequences. See T. Page 1977, 207, 211.

3 A comprehensive list of community theorists is beyond the scope of this study. But for a bibliographic essay deciphering the range of the literature, see Brooks (1995).

4 For instance, H.R. 2099, 104th Cong., 1st Sess. (1995) (appropriations bill rider reducing the Environmental Protection Agency budget by one third); S. 343, 104th Cong., 1st Sess. (1995) (regulatory relief bill); H.R. 479, 104th Cong., 1st Sess. (1995) (repeals Clean Air Act Amendments of 1990); S. 851, 104th Cong., 1st Sess. (1995) (reforms wetland regulatory program); H.R. 198, 104th Cong., 1st Sess. (1995) (wetlands conversion); H.R. 2036, 104th Cong., 1st Sess. (1995) (reduces restrictions for solid waste land disposal).

5 The case studies and lead agencies are Anacostia River Watershed (Army Corps of Engineers), Great Lakes (EPA), Coastal Louisiana Wetlands (EPA), South Florida Everglades (DOI), Southern Appalachian Highlands (U.S. Forest Service), Pacific Northwest Forests (U.S. Forest Service and DOI), and Prince William Sound (National Oceanic and Atmospheric Association).

6 See U.S. EPA 1995c, 11: "Agency sources explain that EPA has used the terms such as 'community,' 'ecosystem,' and 'watershed' to define 'place-based' approaches, but that in any particular situation, the three terms may encompass very different elements."

7 A June 1995 draft Ecosystem Task Force report recommended that federal agencies coordinate local land management responsibilities to protect specific geographic areas (U.S. EPA 1995e, 1).

8 The specific strategies included regulatory flexibility in existing laws to allow pollution prevention incentives, increased research and information, identifying multimedia strategies, incorporating pollution prevention in settlements, exploring cooperative projects with other agencies, and creating the Industrial Toxic Project.

9 Exec. Order 12856, 58 Fed. Reg. 41981 (1993) (Federal Compliance with Right-to-Know Laws and Pollution Prevention Requirements). Specifically, Clinton required all federal agencies to develop internal pollution prevention strategies to reduce use

of toxic waste up to 50 percent, develop pollution prevention criteria for use in acquisitions and procurement, and to comply with all toxics reporting requirements.

10 For a history of the split between natural resources law and pollution abatement law, see Campbell-Mohn, Breen, and Futrell (1993, ch. 1).

11 The material in this section is directly from ibid.

12 Indeed, the hotly debated but relatively successful community-right-to-know laws begin with this premise. See, e.g., 42 U.S.C. §§ 11001–11050.

13 With the exception of the Endangered Species Act, the Surface Mining Control and Reclamation Act, and the National Environmental Policy Act, if a federal action is present.

14 This is not to say that communities should be the only level to address these issues. Environmental impacts transcend community boundaries and communities often fail to account for wider impacts, leading to the "not-in-my-backyard" syndrome.

15 There is a rich and growing field of legal literature on the intersection of environmental law and private property. See, e.g., Rose (1994); Poirier (1997); and Huffman (1998).

Laitos, Sustainability and the Use of Public Lands

1 The amount of land that is federally owned has remained relatively constant over the past two decades, decreasing slightly from 761 million acres in 1973 to 657 million acres in 1994 (see U.S. Dept. of Interior, BLM 1974, 1996). The agencies with the most acreage under their control, the BLM and the Forest Service, have, by tradition and statutory mandate, imposed a multiple-use management philosophy on their lands (see Wilkinson (1992, 20–21, 75–218)). Lands under the control of the BLM and Forest Service constitute 463 million acres: 272 million acres for BLM and 191 million acres for the Forest Service. These "multiple-use" lands constitute over 73 percent of the federal land base.

2 The standard statutory definition of *multiple use* is found in the Multiple Use–Sustained Yield Act of 1960: "Multiple Use means: The management of the various renewable surface resources . . . so that they are utilized in the combination that will best meet the needs of the American people; . . . and harmonious and coordinated management of the various resources, each with the other, without impairment of productivity of the land, with consideration being given to the relative values of the various resources" (16 U.S.C. § 531(a)). Similar definitions appear in the organic acts for the two primary federal multiple agencies. See, for the BLM, 43 U.S.C. § 1702(c) of the Federal Land Management Policy Act, and for the Forest Service, 16 U.S.C. § 1600 of the National Forest Management Act.

3 Another commodity resource found on public lands, water, is not discussed in this essay because of the unique nature of the legal relationship that exists between private parties and water "created" for private use through federal reclamation projects.

4 Other attempts at formulating an alternative to multiple use as a preferred public lands management standard include Glicksman (1997, 647); Jeffery (1996, 79); Blumm (1994, 405).

5 Some commentators have acknowledged the transformation on public lands from

a commodity- to a recreation/preservation-based use. See Power (1996); Wilkinson (1992).

6 Applying a multiple-use strategy to only two dominant uses is both internally contradictory and aggravated by the reality that multiple use was historically grounded in commodity exploitation, the complete opposite of recreation and preservation. See generally Blumm (1994).

7 *United States v. Town of Plymouth*, 6 F. Supp. 2d 81 (D. Mass. 1998) (Fish and Wildlife Service entitled to preliminary injunction banning off-road vehicles from beach to protect endangered species); *Southern Utah Wilderness Alliance v. Dabney*, 7 F. Supp. 2d 1205 (D. Utah 1998) (challenge to National Park Service decisions to permit off-road motorized vehicles in national parks).

8 Some types of recreation may be compatible with extractive uses. Timber cuts create open areas that attract wildlife and thereby benefit hunters. Building roads in a forest for timber also increases access for recreational hikers.

9 Economic efficiency embodies a number of important assumptions, including the following: (1) economic value reflects the full social benefits and costs of all resources; (2) the benefits and costs over different time periods must be adjusted by the appropriate discount rate; (3) economic value is ultimately derived from human preferences, a philosophical assumption that is both utilitarian and anthropocentric; and (4) all economic valuations reflect the given distribution of income; changes in the initial distribution of income would lead to different valuations of resources.

10 See, e.g., Talbot Page (1977). Efficiency has also been invoked by both the proponents and opponents of privatizing federal lands. Compare Stroup and Baden (1983), Anderson and Leal (1991), and J. V. Krutilla and Haigh (1978, 377–81).

11 Costs are usually measured in terms of opportunity costs—the social value forgone when an allocation moves away from one use (e.g., commodities) to another (e.g., recreation).

12 Proposed Rules: Natural Resource Damage Assessments under the Oil Pollution Act of 1990, 58 Fed. Reg. 4601, 4610 (1993). The panel guidelines for conducting CVM studies include the use of personal interviews, use of a future-based willingness to pay measure rather than a willingness to accept measure, use of a referendum format question, and certain reminders to respondents during the interviews (4608–10).

13 The 1990 RPA Program, appendix B identifies several literature reviews that served as the basis for recreation prices: Song and Loomis (1984); Walsh, Johnson, and McKean (1988); Connelly and Brown (1988).

14 The Forest Service collects data on nine different categories of recreation: (1) mechanized travel and viewing scenery; (2) camping, picnicking, and swimming; (3) hiking, horseback riding, and water travel; (4) winter sports; (5) hunting; (6) resorts, cabins, and organization camps; (7) fishing; (8) nature studies; (9) "other" includes team sports, gathering forest products and attending talks and programs (U.S. Department of Agriculture 1997, XII-30). BLM identifies twelve different types of recreational activities: (1) camping, (2) fishing, (3) hunting, (4) miscellaneous land-based activities, (5) miscellaneous water-based activities, (6) motorized boating, (7) off-highway vehicle travel, (8) motorized winter sports, (9) nonmotorized boating, (10) nonmotor-

ized travel, (11) nonmotorized winter sports, (12) driving for pleasure (U.S. Department of Interior, BLM 1994, 243).

15 U.S. Department of Agriculture, Forest Service 1995, section 3; U.S. Department of Interior, BLM 1995, 282.

16 Zinser (1995, 363); BLM Web site (www.blm.gov/nhp/BLMinfo/stratplan/1997/publics).

17 Forest Service Web site (www.fs.fed.us/outdoors/wildlife/fish); BLM Web site (www.blm.gov/nhp/facts/facts).

18 Forest Service Web site (www.fs.fed.us/outdoors/wildlife/get); BLM Web site (www.blm.gov/nhp/facts/facts).

19 Forest Service Web site (www.fs.fed.us/outdoors/wildlife/TES/tes); BLM Web site (www.blm.gov/nhp/BLMinfo/stratplan/1997/3publics).

20 Forest Service Web site (www.fs.fed.us/outdoors/wildlife/TES/tes); BLM Web site (www.blm.gov/nhp/BLMinfo/stratplan/1997/3publics).

21 The Forest Service RPA Program reports utilize three different accounting stances to measure benefits: existing fees, market clearing prices, and market clearing prices plus consumer surplus. The analysis here applies to the first two accounting stances. See U.S. Department of Agriculture (1997) and U.S. Department of Agriculture, Forest Service (1995), respectively. NFS grazing benefits were calculated using an appraised fair market rental valuation figure derived by the Forest Service. See U.S. Department of Agriculture, Forest Service 1990 (appendix B utilizes a market appraisal of grazing lands to obtain a clearing price on forage); Tittman and Brownell (1984) *Appraisal Report Estimating Fair Market Rental Value of Public Rangelands in . . . the Western United States.* BLM market values of timber, range, and minerals benefits were obtained from BLM's *1997 Strategic Plan* (www.blm.gov/nhp/BLMinfo/stratplan/1997).

22 The objective of this exercise is to illustrate some of the innovative methods that can be used to estimate the benefits of nonmarket goods and services on public lands. These calculations rely on aggregated data and should be viewed as only preliminary, illustrative calculations. Further research in this area should be able to refine the techniques and improve the level of confidence about such estimates. Land managers seeking to implement an economic efficiency-based policy will obviously need to address quantification issues.

23 The change in population in the mountain region was 37.2 percent for 1970–80, 20.1 percent for 1980–90, and 14.5 percent for 1990–95. The corresponding increases for the entire United States was 11.4 percent, 9.8 percent, and 5.6 percent, respectively (U.S. Department of Commerce 1996, 29).

24 Other commentators have endorsed user fees as a means of promoting an efficient land management policy. See Clawson (1984, 204); Anderson and Leal (1991, 76).

25 Federal land management statutes usually pair multiple use with a companion concept: sustained yield. The term sustained yield means "the achievement and maintenance in perpetuity of a high-level annual or regular periodic output of the various renewable resources" (16 U.S.C. §531(b); 43 U.S.C. § 1702(h); see also 16 U.S.C. §§ 529, 531(h); 43 U.S.C. § 1732(a)).

26 See Executive Order 12,852,58 Fed. Reg. 35,841 (193), which commissioned the Council that issued its report in February 1997: President's Council on Sustainable Development 1996.

Lowry, The Impact of Political Institutions on Preservation of U.S. and Canadian National Parks

1 National Park Service Organic Act, *U.S. Code Annotated,* title 16, sec. 1:66; National Parks of Canada Act, 1930.
2 See testimony of Rick Gale (15), Paul Pritchard (17), and Robin Winks (40) in U.S. Congress (1995).
3 See testimony of Ridenour in U.S. Congress (1995, 10).

Percival, Global Environmental Accountability: The Missing Link in the Pursuit of Sustainable Development?

1 *Jacobellis v. Ohio,* 378 U.S. 184, 197 (1964) (Stewart, J., concurring).
2 See principles 1, 3, and 4 of United Nations Conference on Environment and Development, 1992.
3 Regulators must be given advance notice before a new chemical is marketed. See Toxic Substances Control Act Sec. 5, 15 U.S.C. 2604.
4 For example, section II 2(i)(5) of the Clean Air Act allows sources of hazardous air pollutants to delay installation of MACT (Maximum Achievable Control Technology) for five years if they voluntarily reduce emissions by 90 percent or more prior to proposal of MACT standards for their source category. Within two years after enactment of the 1990 Amendments, EPA had received seventy-six "enforceable commitments" for early reductions from sources. See 57 Fed. Reg. 58,203 (1992).

Buck, Saving All the Parts: Science and Sustainability

1 Of course, all environmental policies are concerned with scientific issues at their center. However, in this essay I am concerned with *direct* application of scientific knowledge to the policy formation process, that is, relying on scientific input that has not first been filtered through an economic or political screen.
2 An excellent paper that examines the role of the scientific community in the acid deposition arena is Alm and Simon (1999).
3 Twelve nations participated: Austria/Hungary, Denmark, Finland, France, Germany, the Netherlands, Norway, Russia, Sweden, the United Kingdom, Canada, and the United States.
4 Unless otherwise noted, this material is from Zumberge (1986).
5 Unless otherwise noted, this discussion is from Paterson (1996, 78–110).

Kassiola, Why Environmental Public Policy Analyses Must Include Explicit Normative Considerations: Reflections on Seven Illustrations

1 For my discussion of such cultures, and the crucial and unique role the environmental crisis can play in the social transformation of such social orders, see Kassiola (1990).

2 For the path-breaking analysis of the limitless desires created by the current liberal, neoclassical economic structure necessarily constrained by the finite environment, by an economist, see the works of Herman E. Daly listed in the references.

3 For an excellent collection of essays on the nature of limitless consumption and its adverse moral, political, psychological, and environmental consequences, see Crocker and Linden (1998).

Bowersox, Sustainability and Environmental Justice: A Necessary Connection

1 Here I am reminded of and persuaded by Dewey's historicist conception of liberalism: "Liberty in the concrete signifies the release from the impact of *particular* oppressive forces; emancipation from something once taken as a normal part of human life but now experienced as bondage" (1963, 48).

2 There are, of course, exceptions, most notably Herman Daly. See Daly, "Moving to a Steady-State Economy," Paul Ehrlich and Holdren (1988, 271–85).

Sagoff, The Hedgehog, the Fox, and the Environment

1 As the concept of entropy occurs in thermodynamics, it has to do with energy. But Kenneth Boulding proposes that "material entropy can be taken as a measure of the uniformity of the distribution of elements and, more uncertainly, compounds and other structures on the earth's surface" (1993, 301).

Conclusion

1 "When he is told that some pleasures should be sought and valued as arising from desires of a high order, others chastised and enslaved because the desires are base, he will shut the gates of the citadel against the messengers of truth, shaking his head and declaring that one appetite is as good as another and all must have their equal rights. So he spends his days indulging the pleasure of the moment, now intoxicated with wine and music, and then taking to a spare diet. . . . Every now and then he takes a part in politics, leaping to his feet to say or do whatever comes into his head. . . . His life is subject to no order or restraint, and he has no wish to change an existence which he calls pleasant, free, and happy " (Plato 1941, book 8, 561c–d, p. 286).

2 See Gillroy (2000) for a detailed argument about risk perception and the difference between responsible and responsive government.

3 For discussions of stakeholder processes, see, e.g., Andranovich (1995); Finney and Polk (1995); Painter (1988).

4 See Gillroy (2000) for an argument about the critical nature of autonomy in environmental policy decision making.

5 In essence, then, our discussants agree at least in part with the diagnosis of Jonathan Baert Wiener (this volume), if not his prescription.

References

Agee, James K. 1996. "Ecosystem Management: An Appropriate Concept for Parks?" In *National Parks and Protected Areas,* ed. R. G. Wright, 31–44. Cambridge, MA: Blackwell Science.

Alm, Leslie, and Marc Simon. 1999. "Scientists, Advocacy, and Environmental Policy Making: The Case of Acid Rain." Prepared for the Southwestern Political Science Association meeting, San Antonio, TX, 31 March–3 April.

Alper, J. S. 1981. "Facts, Values, and Biology." *Philosophical Forum* 13:99.

Anderson, Terry L., and Donald R. Leal. 1991. *Free Market Environmentalism.* San Francisco: Pacific Research Institute for Public Policy; Boulder, CO: Westview Press.

Andranovich, Greg. 1995. "Achieving Consensus in Public Decision Making." *Journal of Applied Behavioral Science* 31, no. 4: 429–45.

Andrews, Richard N. L. 1999. *Managing the Environment, Managing Ourselves: A History of American Environmental Policy.* New Haven: Yale University Press.

Apel, Karl Otto. 1980. *Towards a Transformation of Philosophy.* London: Routledge and Kegan Paul.

"Appliance Standards Are Getting Results." 1995. *Energy Conservation News* 18, no. 2 (September 1). Available at http://web.lexis-nexis.com/universe.

Arendt, Hannah. 1963. *On Revolution.* New York: Viking.

Argyle, Michael. 1987. *The Psychology of Happiness.* New York: Methuen.

Ariansen, Per. 1997. "The Non-utility Value of Nature: An Investigation into Biodiversity and the Value of Natural Wholes." In *Communications of the Norwegian Forest Research Institute,* ed. Meddelelser fra Skogforsk, 47. Aas, Norway: Agricultural University of Norway.

Arrhenius, Svante. 1908. *Worlds in the Making.* New York: Harper.

Arrow, K., B. Bolin, R. Costanza, P. Dasgupta, C. Folke, C. S. Holling, B.-O Jansson, S. Levin, K.-G. Maler, C. Perrings, and D. Pimentel. 1995. "Economic Growth, Carrying Capacity, and the Environment." *Science* 268: 520–21.

"'Ask a Silly Question . . .': Contingent Valuation of Natural Resource Damages." *Harvard Law Review* 105: 1981–2000.

Atcheson, John. 1991. "The Department of Risk Reduction of Risky Business." *Environmental Law* 21: 1375.

Athanasiou, Tom. 1996. *Divided Planet: The Ecology of Rich and Poor.* Boston: Little, Brown.

Audi, Robert, ed. 1995. *The Cambridge Dictionary of Philosophy.* Cambridge, England: Cambridge University Press.

Austin, J. L. 1962. *How to Do Things with Words.* New York: Oxford University Press.

Ausubel, Jesse. 1996. "Can Technology Spare the Earth?" *American Scientist* 84: 164–70.

Bardach, Eugene. 1982. *The Implementation Game.* New York: Harper.

Barnett, H. J., and C. Morse. 1963. *Scarcity and Growth: The Economics of Natural Resource Availability.* Baltimore: Johns Hopkins University Press.

Bartholomew, Joy A., Robert Costanza, and Herman E. Daly. 1991. "Goals, Agenda, and Policy Recommendations for Ecological Economics." In *Ecological Economics: The Science and Management of Sustainability,* ed. Robert Costanza, 1–20. New York: Columbia University Press.

Baxter, Brian. 1999. *Ecologism: An Introduction.* Washington, DC: Georgetown University Press.

Bearak, Barry. 1998. "Nature-Loving Indians Turn Poachers into Prey." *New York Times* November 29, A26.

Bella, Leslie. 1987. *Parks for Profit.* Montreal: Harvest House.

Bergman, Brian. 1986. "National Parks Going Downhill." *Equinox* 29: 115–16.

Berkes, Fikret, and Carl Folke. 1994. "Investing in Cultural Capital for Sustainable Use of Natural Capital." In *Investing in Natural Capital,* ed. AnnMari Jansson et al. Washington, DC: Island Press.

Berlin, Isaiah, Sir. 1957. *The Hedgehog and the Fox: An Essay on Tolstoy's View of History.* New York: New American Library.

Binger, Brian R., Robert Copple, and Elizabeth Hoffman. 1995. "The Use of Contingent Valuation Methodology in Natural Resource Damage Assessments: Legal Fact and Economic Fiction?" *Northwestern Law Review* 89: 1029.

Birkland, Thomas A. 1997. *After Disaster: Agenda Setting, Public Policy, and Focusing Events.* Washington, DC: Georgetown University Press.

Blamey, R. M. Common, and J. Quiggen. 1993. "Respondents to Contingent Valuation Surveys: Consumers or Citizens?" (Manuscript). Canberra, Australia: Centre for Resource and Environmental Studies and Centre for Economic Policy Research, Australian National University.

—. 1994. "Respondents to Contingent Valuation Surveys: Consumers or Citizens?" (Manuscript). Canberra, Australia: Center for Economic Policy Research, Australian National University.

Blumm, Michael C. 1994. "Public Choice Theory and the Public Lands: Why 'Multiple Use' Failed." *Harvard Environmental Law Review* 18: 405.

Bookchin, Murray. 1981. "What Is Social Ecology?" In *Environmental Philosophy: From Animal Rights to Radical Ecology,* ed. Michael Zimmerman et al. Englewood Cliffs, NJ: Prentice Hall.

Boulding, Kenneth. 1993. "The Economics of Coming Spaceship Earth." In *Valuing the Earth,* ed. Herman Daly and Kenneth Townsend. Cambridge, MA: MIT Press.

Bowers, C. A. 1995. *Educating for an Ecologically Sustainable Culture.* Albany: State University of New York Press.

—. 1997. *The Culture of Denial.* Albany: State University of New York Press.

Bowers, Michael D., and John V. Krutilla. 1989. *Multiple-Use Management: The Economics of Public Forestlands.* Washington, DC: Resources for the Future.

Bowersox, Joe. 1995. "The Public Space of Environmentalism: Reason, Values, and Legitimacy in Environmental Ethics and Politics." Ph.D. diss., University of Wisconsin–Madison.

—. 1998. "What Is Scarce in the Politics of Scarcity?" Paper presented at the Western Political Science Association annual meeting, Los Angeles, March 19–21.

—. 2000. "From Water Development to Water Management." *American Behavioral Scientist* 44, no. 4: 599–613.

Brecht, Arnold. 1959. *Political Theory.* Princeton, NJ: Princeton University Press.

Breen, Barry. 1993. "Objectives and Tools." In *Sustainable Environmental Law,* ed. Celia Campbell-Mohn et al. Eagan, MN: West.

Brooks, Richard. 1995. *Community and the Law: A Post-Modern Perspective.* Unpublished manuscript.

Brower, Michael. 1990. *Cool Energy: The Renewable Solution to Global Warming: A Report by the Union of Concerned Scientists.* Cambridge, MA: Union of Concerned Scientists.

Brown, Beverly. 1995. *In Timber Country: Working People's Stories of Environmental Conflict and Urban Flight.* Philadelphia: Temple University Press.

Brown, Lester R., Christopher Flavin, and Hal Kane. 1996. *Vital Signs 1996.* New York: Norton.

Brown, Lester R., Hal Kane, and David M. Roodman. 1994. *Vital Signs 1994.* New York: Norton.

Browner, Carol. 1994. "The Priorities and Key Activities of the Office of Pollution Prevention." March 11. TTN Doc. G-0001-D.ZIP.

Browner, Carol, and Environmental Protection Agency (EPA). 1993. "Pollution Prevention Policy Statement: New Directions for Environmental Protection." Reprinted in *Environmental Law Reporter* 23: 35555.

Bryan, Rorke. 1973. *Much Is Taken, Much Remains.* North Scituate, MA: Duxbury Press.

Bryant, Bunyan, ed. 1995. *Environmental Justice: Issues, Policies, and Solutions.* Washington, DC: Island Press.

Buchanan, James, and Gordon Tullock. 1962. *The Calculus of Consent.* Ann Arbor: University of Michigan Press.

Buck, Susan J. 1998. *The Global Commons.* Covelo, CA: Island Press.

Buelow, Kristine M. 1997. "Barriers to Regulatory Reform as Experienced in the 3M Project XL Pilot." Master's thesis, Duke University, Durham, NC.

Bullard, Robert. 1993. *Confronting Environmental Racism: Voices from the Grassroots.* Boston: South End Press.

Bureau of the Census, U.S. Department of Commerce. 1996. *Statistical Abstract of the United States 1996.* Available at http://www.census.gov/prod/2/gen/96statab/96statab.html.

Busenberg, George. 1999. "Collaborative and Adversarial Analysis in Environmental Policy." *Policy Sciences* 32: 1–11.

Cable, Sherry, and Charles Cable. 1995. *Environmental Problems, Grassroots Solutions: The Politics of Grassroots Environmental Conflict.* New York: St. Martin's Press.

Callicott, J. Baird. 1989. *In Defense of the Land Ethic.* Albany: State University of New York Press.

—. 1995. "Environmental Philosophy Is Environmental Activism: The Most Radical and Effective Kind." In *Environmental Philosophy and Environmental Activism,* ed. Don E. Marietta Jr. and Lester Embree, 19–35. Lanham, MD: Rowman and Littlefield.

Callicott, J. Baird, and Karen Mumford. 1997. "Ecological Sustainability as a Conservation Concept." *Conservation Biology* 11: 32.

Cameron, James, and Juli Abouchar. 1991. "The Precautionary Principle: A Fundamental Principle of Law and Policy for the Protection of the Global Environment," *Boston College International and Comparative Law Review* 14: 1.

Campbell, A., P. E. Converse, and W. Rodgers. 1976. *The Quality of American Life: Perceptions, Evaluations, Satisfactions.* New York: Russell Sage Foundation.

Campbell-Mohn, Celia. 1993. *Sustainable Environmental Law: Integrating Natural Resource and Pollution Abatement Law from Resources to Recovery.* St. Paul, MN: West.

Canadian Environmental Advisory Council (CEAC). 1991. *A Protected Areas Vision for Canada.* Ottawa: CEAC.

Canadian Nature Federation. 1991. "National Parks Policy Review." *Nature Alert* 1:3.

Canadian Parks Service (CPS). 1988. *Banff National Park Management Plan.* Ottawa: Environment Canada.

Carley, Michael, and Philippe Spapens. 1998. *Sharing the World: Sustainable Living and Global Equity in the 21st Century.* London: Earthscan.

Carruthers, J. A. 1979. "Planning a Canadian National Park and Related Reserve System." In *The Canadian National Parks: Today and Tomorrow Conference II,* ed. J. G. Nelson, 645–70. Waterloo: University of Waterloo.

Carson, Rachel. 1964. *Silent Spring.* Boston: Houghton Mifflin.

Catton, William R., Jr. 1980. *Overshoot: The Ecological Basis of Revolutionary Change.* Urbana: University of Illinois Press.

Chase, Alston. 1987. *Playing God in Yellowstone.* San Diego: Harcourt Brace Jovanovich.

Chavis, Benjamin F., Jr. 1993. Foreword to *Confronting Environmental Racism: Voices from the Grassroots,* ed. Robert D. Bullard, 3–5. Boston: South End Press.

Chiras, Daniel. 1994. *Environmental Science: Action for a Sustainable Future,* 4th ed. New York: Benjamin/Cummings.

Christensen, Paul. 1991. "Driving Forces, Increasing Returns, and Ecological Sustainability." In *Ecological Economics: The Science and Management of Sustainability,* ed. Robert Costanza, 75–87. New York: Columbia University Press.

Clark, Edward R. 1994. "Cumulative Effects Analysis in the NEPA Process." *ALI-ABA* C933: 655.

Clawson, Marion. 1978. "The Concept of Multiple Use Forestry." Environmental Law 8, no. 2: 281–308.

—. 1984. "Major Alternatives for the Future Management of the Federal Lands." In *Re-*

thinking the Federal Lands, ed. Sterling Brubaker. Baltimore: Johns Hopkins University Press.

Clawson, Marion, and Jack L. Knetsch. 1966. *Economics of Outdoor Recreation.* Baltimore: Published for Resources for the Future by Johns Hopkins University Press.

Coase, Ronald. 1960. "The Problem of Social Cost." *Journal of Law and Economics* 3:1.

Collin, Robert, and Robin Morris Collin. 1994. "Essays on Environmental Justice: Equity as the Basis of Implementing Sustainability: An Exploratory Essay." *West Virginia Law Review* 96: 1173.

Congressional Research Service (CRS). 1994. *Ecosystem Management: Federal Agency Actions.* Washington, DC: Government Printing Office.

Connelly, James, and Graham Smith. 1999. *Politics and the Environment: From Theory to Practice.* New York: Routledge.

Connelly, N. A., and T. L. Brown. 1988. *Estimates of Nonconsumptive Wildlife Use on Forest Service and BLM Lands.* Washington, DC: Human Dimensions Research Unit, Dept. of Natural Resources, New York State College of Agriculture and Life Sciences, Cornell University.

Costanza, Robert. 1995. "Ecological Economics: Toward a New Transdisciplinary Science." In *A New Century for Natural Resources Management*, ed. Richard Knight and Sarah Bates, 323–48. Washington, DC: Island Press.

Costanza, Robert, and Herman E. Daly, 1992. "Natural Capital and Sustainable Development." *Conservation Biology* 6: 37–46.

Costanza, Robert, Herman Daly, and Joy Bartholomew. 1991. "Goals, Agenda, and Policy Recommendations for Ecological Economics." In *Ecological Economics: The Science and Management of Sustainability*, ed. Robert Costanza. New York: Columbia University Press.

Costanza, Robert, and the International Society for Ecological Economics. 1997. *An Introduction to Ecological Economics.* Boca Raton, FL: St. Lucie Press.

Costanza, Robert, et al. 1997. "The Value of the World's Ecosystem Services and Natural Capital." *Nature* 387: 253–260.

Costanza, Robert, Bryan Norton, and Benjamin Haskell, eds. 1992. *Ecosystem Health: New Goals for Environmental Management.* Covelo, CA: Island Press.

Council of State Governments. 1999. *Resource Guide to State Environmental Management*, 5th ed. Lexington, KY: Council of State Governments.

Cowell, Alan. 1999. "Annan Fears Backlash over Global Crisis." *New York Times* February 1, A12.

Crocker, David A., and Toby Linden, eds. 1998. *Ethics of Consumption: The Good Life, Justice, and Global Stewardship.* Lanham, MD: Rowman and Littlefield.

Cross, Frank B. 1996. "Paradoxical Perils of the Precautionary Principle." *Washington and Lee Law Review* 53:851.

Cushman, John H., Jr. 1999. "White House Calls for Ban on Mining in Rockies Habitat: Conserving Public Land." *New York Times* February 4, A1.

"Cyprus Amax Sees Many Reasons for SO_2 Price." 1995. *Air Daily* (25 October): 1.

Daily, Gretchen C., and Paul Ehrlich. 1992. "Population, Sustainability, and Earth's Carrying Capacity." *Bioscience* 42:761.

Daly, Herman E. 1977. *Steady-State Economics: The Economics of Biophysical Equilibrium and Moral Growth.* San Francisco: W. H. Freeman.
—. 1979. "Entropy, Growth, and the Political Economy of Scarcity." In *Scarcity and Growth Revisited,* ed. V. Kerry Smith, 67–94. Baltimore: Johns Hopkins/Resources for the Future.
—. 1990. "Toward Some Operational Principles of Sustainable Development." *Ecological Economics* 2, no. 1: 1–6.
—. 1991. "Elements of Environmental Economics." In *Ecological Economics,* ed. Robert Costanza. New York: Columbia University Press.
—. 1992. "Allocation, Distribution, and Scale: Towards an Economics That Is Efficient, Just, and Sustainable." *Ecological Economics* 6: 185–93.
—. 1993. "The Perils of Free Trade." *Scientific American* November, 50–57.
—. 1994. "Operationalizing Sustainable Development by Investing in Natural Capital." In *Investing in Natural Capital: The Ecological Economics Approach to Sustainability,* ed. AnnMari Jansson et al. Washington, DC: Island Press.
—. 1996. *Beyond Growth: The Economics of Sustainable Development.* Boston: Beacon Press.
—. 1973. *Toward a Steady-State Economy.* San Francisco: W. H. Freeman.
—. 1980. *Economics, Ecology, Ethics: Essays toward a Steady-State Economy.* San Francisco: W. H. Freeman.
Daly, Herman E., and John B. Cobb Jr., with contributions by Clifford W. Cobb. 1989. *For the Common Good: Redirecting the Economy Toward Community, the Environment, and a Sustainable Future.* Boston: Beacon Press.
Daly, Herman E., and Kenneth N. Townsend, eds. 1993. *Valuing the Earth: Economics, Ecology, Ethics.* Cambridge, MA: MIT Press.
Daniels, Steven E. 1987. "Rethinking Dominant Use Management in the Forest Planning Era." *Environmental Law Review* 17: 483.
Davies, J. Clarence. 1989. *The Environmental Protection Act.* Washington, DC: Conservation Foundation.
Davies, J. Clarence, and Jan Mazurek. 1997. *Pollution Control in the United States: Evaluating the System.* Washington, DC: Resources for the Future.
Davis, Charles E. 1993. *The Politics of Hazardous Waste.* Englewood Cliffs, NJ: Prentice Hall.
Davis, Robert K. 1963. "Recreation Planning as an Economic Problem." *Natural Resources Journal* 3: 239.
Deacon, Robert T. 1994. "Deforestation and the Rule of Law in a Cross-Section of Countries." *Land Economics* November, 70.
Derthick, Martha. 1979. *Policymaking for Social Security.* Washington, DC: Brookings.
Deruiter, Darla S., and Glenn E. Haas. 1995. *National Public Opinion Survey on the National Park System.* Washington, DC: National Parks and Conservation Association.
Devall, Bill, and George Sessions. 1985. *Deep Ecology.* Salt Lake City: Peregrine Smith Books.
Dewey, John. 1963, 1935. *Liberalism and Social Action.* New York: Capricorn Books.
Diamond, Jared. 1990. "Playing Dice with Megadeath." *Discover* April, 55–59.

Diamond, Peter A., and Jerry A. Hausman. 1994. "Contingent Valuation: Is Some Number Better Than No Number?" *Journal of Economic Perspectives* 8: 45.
Dobson, Andrew. 1998. *Justice and the Environment: Conceptions of Environmental Sustainability and Theories of Distributive Justice.* Oxford: Oxford University Press.
Dryzek, John. 1987. *Rational Ecology.* New York: Basil Blackwell.
—. 1990. "Green Reason: Communicative Ethics for the Biosphere." *Environmental Ethics* 12:195–210.
—. 1993. "Policy Analysis and Planning: From Science to Argument." In *The Argumentative Turn in Policy Analysis and Planning,* ed. Frank Fischer and John Forester. Durham, NC: Duke University Press.
Dryzek, John. 1996. *Democracy In Capitalist Times: Ideals, Limits, and Struggles.* New York: Oxford University Press.
—. 1997. *The Politics of the Earth: Environmental Discourses.* New York: Oxford University Press.
Duncan, Myrl. 1996. "Property as Public Conversation, Not a Lockean Soliloquy." *Environmental Law* 26:1095.
Durning, Alan Thein, and Yoram Bauman. 1998. *Tax Shift.* Seattle: Northwest Environment Watch.
Eckersley, Robyn. 1992. *Environmentalism and Political Theory: Toward an Ecocentric Approach.* Albany: State University of New York Press.
Ehrenfeld, David. 1993. *Beginning Again.* New York: Oxford University Press.
Ehrlich, Paul. 1968. *The Population Bomb.* New York: Ballantine Books.
—. 1989. "The Limits to Substitution: Meta-Resource Depletion and a New Economic-Ecological Paradigm," *Ecological Economics* 1: 9–16.
—. 1994. "Ecological Economics and the Carrying Capacity of the Earth." In *Investing in Natural Capital: The Ecological Economics Approach to Sustainability,* ed. Ann-Mari Jansson et al., 38–56. Washington, DC: Island Press.
Ehrlich, Paul, Gretchen C. Daily, Scott C. Daily, Norman Myers, and James Salzman. 1997. "No Middle Way on the Environment." *Atlantic Monthly* 280, no. 6: 98–104.
Ehrlich, Paul R., and Anne H. Ehrlich. 1970. *Population, Resources, Environment: Issues in Human Ecology.* San Francisco: W. H. Freeman.
—. 1990. *The Population Explosion.* New York: Simon and Schuster.
—. 1996. *The Betrayal of Science and Reason.* Washington, DC: Island Press.
Ehrlich, Paul R., and John P. Holdren. 1971. "Impact of Population Growth." *Science* 171: 1212–217.
—, eds. 1988. *The Cassandra Conference: Resources and the Human Predicament.* College Station: Texas A&M University Press.
Elgie, Stewart. 1998. "The Harmonization Accord: A Solution in Search of a Problem." *Canada Watch* January, 10–11.
Erskine, Hazel. 1964. "The Polls: Some Thoughts about Life and People." *Public Opinion Quarterly* 28:3.
Farber, D. A., and P. A. Hemmersbaugh. 1993. "The Shadow of the Future: Discount Rates, Later Generations, and the Environment." *Vanderbilt Law Review* 46: 267.

Farmer, Michael C., and Alan Randall. 1997. "Policies for Sustainability: Lessons from an Overlapping Generations Model." *Land Economics* 73: 608.

Feldman, David, and Catherine Wilt. 1998. "Climate Change from a Bioregional Perspective." In *Bioregionalism*, ed. Michael McGinniss. New York: Routledge.

Feldman, Elliot, and Michael A. Goldberg, eds. 1997. *Land Rites and Wrongs*. Cambridge, MA: Lincoln Institute of Land Policy.

Finney, Carolyn, and Ruth E. Polk. 1995. "Developing Stakeholder Understanding, Technical Capability, and Responsibility: The New Bedford Harbor Superfund Forum." *Environmental Impact Assessment Review* 15:517–41.

Fischer-Kowalski, Marina. 1997. "Society's Metabolism: On the Childhood and Adolescence of a Rising Conceptual Star." In *The International Handbook of Environmental Sociology*, ed. Michael Redclift and Graham Woodgate. Cheltenham, England: Edward Elgar.

Fisher, Marie Ramirez. 1994. "Comment: On the Road from Environmental Racism to Environmental Justice." *Villanova Environmental Law Journal* 5: 449.

Fishkin, James S. 1979. *Tyranny and Legitimacy: A Critique of Political Theories*. Baltimore: Johns Hopkins University Press.

Flathman, Richard. 1976. *The Practice of Rights*. Cambridge, England: Cambridge University Press.

Folke, Carl, et al. 1994. "Investing in Natural Capital: Why, What, and How?" In *Investing in Natural Capital: The Ecological Economics Approach to Sustainability*, ed. Ann-Mari Jansson et al. Washington, DC: Island Press, 1994.

Foreman, Christopher H. 1998. *The Promise and Peril of Environmental Justice*. Washington, DC: Brookings Institution.

Foreman, Dave. 1991. *Confessions of an Eco-Warrior*. New York: Harmony.

Foresta, Ron. 1984. *America's National Parks and Their Keepers*. Washington, DC: Resources for the Future.

Fourier, Jean Baptiste Joseph. 1827. *Memoire sur Les Temperatures du Globe Terrestre et des Espace Planetaires*. Paris: Institut de France, Academe des Sciences Memoires.

Fowler, Robert Booth. 1991. *The Dance with Community: The Contemporary Debate in American Political Thought*. Lawrence: University Press of Kansas.

Freeman, A. Myrick, III. 1979. *The Benefits of Environmental Improvement*. Baltimore: Published for Resources for the Future by Johns Hopkins University Press.

—. 1993. *The Measurement of Environmental and Resource Values*. Washington, DC: Resources for the Future.

Freemuth, John. 1989. "The National Parks." *Public Administration Review* 49, no. 3: 278–86.

—. 1991. *Islands under Siege*. Lawrence: University Press of Kansas.

Frege, Gotthold. 1892. "Sinn und Begriff." *Zeitschrift fuer Philosophie und Kritik:* 100.

Frome, Michael. 1992. *Regreening the National Parks*. Tucson: University of Arizona Press.

Fullem, Gregory D. 1995. "The Precautionary Principle: Environmental Protection in the Face of Scientific Uncertainty." *Willamette Law Review* 31:495.

Gale, R. P., and S. M. Corday. 1991. "What Should Forests Sustain? Eight Answers." *Journal of Forestry* 31: 31–36.

Georgescu-Roegen, Nicholas. 1979. "Comments on the Papers by Daly and Stiglitz." In *Scarcity and Growth Reconsidered,* ed. V. Kerry Smith. Baltimore: Published for Resources for the Future by the Johns Hopkins University Press.

Gibbs, Lois Marie. 1995. *Dying from Dioxin.* Boston: South End Press.

Gillroy, John Martin. 1995. "Comparative Risk." In *Conservation and Environmentalism: An Encyclopedia,* ed. Robert Paehlke, 144–47. New York: Garland.

—. 1998. "Beyond Sustainability." Paper presented at the annual meeting of the American Political Science Association, Boston, September 3–6.

—. 2000. *Justice and Nature: Kantian Philosophy, Environmental Policy and the Law.* Washington, DC: Georgetown University Press.

Gillroy, John Martin, ed. 1993. *Environmental Risk, Environmental Values, and Political Choices: Beyond Efficiency Tradeoffs in Policy Analysis.* Boulder, CO: Westview Press.

Gillroy, John Martin, Mark Sagoff, Joel Kassiola, Bob Pepperman Taylor, Robert Paehlke, Joe Bowersox, Susan Buck, and Gary Woller. 1997. "A Forum on the Role of Environmental Ethics in Restructuring Environmental Policy and Law for the Next Century." *Policy Currents* 7, no. 2: 1–13.

Gillroy, John Martin, and Maurice Wade, eds. 1992. *The Moral Dimensions of Public Policy Choice: Beyond the Market Paradigm.* Pittsburgh: University of Pittsburgh Press.

Glicksman, Robert. 1997. "Fear and Loathing on the Federal Lands." *University of Kansas Law Review* 45: 647.

Goodin, Robert. 1976. *The Politics of Rational Man.* London: Wiley.

—. 1992. *Green Political Theory.* Cambridge, England: Polity Press.

Goodstein, Eban S. 1995. *Economics and the Environment.* Englewood Cliffs, NJ: Prentice Hall.

Graham, John D., and Maria Segui-Gomez. 1997. "Airbags: Benefits and Risks." *Risk in Perspective,* Harvard Center for Risk Analysis, July, 7.

Graham, John D., and Jonathan Baert Wiener. 1995. *Risk vs. Risk: Tradeoffs in Protecting Health and the Environment.* Cambridge, MA: Harvard University Press.

Grant, Ruth. 1987. *John Locke's Liberalism.* Chicago: University of Chicago Press.

Greber, Brian J., and K. Norman Johnson. 1991. "What's All This Debate about Overcutting? It Depends upon Your Perspective." *Journal of Forestry* 89, no. 11 (November): 25–30.

Greenstone, David. 1993. *The Lincoln Persuasion.* Princeton, NJ: Princeton University Press.

Grumbie, R. Edward. 1994. "What Is Ecosystem Management?" *Conservation Biology* March, 27.

Gundling, Lothar. 1990. "What Obligation Does Our Generation Owe to the Next?" *American Journal of International Law* 84:207.

Guring, Gerald. 1960. *Americans View Their Mental Health.* New York: Basic Books.

Habermas, Jürgen. 1987. *Philosophical Discourse of Modernity.* Cambridge, MA: MIT Press.

Hagen, Daniel A., James W. Vincent, and Patrick G. Wells. 1992. "Benefits of Preserving Old-Growth Forests and the Spotted Owl." *Contemporary Policy Issues* 10:13.

Hanemann, W. Michael. 1994. "Valuing the Environment through Contingent Valuation." *Journal of Economic Perspectives* 8: 19.
Hardin, Garrett. 1968. "The Tragedy of the Commons." *Science* 162 (December 13): 1243–248.
Hardt, Scott W. 1994. "Federal Land Management in the Twenty First Century: From Wise Use to Wise Stewardship." *Harvard Environmental Law Review* 18: 345.
Hargrove, Eugene. 1992. "Weak Anthropocentric Intrinsic Value." *The Monist* 75:183.
Harper, S. 1995. "Cleaner, Cheaper: It's Common Sense." *Environmental Forum* May–June.
Harrison, Glenn, and James Lesley. 1996. "Must Contingent Valuation Surveys Cost So Much?" *Journal of Environmental Economics and Management* 31:79.
Harrison, Kathryn. 1996. *Passing the Buck: Federalism and Canadian Environmental Policy.* Vancouver: University of British Columbia Press.
Hart, H. L. A. 1961. *The Concept of Law.* Oxford: Clarendon.
Hartz, Louis. 1955. *The Liberal Tradition in America.* New York: Harcourt Brace Jovanovich.
Hartzog, George. 1988. *Battling for the National Parks.* Mt. Kisco, N.Y.: Moyer Bell Limited.
Heclo, Hugh. 1977. *A Government of Strangers.* Washington, DC: Brookings.
Heilbroner, Robert. 1980. *Inquiry into the Human Prospect.* Updated ed. New York: Norton.
Hey, Ellen. 1992. "The Precautionary Concept in Environmental Policy and Law: Institutionalizing Caution." *Georgetown International Environmental Law Review* 4:303.
Hickey, James E., Jr., and Vern R. Walker. 1995. "Refining the Precautionary Principle in International Environmental Law." *Virginia Environmental Law Journal* 14:423.
Hildebrandt, Walter. 1995. "An Historical Analysis of Parks: Canada and Banff National Park, 1968–1995." Unpublished manuscript for Banff–Bow Valley Study.
Hinds, M. D. 1992. "Much Steaming over Steamtown." *New York Times* February 8, A12.
Hodas, David R. 1998. "The Role of Law in Defining Sustainable Development: NEPA Reconsidered." *Widener Law Symposium* 3:1–60.
Hoelting, Rebecca. 1994. "After Rio: The Sustainable Development Concept Following the United Nations Conference on Environment and Development." *Georgetown Journal of International and Comparative Law* 24:117.
Holland, Alan. 1999. "Sustainability: Should We Start from Here?" In *Fairness and Futurity,* ed. Andrew Dobson. Oxford: Oxford University Press.
Holland, Alan, and Kate Rawls. 1993. "Values in Conservation." *Ecos* 14: 14–19.
Holling, C. S. 1995. "What Barriers? What Bridges?" In *Barriers and Bridges to Renewal of Ecosystems and Institutions,* ed. L. H. Gunderson, C. S. Holling, and S. S. Light. New York: Columbia University Press.
—. 1996. "Engineering Resilience versus Ecological Resilience." In *Engineering within Ecological Constraints.* Washington, DC: National Academy Press.
Holmes, Stephen, and Cass R. Sunstein. 1999. *The Cost of Rights: Why Liberty Depends on Taxes.* New York: Norton.

Holsworth, Robert D. 1979. "Recycling Hobbes: The Limits of Political Ecology." *Massachusetts Review* 20:9–40.
Horwitz, Robert Britt. 1989. *The Irony of Regulatory Reform.* New York: Oxford University Press.
Howarth, Richard B. 1995. "Sustainability under Uncertainty: A Deontological Approach." *Land Economics* 71:417.
—. 1997. "Sustainability as Opportunity." *Land Economics* 73:569.
Howe, Charles W. 1997. "Dimensions of Sustainability: Geographical, Temporal, Institutional, and Psychological." *Land Economics* 73:597.
Howse, John. 1989. "An Identity Crisis." *MacLean's* 2, no. 27: 48.
Huber, Peter W. 1999. *Hard Green.* New York: Basic Books.
Huffman, James. 1998. "The Public Interest in Private Property Rights." *Public Land and Resources Digest* 35: 339.
Hume, David. 1888. *A Treatise on Human Nature.* Oxford: Clarendon Press.
International Union for the Conservation of Nature. 1972. *Second World Conference on National Parks,* ed. H. Elliot. Lausanne, Switzerland: IUCN.
Janicke, Martin. 1990. *State Failure: The Impotence of Politics in Industrial Society,* trans. Alan Braley. University Park: Pennsylvania State University Press.
Jansson, AnnMari, et al., eds. 1994. *Investing in Natural Capital: The Ecological Approach to Sustainability.* Washington, DC: Island Press.
Jefferson, Thomas. 1785. (1954) *Notes on the State of Virginia.* Edited with an introduction by William Peden. Chapel Hill: University of North Carolina Press.
—. 1813. (1959) Letter, Jefferson to John Adams, October 28. Reprinted in *The Adams-Jefferson Letters,* ed. Lester J. Cappon. Vol. 2: 387–92. Chapel Hill: University of North Carolina Press.
—. 1816. (1830) "On the Present Need to Promote Manufacturing" (letter to Benjamin Austin). In *Memoirs, Correspondence, and Private Papers of Thomas Jefferson,* ed. Thomas Jefferson Randolph, 4:280–82. Boston: Gray and Bowen.
Jeffery, Michael I. 1996. "Public Lands Reform: A Reluctant Leap into the Abyss." *Virginia Environmental Law Review* 16: 79.
Johnson, Stephen M. 1992. "From Reaction to Proaction: The 1990 Pollution Prevention Act." *Columbia Journal of Environmental Law* 17:153.
Johnsson, Thomas B., et al., eds. 1993. *Renewable Energy: Sources for Fuels and Electricity.* Washington, DC: Island Press.
Kassiola, Joel Jay. 1990. *The Death of Industrial Civilization: The Limits to Economic Growth and the Repoliticization of Advanced Industrial Society.* Albany: State University of New York Press.
—, ed. Forthcoming. *Explorations in Environmental Political Theory.* Albany: State University of New York Press.
Kearney, Anne R., Gordon Bradley, Rachel Kaplan, and Stephen Kaplan. 1999. "Stakeholder Perspectives on Appropriate Forest Management in the Pacific Northwest." *Forest Science* 45, no. 1:62–73.
Keiter, Robert B. 1989. "Taking Account of the Ecosystem on the Public Domain: Law and Ecology in the Greater Yellowstone Region." *Colorado Law Review* 60:923.

—. 1991. "An Introduction to the Ecosystem Debate." In *The Greater Yellowstone Ecosystem,* ed. Robert B. Keiter and M. S. Boyce, 3–18. New Haven: Yale University Press.

—. 1994. "Beyond the Boundary Line: Constructing a Law of Ecosystem Management." *University of Colorado Law Review* 65: 293.

—. 1996. "Ecosystem Management: Exploring the Legal-Political Framework." In *National Parks and Protected Areas,* ed. R. G. Wright, 63–88. Cambridge, MA: Blackwell Science.

—. 1997. "Preserving Nature in the National Parks." *Denver University Law Review* 74, no. 3: 649.

Keiter, Robert B., and Mark S. Boyce. 1991. "Greater Yellowstone's Future." In *The Greater Yellowstone Ecosystem,* ed. Robert B. Keiter and M. S. Boyce, 379–413. New Haven: Yale University Press.

Kelleher, Graeme, Chris Bleakley, and Sue Wells. 1995. *A Global Representative System of Marine Protected Areas.* Washington, DC: World Bank.

Kelly, Erin. 1998. "A Noisy Debate on National Parks." *Denver Post* June 7, A12.

Kempton, Willet, James Boster, and Jennifer Hartly. 1995. *Environmental Values in American Culture.* Cambridge, MA: MIT Press.

Kimber, Cliona. 1995. "A Comparison of Environmental Federalism in the United States and the European Union." *Maryland Law Review* 54:1658.

King, T. F. 1988. "Park Planning, Historic Resources, and the National Historic Preservation Act." In *Our Common Lands,* ed. D. Simon, 275–91. Washington, DC: Island Press.

Kingdon, John W. 1999. *America the Unusual.* New York: St. Martin's.

Kneese, Allen V., and Blair T. Bower, eds. 1972. *Environmental Quality Analysis: Theory and Method in the Social Sciences.* Washington, DC: Resources for the Future.

Knight, Richard L., and Sarah F. Bates, eds. 1995. *A New Century for Natural Resources Management.* Washington, DC: Island Press.

Knight, Richard, and T. Luke George. 1995. "New Approaches, New Tools: Conservation Biology." In *A New Century for Natural Resources Management,* ed. Richard Knight and Sarah Bates, 279–95. Washington, DC: Island Press.

Knight, Richard L., and Peter B. Landres. 1998. *Stewardship across Boundaries.* Washington, DC: Island Press.

Kopp Raymond J., and V. Kerry Smith, eds. 1993. *Valuing Natural Assets.* Washington, DC: Resources for the Future.

Kraft, Michael E., and Denise Scheberle. 1998. "Environmental Federalism at Decade's End: New Approaches and Strategies." *Publius: The Journal of Federalism* 28, no. 1 (winter): 131–46.

Kristof, Nicholas D. 1998. "As Free-Flowing Capital Sinks Nations, Experts Prepare to 'Rethink System.'" *New York Times* September 20, A18.

Krutilla, John C., and Anthony C. Fisher. 1985. *The Economics of Natural Environments.* Washington, DC: Resources for the Future.

Krutilla, John V. 1967. "Conservation Reconsidered." *American Economic Review* 57: 777.

Krutilla, John V., and John A. Haigh. 1978. "An Integrated Approach to National Forest Management." *Environmental Law* 8: 373.
Kuhn, Thomas. 1962, 1996. *Structure of Scientific Revolutions.* Chicago: University of Chicago Press.
Laitos, Jan G., and Thomas A. Carr. 1998. "The New Dominant Use Reality on Multiple Use Lands." *Rocky Mountain Minnesota Law Institution* 44:1.
—. 1999. "The Transformation of Public Lands." *Ecology Law Quarterly* 26:140.
Lancaster, John. 1991. "Weighing the Gain in Oil-Spill Cures: Harm from Aggressive Hot-Water Cleanup May Eclipse the Environmental Benefits." *Washington Post* April 22, A3.
Lane, Robert E. 1998. "The Road Not Taken: Friendship, Consumerism, and Happiness." In *Ethics of Consumption: The Good Life, Justice, and Global Stewardship,* ed. David A. Crocker and Toby Linden, 218–48. Lanham, MD: Rowman and Littlefield.
Lash, Jonathan. 1997. "Toward a Sustainable Future." *Natural Resources and Environment* 12: 83.
Laslett, Peter. 1963. Introduction to *Two Treatises of Government,* by John Locke. New York: Mentor.
Lazarus, Richard. 1992. "Pursuing Environmental Justice: The Distributional Effects of Environmental Protection." *Northwestern University Law Review* 87: 787.
Leeson, Susan. 1978. "Philosophic Implications of the Ecological Crisis: The Authoritarian Challenge to Liberalism." *Polity* II: 303–18.
Leiss, William. 1972. *The Domination of Nature.* New York: George Braziller.
Leman, C. 1987. "The Concepts of Public and Private and Their Applicability to North American Lands." In *Land Rites and Wrongs,* ed. E. Feldman and M. Goldberg, 23–37. Cambridge, MA: Lincoln Institute of Land Policy.
Leopold, Aldo. 1966. *A Sand County Almanac with Essays on Conservation from Round River* [*A Sand County Almanac,* 1949; *Round River,* 1953]. Oxford: Oxford University Press.
Leshy, John D. 1984. "Sharing Multiple Use Lands: Historic Lessons and Speculations for the Future." In *Rethinking the Federal Lands,* ed. Sterling Brubaker. Washington, DC: Resources for the Future.
Lester, James P. 1994. "A New Federalism? Environmental Policy in the States." In *Environmental Policy in the 1990s,* ed. Norman J. Vig and Michael E. Kraft. Washington, DC: Congressional Quarterly Press.
Lindblom, Charles E. 1959. "The 'Science' of Muddling Through." *Public Administration Review* 19: 79–88.
Lipschutz, Ronnie D. 1991. "Wasn't the Future Wonderful? Resources, Environment, and the Emerging Myth of Global Sustainable Development." *Colorado Journal of International Environmental Law* 2:35.
Locke, John. 1955 (1689). *A Letter Concerning Toleration.* New York: Bobbs-Merrill.
—. 1959 (1690). *Essay Concerning Human Understanding.* Ed. Alexander C. Fraser. New York: Dover.
Lofblom, Nancy. 1998. "Forest Users Face New Rules." *Denver Post* December 8, A9.
Loomis, John B. 1993. *Integrated Public Lands Management: Principles and Applications*

to National Forests, Parks, Wildlife Refuges, and BLM Lands. New York: Columbia University Press.
Lopez, Rigoberto A., Farhed A. Shah, and Marilyn A. Altobello. 1994. "Amenity Benefits and the Optimal Allocation of Land." *Land Economics* 70: 53.
Lovins, Amory B. 1991. "Energy, People, and Industrialization." In *Resources, Environment, and Population: Present Knowledge, Future Options,* ed. Kingsley Davis and Nikhail S. Bernstam. New York: Oxford University Press.
Lowey, Mark. 1985. "National Park Jewels Are Losing Their Shine." *Calgary Herald,* July 13, A5.
Lowry, William R. 1992. *The Dimensions of Federalism: State Governments and Pollution Control Policies.* Durham, NC: Duke University Press.
—. 1994. *The Capacity for Wonder.* Washington, DC: Brookings Institution.
Lucas, P. H. C. 1991. Foreword to *Protected Areas of the World.* Lausanne, Switzerland: IUCN. International Union for the Conservation of Nature.
Lukes, Steven. 1974. *Power: A Radical View.* New York: Macmillan.
Lutter, Randall, and Christopher Wolz. 1997. "UV-B Screening by Tropospheric Ozone: Implications for the National Ambient Air Quality Standard." *Environmental Science and Technology* 31, 142A–46A.
Majone, Giandomenico. 1989. *Evidence, Argument and Persuasion in the Policy Process.* New Haven: Yale University Press.
Manaster, Kenneth A. 1995. *Environmental Protection and Justice: Readings and Commentary on Environmental Law and Practice.* Cincinnati, OH: Anderson Publishing.
Marx, Karl. 1969. "The Class Struggle in France." In *Marx/Engels Selected Works.* Moscow: Progress Publishers.
Maser, Chris, Russ Beaton, and Kevin Smith. 1998. *Setting the Stage for Sustainability: A Citizen's Handbook.* Boca Raton, FL: Lewis Publishers.
Mazmanian, Daniel, and Paul Sabatier. 1983. *Implementation and Public Policy.* Glenview, IL: Scott, Foresman.
McConnell, K. E. 1989. "The Optimal Quantity of Land in Agriculture." *Northeastern Journal of Agriculture and Resource Economics* 18: 63.
McLaren, C. 1986. "Creation of More National Parks Urged." *Toronto Globe and Mail* June 2, A10.
McManus, Reed. 1998. "What Money Can Buy." *Sierra* 83, no. 1:35.
McNamee, Thomas. 1997. *The Return of the Wolf to Yellowstone.* New York: Henry Holt.
McPherson, C. B. 1962. *The Political Theory of Possessive Individualism.* Oxford: Clarendon.
Meadows, Donella, Dennis Meadows, and Jorgen Randers. 1992. *Beyond the Limits: Confronting Global Collapse, Envisioning a Sustainable Future.* Post Mills, VT: Chelsea Green Publishing.
Meine, Curt. 1988. *Aldo Leopold: His Life and Work.* Madison: University of Wisconsin Press.
Menell, Peter S., and Richard B. Stewart. 1994. *Environmental Law and Policy.* Boston: Little, Brown.
Mill, John Stuart. 1904 (1884). *Principles of Political Economy.* New York: Appleton.
—. 1986 (1849). *On Liberty.* Buffalo, NY: Prometheus Books.

Miller, Alan S. 1991. *Gaia Connections.* Savage, MD: Rowman and Littlefield.
—. 1998. "Environmental Policy in the New World Economy." *Widener Law Symposium Journal* 3:287.
Miller, Gary J. 1992. *Managerial Dilemmas: The Political Hierarchy of Political Economy.* Cambridge, England: Cambridge University Press.
Minnesota Office of Environmental Assistance (MOEA). 1996. *1996 Pollution Prevention Evaluation Report.* St. Paul: OEA.
—. 1997. *Strategies for Improved Resource Conservation.* St. Paul: OEA.
—. 1998. *1998 Pollution Prevention Evaluation Report.* St. Paul: OEA.
Minnesota Pollution Control Agency (MPCA). 1996. *Protecting Minnesota's Environment: A Progress Report.* St. Paul: MPCA.
Minnesota Pollution Control Agency, Environmental Protection Agency, and 3M. 1996. *Minnesota XL: Final Project Agreement (FPA).* St. Paul: MPCA.
Mintrom, Michael. 1997. "Policy Entrepreneurs and School Choice." *American Journal of Political Science.* 41, no. 3 (July): 738–70.
Mintrom, Michael, and Sandra Vergari. 1998. "Policy Networks and Innovation Diffusion: The Case of State Education Reforms." *Journal of Politics* 60, no. 1 (February): 126–48.
Mishan, Edward J. 1976. *Cost-Benefit Analysis.* New York: Praeger.
Mitchell, Robert Cameron, and Richard T. Carson. 1989. *Using Surveys to Value Public Goods: The Contingent Value Method.* Washington, DC: Resources for the Future.
Moe, Terry M. 1985. "The Politicized Presidency." In *The New Direction in American Politics,* ed. John Chubb and Paul Peterson, 235–71. Washington, DC: Brookings.
Moynihan, Daniel Patrick. 1993. "Iatrogenic Government: Social Policy and Drug Research." *American Scholar* (summer): 351.
Munro, M. 1987. "National Parks: Greed Triumphs over Pride." *Montreal Gazette* December 27, A7.
Myers, Norman. 1984, 1992. *The Primary Source: Tropical Forests and Our Future.* New York: Norton.
—. 1998. "Lifting the Veil on Perverse Subsidies." *Nature* 392:327–28.
Nash, Roderick. 1989. *The Rights of Nature.* Madison: University of Wisconsin Press.
Nassauer, Joan Iverson, ed. 1997. *Placing Nature: Culture and Landscape Ecology.* Washington, DC: Island Press.
Nathan, Richard P. 1983. *The Administrative Presidency.* New York: Wiley.
National Parks and Conservation Association (NPCA). 1988. *Investing in Park Futures.* Washington, DC: NPCA.
Newman, Lawrence W. 1999. "Latin America Non Conveniens Dismissals." *New York Law Journal* 21, no. 3.
Nicholson, E. Max. 1972. "What Is Wrong with the National Park Movement?" In *Second World Conference on National Parks,* ed. H. Elliot, 32–37. Lausanne, Switzerland: International Union for the Conservation of Nature.
Nietzsche, Friedreich Wilhelm. 1967. *On the Genealogy of Morals.* Trans. and ed. Walter Kaufmann. New York: Vintage Books.
Norton, Bryan G. 1986. "Conservation and Preservation: A Conceptual Rehabilitation." *Environmental Ethics* 8: 195–220.

—. 1992. "A New Paradigm for Environmental Management." In *Ecosystem Health: New Goals for Environmental Management,* ed. Robert Costanza et al. Washington, DC: Island Press.

—. 1995. "Why I Am Not a Nonanthropocentrist: Callicott and the Failure of Monistic Inherentism." *Environmental Ethics* 17:355.

Norton, Bryan G., and Michael A. Toman. 1997. "Sustainability: Ecological and Economic Perspectives." *Land Economics* 73:553.

Noss, Reed F. 1996. "Protected Areas: How Much Is Enough?" In *National Parks and Protected Areas,* ed. R. G. Wright, 91–120. Cambridge, MA: Blackwell Science.

Okin, Susan Moller. 1989. "Justice as Fairness—For Whom?" In *Justice, Gender, and the Family.* New York: Basic Books.

Olson, Mancur. 1965. *The Logic of Collective Action.* Cambridge, MA: Harvard University Press.

Ophuls, William. 1977. *Ecology and the Politics of Scarcity.* San Francisco: W. H. Freeman.

—. 1997. *Requiem for Modern Politics.* Boulder, CO: Westview Press.

Ophuls, William, and A. Stephan Boyan. 1992. *Ecology and the Politics of Scarcity Revisited: The Unraveling of the American Dream.* New York: Freeman.

O'Riordan, Timothy, and J. Cameron, eds. 1994. *Interpreting the Precautionary Principle.* London: Earthscan.

Orr, David. 1994. *Earth in Mind: On Education, the Environment, and the Human Prospect.* Washington, DC: Island Press.

Ostrom, Elinor. 1990. *Governing the Commons.* Cambridge, England: Cambridge University Press.

Paehlke, Robert. 1979. "Occupational Health Policy in Canada." In *Ecology versus Politics in Canada,* ed. William Leiss, 97–129. Toronto: University of Toronto Press.

—. 1989. *Environmentalism and the Future of Progressive Politics.* New Haven: Yale University Press.

—. 1996. "Environmental Challenges to Democratic Practice." In *Democracy and the Environment: Problems and Prospects,* ed. William Lafferty and James Meadowcroft, 18–38. Cheltenham, England: Edward Elgar.

—. 1999. "Toward Defining, Measuring and Achieving Sustainability." In *Sustainability and the Social Sciences,* ed. Egon Becker and Thomas Jahn. London: ZED Books.

Page, Robert. 1996. *Banff–Bow Valley: At the Crossroads.* Ottawa: Canadian Heritage.

Page, Talbot. 1977. *Conservation and Economic Efficiency: An Approach to Materials Policy.* Baltimore: Published for Resources of the Future by the Johns Hopkins University Press.

—. 1978. "A Generic View of Toxic Chemicals and Similar Risks." *Ecology Law Quarterly* 7:207.

Paine, Thomas. 1953 (1776). *Common Sense.* Indianapolis: Bobbs-Merrill.

Painter, An. 1988. "The Future of Environmental Dispute Resolution." *Natural Resources Journal* 28:145.

Parks Canada (PC). 1979. *Parks Canada Policy.* Ottawa: PC.

—. 1990. *National Parks System Plan.* Ottawa: PC.

—. 1996. *Banff–Bow Valley Technical Report.* Ottawa: PC.

Passell, Peter. 1998. "Economists Point to Values beyond Price." *New York Times,* June 2, D5.

Passmore, John. 1974. *Man's Responsibility for Nature.* New York: Scribner's.

Paterson, Matthew. 1996. *Global Warming and Global Politics.* New York: Routledge.

Pearce, David, Anil Markandya, and Edward Barbier. 1989. *Blueprint for a Green Economy.* London: Earthscan.

Percival, Robert, Alan S. Miller, Christopher Schroeder, and James Leape. 1992. *Environmental Regulation: Law, Science, and Policy.* Boston: Little, Brown.

Pezzey, John C. V. 1997. "Sustainability Constraints versus 'Optimality' versus Intertemporal Concern, and Axioms versus Data." *Land Economics* 448: 73.

"Philippine Banana Workers Sue Foreign Chemical Firms." 1998. *Asia Pulse* (22 October) in Nationwide Financial News Section.

Pickett, S. T. A., and Richard Ostfeld. 1995. "The Shifting Paradigm in Ecology." In *A New Century for Natural Resources Management,* ed. Richard Knight and Sarah Bates, 261–78. Covelo, CA: Island Press.

Pigou, Arthur C. 1932. *The Economics of Welfare,* 4th ed. London: Macmillan.

Pinchot, Gifford. 1947. *Breaking New Ground.* New York: Harcourt, Brace.

—. 1967 (1910). *The Fight for Conservation.* Seattle: University of Washington Press.

Plato. 1941. *The Republic.* Trans. Francis MacDonald Cornford. Oxford: Oxford University Press.

Poirier, John. 1997. "Property, Environment, and Community." *Journal of Environmental Law and Literature* 12:43.

Portney, Paul R. 1994. "The Contingent Valuation Debate: Why Economists Should Care." *Journal of Economic Perspectives* 8: 3.

Postel, Sandra. 1994. "Carrying Capacity: Earth's Bottom Line." In *State of the World,* ed. Lester Brown et al. Washington, DC: World Resources Institute.

Power, Thomas M. 1996. *Landscapes and Failed Economies: The Search for the Value of Place.* Washington, DC: Island Press.

Press, Daniel. 1994. *Democratic Dilemmas in the Age of Ecology.* Durham, NC: Duke University Press.

President's Council on Sustainable Development (PCSD). 1996. *Sustainable America: A New Consensus.* Washington, DC: PCSD.

Pressman, J. L., and Aaron Wildavsky. 1984. *Implementation,* 3d ed. Berkeley: University of California Press.

Quirk, James, and Rubin Saponsik. 1968. *Introduction to General Equilibrium Theory and Welfare Economics.* New York: McGraw Hill.

Raab, Jonathan. 1995. "Consensus Building Can Improve Utility/Regulatory/User Relations." *Strategic Planning for Energy and the Environment* 14, no. 3:6–14.

Rabe, Barry G. 1995a. "Integrated Environmental Permitting: Experience and Innovation at the State Level." *State and Local Government Review* 27, no. 3 (fall): 209–20.

—. 1995b. "Integrating Environmental Regulation: Permitting Innovation at the State Level." *Journal of Policy Analysis and Management* 14, no. 3 (summer): 467–72.

—. 1999. "Federalism and Entrepreneurship: Explaining American and Canadian Innovation in Pollution Prevention and Regulatory Integration." *Policy Studies Journal.* 27, no. 2: 288–307.

Rabe, Barry G., and John Martin Gillroy. 1993. "Intrinsic Value and Public Policy Choice: The Alberta Case." In *Environmental Risk, Environmental Values, and Political Choices: Beyond Efficiency Tradeoffs in Public Policy Analysis,* ed. John Martin Gillroy. Boulder, CO: Westview Press.

Randall, Alan. 1981. *Resource Economics: An Economic Approach to Natural Resources and Environmental Policy.* Columbus, OH: Grid Publishing.

Rapaczynski, Andrzej. 1987. *Nature and Politics: Liberalism in the Philosophies of Hobbes, Locke, and Rousseau.* Ithaca, NY: Cornell University Press.

Rawls, John. 1985. "Justice as Fairness: Political, Not Metaphysical." *Philosophy and Public Affairs* 14(3): 223–51.

Reid, T. R. 1989. "Passive Policy on Natural Forest Fires Reaffirmed." *Washington Post* June 2, A3.

Repetto, Robert, ed. 1985. *The Global Possible.* New Haven: Yale University Press.

Rescher, Nicholas. 1980. *Unpopular Essays on Technological Progress.* Pittsburgh: University of Pittsburgh Press.

Rettie, Dwight. 1995. *Our National Park System.* Urbana: University of Illinois Press.

Rice, Vernon C. 1994. "Regulating Reasonably." *Environmental Forum* 11: (May–June) 16–23.

Ridenour, J. M. 1994. *The National Parks Compromised.* Merrillville, IN: ICS Books.

Riley, Patrick. 1982. *Will and Political Legitimacy.* Cambridge, MA: Harvard University Press.

Robinson, John, and Jon Tinker. 1997. "Reconciling Ecological, Economic, and Social Imperatives: A New Conceptual Framework," in *Surviving Globalism: The Social and Environmental Challenges,* ed. Ted Schrecker. London: Macmillan.

Rodman, John. 1977. "The Liberation of Nature?" *Inquiry* 20: 83–131.

Rogers, Raymond A. 1998. *Solving History: The Challenge of Environmental Activism.* Montreal: Black Rose Books.

Rohr, John A. 1989. *Ethics for Bureaucrats,* 2d ed. New York: Marcel Dekker.

Roots, E. Fred. 1986. "The Role of Science in the Antarctic Treaty System." In Polar Research Board, *Antarctic Treaty System: An Assessment.* Washington, DC: National Academy Press.

Rose, Carol M. 1994. *Property and Persuasion.* Boulder, CO: Westview Press.

Rosenbaum, Ken. 1993. "Timber." In *Sustainable Environmental Law,* ed. Celia Campbell-Mohn et al. St. Paul: West.

Rosenbaum, Walter. 1998. *Environmental Politics and Policy,* 4th ed. Washington, DC: CQ Press.

Rosenthal, Alan. 1998. *The Decline of Representative Democracy: Process, Participation, and Power in the State Legislatures.* Washington, DC: Congressional Quarterly Press.

Rourke, Francis E. 1984. *Bureaucracy, Politics, and Public Policy,* 3d ed. Boston: Little, Brown.

Rowlands, Ian. 1995. *The Politics of Global Atmospheric Change.* Manchester, England: Manchester University Press.

Rudzitis, Gundars. 1996. *Wilderness and the Changing American West.* New York: Riley.

Ruhl, J. B. 1998. "The Seven Degrees of Relevance: Why Should Real-World Environmen-

tal Law Attorneys Care about Sustainable Development Policy?" *Duke Environmental Law and Policy Forum* 8: 273.
Runte, Alfred. 1987. *National Parks: The American Experience,* 2d ed. Lincoln: University of Nebraska Press.
Sachs, Jeffrey D. N.d. "Globalization and the Rule of Law." *Yale Law School Occasional Papers, Second Series* No. 4.
Sagoff, Mark. 1985. *Risk-Benefit Analysis in Decisions Concerning Public Safety and Health.* Dubuque, IA: Kendall Hunt.
—. 1992. "Has Nature a Good of Its Own?" In *Ecosystem Health,* ed. Robert Costanza et al. Washington, DC: Island Press.
Salzman, James. 1997. "Nature's Services." *Ecology Law Quarterly* 24:887.
Samuelson, Paul A. 1953. *Foundation of Economic Analysis.* Cambridge, MA: Harvard University Press.
Sandel, Michael. 1984. "Justice and the Good." In *Liberalism and Its Critics,* ed. Michael Sandel. New York: New York University Press.
Sax, Joseph. 1972. *Defending the Environment.* New York: Vintage Books.
—. 1995. "Perspectives on Takings." Speech before the Federal Circuit Bar Association, May 25–26.
Schkade, D. A., and J. W. Payne. 1994. "How People Respond to Contingent Valuation Questions: A Verbal Protocol Analysis of Willingness to Pay for an Environmental Regulation." *Journal of Environmental Economics and Management* 26: 88–109.
Schmidt, James, ed. 1996. *What Is Enlightenment?* Berkeley: University of California Press.
Schmidt-Bleek, F. 1994. *Wieviel Umwelt Braucht der Mensch? MIPS: Das Masz für ökologisches Wirtschaften.* Basel, Switzerland: Birkhäuser.
Schneider, Mark, and Paul Teske, with Michael Mintrom. 1995. *Public Entrepreneurs: Agents for Change in American Government.* Princeton, NJ: Princeton University Press.
Sedjo, Roger A. 1995. "Forests: Conflicting Signals." In *The True State of the Planet,* ed. Ronald Bailey. New York: Free Press.
Sen, A. K. 1977. "Rational Fools: A Critique of the Behavioral Foundations of Economic Theory." *Philosophy and Public Affairs* 16: 317–44.
Sellars, Richard W. 1997. *Preserving Nature in the National Parks.* New Haven: Yale University Press.
Sessions, Kathy. 1993. "Building a Capacity to Change." *EPA Journal* 19, no. 2 (April–June): 15.
Shanklin, Pathfinder. 1997. "Pathfinder: Environmental Justice." *Ecology Law Quarterly.* 24:333.
Shannon, Margaret, and Alexios Antypas. 1997. "Open Institutions: Uncertainty and Ambiguity in 21st Century Forestry." In *Creating a Forestry for the 21st Century,* ed. Kathryn Kohm and Jerry F. Franklin, 437–45. Washington, DC: Island Press.
Shapley, Deborah. 1985. *The Seventh Continent: Antarctica in a Resource Age.* Washington, DC: Resources for the Future.
Shleifer, Andrei, and Robert W. Vishny. 1998. *The Grabbing Hand: Government Pathologies and Their Cures.* Cambridge, MA: Harvard University Press.

Slocombe, D. Scott. 1993. "Environmental Planning, Ecosystem Science, and Ecosystem Approaches for Integrating Environment and Development." *Environmental Management* 17:289–303.

Smeloff, Ed, and Peter Asmus. 1997. *Reinventing Electric Utilities.* Washington, DC: Island Press.

Smith, Adam. 1761. *The Theory of Moral Sentiments.* London: Millar, Kincaid, and Bell.

—. 1828. *An Inquiry into the Nature and Causes of the Wealth of Nations.* Edinburgh: A. Black and W. Tait.

Smith, Adfian. 1997. *Integrated Pollution Control: Change and Continuity in the UK Industrial Pollution Policy Network.* Brookfield, VT: Ashgate.

Smith, Fred L. 1994. "Sustainable Development: A Free Market Perspective." *B.C. Environmental Affairs Law Review* 21:297.

Smith, V. Kerry, ed. 1979. *Scarcity and Growth Revisited.* Baltimore: Johns Hopkins/ Resources for the Future.

Snow, Donald, ed. 1992. *Inside the Environmental Movement: Meeting the Leadership Challenge.* Washington, DC: Island Press.

Solow, Robert. 1992, 1998. "An Almost Practical Step toward Sustainability." Lecture on the occasion of the 40th anniversary of Resources for the Future. In *Natural Resources and the Environment: The RFF Reader,* ed. Wallace Oates. Washington, DC: Resources for the Future.

—. 1993. "Sustainability: An Economist's Perspective." In *Economics of the Environment,* 3d ed., ed. Robert Dorfman and Nancy Dorfman, 179–87. New York: Norton.

Song, C. F., and J. B. Loomis. 1984. "Empirical Estimates of Amenity Forest Values: A Comparative Review." General Technical Report RM-Rocky Mountain Forest and Range Experiment Station, United States Forest Service. Fort Collins, Colorado: USDA.

State of Minnesota, Office of Administrative Hearings for the Minnesota Public Utilities Commission. 1996. *In the Matter of the Quantification of Environmental Costs Pursuant to Laws of Minnesota 1993.* Chapter 356, Section 3. St. Paul: Minnesota Public Utilities Commission.

Steinzor, Rena. 1996. "Regulatory Reinvention and Project XL: Does the Emperor Have Any Clothes?" *Environmental Law Reporter* 26 (October): 10527.

Stevens, Thomas H., Jaime Echeverria, Ronald J. Glass, Tim Hager, and Thomas A. More. 1991. "Measuring the Existence Value of Wildlife: What do CVM Estimates Really Show?" *Land Economics* 67: 390.

Stevens, William. 1997. "Study of Ocean Currents Offers Clues to Global Climate Shifts." *New York Times* March 18.

Stewart, Richard, ed. 1995. *Natural Resource Damages: A Legal, Economic, and Policy Analysis.* Washington, DC: National Legal Center for the Public Interest.

Stiglitz, Joseph E. 1989. "On the Economic Role of the State." In *The Economic Role of the State,* ed. A. Heertje. Oxford: Basil Blackwell.

Stokey, Edith, and Richard Zeckhauser. 1978. *A Primer for Policy Analysis.* New York: Norton.

Stone, Christopher D. 1994. "Deciphering 'Sustainable Development.'" *Chicago Kent Law Review* 69:977.

—. 1995. "Reflections on the Moral and Institutional Challenges of 'Sustainable Development.'" SOAS Law Department Working Paper no. 5, School of Oriental and African Studies, University of London, February.

Stroup, Richard L., and John A. Baden. 1983. *Natural Resources: Bureaucratic Myths and Environmental Management.* Cambridge, MA: Ballinger.

Sunstein, Cass R. 1990. *After the Rights Revolution: Reconceiving the Regulatory State.* Cambridge, MA: Harvard University Press.

—. 1997. *Free Markets and Social Justice.* New York: Oxford University Press.

Tarlock, Dan. 1992. "Environmental Law: the Role of Non-Governmental Organizations in the Development of International Environmental Law." *Chicago-Kent Law Review* 68: 61.

Thibodeau, Francis R., and Bart D. Ostro. 1981. "An Economic Analysis of Wetland Protection." *Journal of Environmental Management* 12, no. 1: 19–30.

Thomas, John Clayton. 1995. *Public Participation In Public Decisions.* San Francisco: Jossey-Bass.

Thompson, Janna L. 1983. "Preservation of Wilderness and the Good Life." In *Environmental Philosophy,* ed. Robert Elliot and Arran Gare. University Park: Pennsylvania State University Press.

—. 1990. "A Refutation of Environmental Ethics." *Environmental Ethics* 12:159.

Thomson, Rebecca W. 1995. "Ecosystem Management: Great Idea, But What Is It, Will It Work, and Who Will Pay?" *Natural Resources and Environment* 9 (winter): 3.

Thoreau, Henry David. 1966 (1849). *Walden and Civil Disobedience.* Ed. Owen Thomas. New York: Norton.

Tietenberg, Tom. 1996. *Environmental and Natural Resource Economics,* 4th ed. New York: HarperCollins College Publishers.

Tittman, Paul, and Clifton Brownell. 1984. *Appraisal Report Estimating Fair Market Rental Value of Public Rangelands in the Western United States.* Washington, DC: U.S. Forest Service.

Torgerson, Douglas. 1999. *The Promise of Green Politics.* Durham, NC: Duke University Press.

Torry, Saundra. 1999. "Cigarette Firms Sued by Foreign Governments." *Washington Post* January 17, A12.

Tully, James. 1980. *A Discourse on Property.* New York: Cambridge University Press.

—. 1993. *An Approach to Political Philosophy: Locke in Contexts.* Cambridge: Cambridge University Press.

Turner, R. Kerry, David Pearce, and Ian Bateman. 1993. *Environmental Economics.* Baltimore: Johns Hopkins University Press.

United Nations. 1989. *General Assembly Resolution on Protection of the Global Climate for Present and Future Generations of Mankind,* A/Res/44/207.

—. 1996. *World Economic and Social Survey.* New York: United Nations.

—. Conference on Environment and Development. 1992. *Declaration of Principles.* Available at gopher://infoserver.ciesin.org/00/human/domains/political-policy/intl/confs/UNCED/unced-finals/rio-dec.

—. N.d. *Indicators of Sustainable Development.* (www.un.org/esa/sustdev/isd.htm).

U.S. Congress. 1995. *Hearings before the House Subcommittee on National Parks for the National Park System Reform Act.* Washington, DC: GPO.

U.S. Congress, House Committee on Interior and Insular Affairs. 1985. *The Greater Yellowstone Ecosystem.* Washington, DC: GPO.

U.S. Department of Agriculture. 1997. *Agricultural Statistics.* Washington, DC: USDA.

U.S. Department of Agriculture, Forest Service. 1990. *Forest Service Program for Forest and Rangeland Resources.* Washington, DC: USFS.

—. 1991. *Resource Planning Assessment Program.* Washington, DC: USFS.

—. 1995. *Draft Resource Planning Assessment Program. http://www.fs.fed.us/pl/rpa/95rpa.*

—. 1999. *Forest Service Web Page: Biology. http://www.fs.fed.us/biology.*

U.S. Department of Interior, Bureau of Land Management. 1974. *Public Land Statistics.* Washington, DC: USDOIBLM.

—. 1991. *Public Land Statistics.* Washington, DC: USDOIBLM.

—. 1994. *Public Land Statistics.* Washington, DC: USDOIBLM.

—. 1995. *Public Land Statistics.* Washington, DC: USDOIBLM.

—. 1996. *Public Land Statistics.* Washington, DC: USDOIBLM.

—. 2000a. *BLM Facts. http://www.blm.gov/nhp/facts.*

—. 2000b. *1997 Strategic Plan. http://www.blm.gov/nhp/info/stratplan/.*

U.S. Environmental Protection Agency (EPA). 1994. *Inside EPA Weekly Report.* November 18. Washington, DC: USEPA.

—. 1995a. *Implementation Strategy for the Clean Air Act Amendments of 1990.* Washington, DC: USEPA.

—. 1995b. *Inside EPA Weekly Report.* April 28. Washington, DC: USEPA.

—. 1995c. *Inside EPA Weekly Report.* May 5. Washington, DC: USEPA.

—. 1995d. *Inside EPA Weekly Report.* June 30. Washington, DC: USEPA.

—. 1995e. *Inside EPA Weekly Report.* July 7. Washington, DC: USEPA.

—. 1997a. *Environmental Performance Partnership Agreement: Minnesota Pollution Control Agency and U.S. Environmental Protection Agency Region 5.* Washington, DC: USEPA.

—. 1997b. *Multimedia Pollution Prevention Permitting Project.* Washington, DC: USEPA.

U.S. Environmental Protection Agency, Office of Policy Planning and Evaluation, and Industrial Economics. 1994. "Planning and Evaluation and Industrial Economics Inc." *Sustainable Industry Phase 1 Report.* June. Washington, DC: USEPA.

U.S. General Accounting Office (GAO). 1991a. *Recreation Concessionaires Operating on Federal Lands.* Washington, DC: GPO.

—. 1991b. *Status of Development at the Steamtown National Historic Site.* Washington, DC: GAO.

—. 1993. *Transfer of the Presidio from the Army to the NPS.* Washington, DC: GAO.

—. 1994a. *Ecosystem Management: Additional Actions Needed to Adequately Test a Promising Approach.* Washington, DC: GAO.

—. 1994b. *Pollution Prevention: EPA Should Reexamine the Objectives and Sustainability of State Programs.* Washington, DC: GAO.

U.S. National Park Service (NPS). 1972. *National Park System Plan.* Washington, DC: NPS.

—. 1974. *Yellowstone Master Plan.* Washington, DC: NPS.

—. 1991. *Yellowstone Statement for Management.* Washington, DC: NPS.

—. 1992. *National Parks for the 21st Century: The Vail Agenda.* Washington, DC: NPS.
—. 1995. *Denver Service Center Annual Report.* Denver, CO: NPS.
U.S. Office of Technology Assessment (OTA). 1995. *Environmental Policy Tools.* Washington, DC: OTA.
von Weizsäcker, Ernst, Amory B. Lovins, and L. Hunter Lovins. 1998. *Factor Four: Doubling Wealth, Halving Resource Use.* London: Earthscan.
Wackernagel, Mathis, and William Rees. 1996. *Our Ecological Footprint: Reducing Human Impact on the Earth.* Gabriola Island, BC: New Society.
Walker, W. 1992. "Feds Falling Behind on National Parks Program." *Vancouver Sun* January 18, A13.
Walsh, Richard G., John B. Loomis, and Richard A. Gillman. 1984. "Valuing Option, Existence, and Bequest Demands for Wilderness." *Land Economics* 60: 14.
Walsh, Richard G., Donn M. Johnson, and John R. McKean. 1988. *Review of Outdoor Recreation Economic Demand Studies with Nonmarket Benefit Estimates 1968–1988.* Fort Collins, Colorado: Colorado Water Resources Research Institute.
Walzer, Michael. 1983. *Spheres of Justice: A Defense of Pluralism and Equality.* New York: Basic Books.
Weaver, R. K. 1992. "Political Institutions and Canada's Constitutional Crisis." In *The Collapse of Canada?,* ed. Washington Weaver, 7–75. Washington, DC: Brookings Institution.
Weiler, Paul, et al. 1993. *A Measure of Malpractice: Medical Injury, Malpractice Litigation, and Patient Compensation.* Cambridge, MA: Harvard University Press.
Weisbrod, Burton A. 1964. "Collective-Consumption Services of Individual-Consumption Goods." *Quarterly Journal of Economics* 78: 471.
—. 1978. *Public Interest Law.* Berkeley: University of California Press.
Weiss, Edith B. 1990. "Our Rights and Obligations to Future Generations." *American Journal of International Law* 84:198.
Wenz, Peter. 1988. *Environmental Justice.* Albany: State University of New York Press.
—. 1996. *Nature's Keeper.* Philadelphia: Temple University Press.
—. 1997. "Philosophy Class as Commercial. *Environmental Ethics* 19:204–16.
Weston, Anthony. 1992. *Toward Better Problems.* Philadelphia: Temple University Press.
—. 1996. "Beyond Intrinsic Value: Pragmatism and Environmental Ethics." In *Environmental Pragmatism,* ed. Andrew Light and Eric Katz, 292–98. New York: Routledge.
Wiener, Jonathan Baert. 1998. "Managing the Iatrogenic Risks of Risk Management." *Risk: Health, Safety and Environment* 9:39.
—. 1999. "Global Environmental Regulation: Instrument Choice in Legal Context." *Yale Law Journal* 108:677.
Wilkinson, Charles F. 1992. *Crossing the Next Meridian: Land, Water, and the Future of the West.* Washington, DC: Island Press.
Williams, Douglas. 1995. "Valuing Natural Environments." *Connecticut Law Review* 27: 365.
Wolf, Charles. 1988. *Markets or Government?* Cambridge, MA: MIT Press.
Wood, Gordon S. 1969. *The Creation of the American Republic 1776–1787.* New York: Norton.

Woodwell, George. 1985. "On the Limits of Nature." In *The Global Possible*, ed. Robert Repetto. New Haven: Yale University Press.

World Bank. 1992. *World Development Report 1992: Environment and Development.* New York: Oxford University Press.

—. 1997. *World Development Report 1997: The State in a Changing World.* New York: Oxford University Press.

World Commission on Environment and Development (WCED). 1987. *Our Common Future.* Oxford: Oxford University Press.

World Resources Institute in collaboration with the United Nations Environmental Programme and the United Nations Development Programme. 1994. *World Resources 1994–95.* New York: Oxford University Press.

—. 1996. *World Resources 1996–1997.* New York: Oxford University Press.

World Wildlife Fund (WWF). 1995. *Endangered Species Progress Report.* Ottawa: WWF.

Wright, George M., J. Dixon, and B. Thompson. 1932. *Fauna of the National Parks of the United States.* Washington, DC: GPO.

Wright, R. G., and J. M. Scott. 1996. "Evaluating the Ecological Suitability of Lands." In *National Parks and Protected Areas*, ed. R. G. Wright, 121–32. Cambridge, MA: Blackwell Science.

Zinser, Charles L. 1995. *Outdoor Recreation: U.S. National Parks, Forests, and Public Lands.* New York: Wiley.

Zumberge, James. 1986. "The Antarctic Treaty as a Scientific Mechanism: The Scientific Committee on Antarctic Research and the Antarctic Treaty System." In Polar Research Board, *Antarctic Treaty System: An Assessment*, 153–184. Washington, DC: National Academy Press.

Contributors

Joe Bowersox is Associate Professor of Political Science at Willamette University in Oregon. His interests lie in the conjunction of political philosophy/environmental theory. He has participated in the Fulbright-Hays Group Project to Jordan, and has been Jacob K. Javits Fellow in Political Science at the University of Wisconsin–Madison.

David J. Brower is Research Professor in the Department of City and Regional Planning, University of North Carolina at Chapel Hill. He specializes in land use and environmental and coastal management. Anna K. Schwab is a Research Associate in the Center for Urban and Regional Studies at the University of North Carolina at Chapel Hill. Brower and Schwab are co-authors with Timothy Beatley of *An Introduction to Coastal Zone Management.*

Susan Buck is Associate Professor of Political Science at the University of North Carolina, Greensboro. She works in the areas of public administration and environmental policy and her most recent book, *The Global Commons,* examines the major issues of current international environmental law as a collective action problem.

Celia Campbell-Mohn is Professor of Law at the Environmental Law Center, Vermont Law School. She specializes in sustainable environmental/natural resources law and the resources-to-recovery model, which is the subject of her co-authored hornbook, *Environmental Law: From Resources to Recovery.*

John Martin Gillroy is John D. MacArthur Professor of Environmental Policy and Law and Director of the Environmental Studies Program at Bucknell University. His newest book, *Justice and Nature,* constructs an alternative to the market paradigm for setting the standards of environmental policy and law based on the moral and political thought of Immanuel Kant.

Joel J. Kassiola is Dean of the College of Behavioral and Social Sciences at San Francisco State University. A long-time advocate for the connection between social justice issues and environmental policy, his seminal book *The Death of Industrial Civilization* charts the roots of our ecological dilemmas to the values of industrialism.

Jan Laitos is Professor of Law at the University of Denver College of Law. Laitos specializes in public lands and natural resources law and is currently writing the new edition of his seminal casebook *Natural Resources Law.*

William Lowry is Associate Professor in the Department of Political Science at Washington University in St. Louis. He specializes in comparative environmental policy, national parks, and environmental federalism. His latest book is *Preserving Public Lands for the Future.*

Bryan Norton is Professor of Public Policy and Philosophy at the School of Public Affairs, Georgia Institute of Technology. He has widely written in the areas of environmental ethics, politics, and philosophy. His classic book *Why Preserve Natural Variety?* helped to define the field of modern environmental ethics.

Robert Paehlke is Professor of Political Science and Environmental and Resource Studies at Trent University in Ontario, Canada. His work in the ascension of environmentalism as a competitor for both capitalism and socialism in twenty-first-century politics has won widespread praise and resulted in a book, *Environmentalism and the Future of Progressive Politics.*

Robert Percival is Director of the Environmental Law Program, Robert Stanton Scholar, and Professor of Law at the University of Maryland School of Law. He is principal author of the most widely used environmental law casebook in the country: *Environmental Regulation: Law, Science, and Policy.*

Barry Rabe is Professor and Dean of the School of Natural Resources and the Environment at the University of Michigan, Ann Arbor. He specializes in comparative environmental policy, comparative federalism, and environmental politics. His latest book is titled *Beyond NIMBY.*

Mark Sagoff is Director of the Institute for Philosophy and Public Policy at the University of Maryland and Visiting Scholar at the Wilson Institute, Washington, D.C. His eclectic writings have examined subjects as diverse as markets for risk, animal rights, and even the absence of environmental crisis. His seminal book, *The Economy of the Earth,* is essential reading.

Bob Pepperman Taylor is Professor of Political Science and directs the John Dewey Honors Program at the University of Vermont. His excellent survey of the history of environmental thought in America, *Our Limits Transgressed: Environmental Political Thought In America* is the authoritative text in the origins and persistence of both conservationist and preservationist movements in this country.

Jonathan Wiener is Associate Professor at the Law School and the Nicholas School of the Environment at Duke University. His published writing includes *Risk vs. Risk: Tradeoffs in Protecting Health and the Environment.*

Index

Names

Cases Cited

Library of Congress Cataloging-in-Publication Data
The moral austerity of environmental decision making : sustainability, democracy, and normative argument in policy and law / edited by John Martin Gillroy and Joe Bowersox.
Includes bibliographical references and index.
ISBN 0-8223-2850-x (cloth : alk. paper)
ISBN 0-8223-2865-8 (pbk. : alk. paper)
1. Environmental policy—Moral and ethical aspects.
2. Environmental ethics. I. Gillroy, John Martin. II. Bowersox, Joe.
GE170 .M66 2002 179′.1—dc21 2002000382